America's First Eclipse Chasers

Stories of Science, Planet Vulcan, Quicksand, and the Railroad Boom

Thomas Hockey

America's First Eclipse Chasers

Stories of Science, Planet Vulcan, Quicksand, and the Railroad Boom

 Springer

Published in association with
Praxis Publishing
Chichester, UK

Thomas Hockey
Department of Earth and Environmental Sciences
University of Northern Iowa
Cedar Falls, IA, USA

SPRINGER-PRAXIS BOOKS IN POPULAR ASTRONOMY

Popular Astronomy
ISSN 2626-8760   ISSN 2626-8779 (electronic)
Springer Praxis Books
ISBN 978-3-031-24123-9   ISBN 978-3-031-24124-6   (eBook)
https://doi.org/10.1007/978-3-031-24124-6

This Springer imprint is published by the registered company Springer Nature Switzerland AG
The registered company address is: Gewerbestrasse 11, 6330 Cham, Switzerland

To Alexander Leo Ivakh.
I so look forward to showing you your first eclipse.

Contents

Acknowledgements

This project took me from the Grand Canyon to the gates of the Old Naval Observatory in Washington, DC (now guarded by friendly but still-intimidating armed soldiers since it was re-purposed to the Department of State). Along the way, I became beholden to many kind people.

Thank you to Lara Szypszak and Joshua Levy, of the Library of Congress; Morgan Aaronson, of the United States Naval Observatory; Katie Moss, of the State Historical Museum of Iowa; and Christine Chandler of the Putnam Museum, who facilitated my visits just as soon as such a thing again became possible after COVID-19 shutdowns.

I appreciate those who took the time to act as sounding boards over coffee, as I planned this book. They include Jennifer Bartlett[1] (United States Naval Academy), David DeVorkin (Smithsonian Institution), Steven Dick (United States Naval Observatory), Jay Pasachoff (Williams College), and Marc Rothenberg (National Science Foundation).

Researching a book during a pandemic was a (hopefully) unique experience. Doing so was possible only through the efforts of sympathetic librarians, archivists, and curators, a partial list of whom includes Joseph Ahern (University of Pennsylvania); Silvia Brown (Williams College); Jennifer Bryan (United States Naval Academy); Lisa Burkhart (Sanford Museum); Susannah Carroll (Franklin Institute); Heather Cecil (Shelbyville History Museum); Elizabeth Cisar (Augustana University); Ilhan Citak and Brian Simboli (Lehigh University); Lisa Clarke (National Oceanographic and Atmospheric Administration); James Cleveland (Shelby County Historical Society); Catherine Cotter, Leonard Levin, and Carla Townsend (Northwestern University); Myles Crowley (Massachusetts Institute of Technology); Elizabeth Engel (State Historical Society of Missouri);

---

[1] Formerly (modern-day equivalent of) Nautical Almanac Office

Gail Fleenor (Bristol Historical Association); Jim Gerencser (Dickenson College); Julie Huffman-Klinkowitz (Cedar Falls Historical Society); Anthony Jahn and Julie Martineau (Des Moines County Historical Society); Dave Joens (Office of the Illinois Secretary of State); Elizabeth Jones-Minsinger (Haverford College); Jerry Jordak (Penn Central Rail Road Historical Society); Bill Kemp (McLean County Museum of History); Leah Loscutoff (Stevens Institute of Technology); David Lotz (Burlington Route Historical Society); Maria McEachern (Harvard University); Susan McElrath (University of California-Berkeley); Rosemary Meany (University of Northern Iowa); Amanda Nelson (Wesleyan University); Hang Nguyen (State Historical Society of Iowa); Scout Noffke (Dartmouth University); Adam Nyhaug (Siouxland Heritage Museums); Dean Rogers (Vassar College); Danielle Rougeau (Middlebury College); Joyce Shepherd (Lawrence County Museum of History); Trenton Streck-Havill (National Museum of Health and Medicine); Pat White (Henry County Trust); and the staff of the Ames Public Library, Atlantic Public Library, Boone County Historical Society, Burlington Public Library, Cedar Falls Public Library, Cedar Rapids Public Library, Cherokee Public Library, Davenport Public Library, Des Moines Public Library, Fairfield Public Library, Fort Dodge Public Library, Grinnell Public Library, Hampton Public Library, Ida Grove Public Library, Iowa City Public Library, Iowa State University Library, James H. Andrew Railroad Museum, Jasper County Historical Society, Jasper County Public Library, Jefferson Public Library, Knoxville Public Library, Le Mars Public Library, Marion Public Library, Marshalltown Public Library, Mount Pleasant Public Library, Muscatine Public Library, Newton Public Library, Orange City Public Library, Osceola Public Library, Oskaloosa Public Library, Pella Public Library, Perry Public Library, Sergeant Bluff Public Library, Sheffield Public Library, Sioux City Public Library, Toledo Public Library, University of Iowa Library, and Vinton Public Library.

Mack Greer and Professor Annette Lynch added context.

The following kindly agreed to read and comment upon early draft chapters for me: Prof. Jennifer Bartlett, Trudy Bell, Brenda Corbin, Dr. Stella Cottam, Dr. David DeVorkin, Dr. Steven Dick, Dr. Richard Feinberg, Prof. Jarita Holbrook, Prof. Philip Nicholson, Prof. James O'Laughlin, Prof. Wayne Osborn, Prof. Jay Pasachoff, Dr. Marc Rothenberg, Prof. Kenneth Rumstay, Dr. Steven Ruskin, Prof. Bradley Schaefer, Dr. William Sheehan, and Prof. Joseph Tenn. Remaining errors are my own.

I am proud to say that I my research was assisted by these University of Northern Iowa undergraduates: Samuel Lala traversed Iowa, even driving through the infamous derecho of 2020, in order to make site visits on my behalf. Trishyan Anthony (aided by Jacob Helmuth) searched the country *via* internet to bring me local resources from the length of the total-eclipse path. Morgan Kaiser sorted documents for me. Sierra Ameen helped populate my census of those who observed the

1869 total eclipse of the Sun. Ethan Ahrens kept me metric. Katelyn Hoff facilitated organization of the citations. Dan Wilkinson made sure my quotes were *verbatim*.

I benefited from i-conversations with members of the Antique Telescope Society. Postings on the Solar Eclipse Mailing List were similarly illuminating. I also want to acknowledge the very useful interactive eclipse tool created by Xavier Jubier and NASA's Astrophysics Data System.

A 'shout out' goes to Hannah Kaufman, recently of Springer, with whom I partnered on several projects including this one.

Finally, I am grateful to the citizens of various towns and cities—for example, those of Ames, Boone, Burlington, Cedar Falls, Cedar Rapids, Cherokee, Iowa City, Jefferson, Keokuk, Kewanee, Marion, Mount Pleasant, Oskaloosa, Ottumwa, Rock Island, Waterloo, and Waverly—who pointed out, and put up with a stranger and his camera poking around, their importance places of long ago.

Preface

I sit at my desk and find myself in an odd state of mind. I realize that I have just finished writing a book different from those that I have ever read before.

Most histories follow a narrative; they have a timeline. This book describes what is nearly but a moment in time. It is not a near-moment of great cosmological significance, such as the immediate aftermath of the Big Bang. Anywhere else in the Universe, this 'moment,' with which I have spent so much real-time, took place and passed without comment on a typical weekend afternoon—if such a time reckoning made any sense to hypothetical sentient beings elsewhere! Yet this moment (actually, a few minutes) was, I contend, of great importance to those who watched the sky of Earth during its tiny span of time.

Because the event herein described was brief, and those who experienced it were mostly separated by great distance from others who did so, my story is dimensionless. Time (the duration of the event) and place (the sky) are a constant. Therefore, as I deconstruct the 1869 total eclipse of the Sun, my audience is not necessarily bound to linear reading.

A chapter titled Introduction (and the one that follows it) is usually always a good start. Those already well-versed in the mechanism and phenomena of a solar eclipse may wish to parse Chapter ("Some Light Upon a Dark Subject") lightly. For the science and adventure of nineteenth-century eclipse expeditions, there are Chapters 4–12 (Navy Astronomers 2,000 Kilometers Ashore; "The Vast Black Orb"; New Astronomy in the Old West; Observing in Style; Meeting of the Gray Hairs; "Overhanging Monster Wings:" The Philadelphia Photographic Corp; Surveying a Solar Eclipse; The Canadians: Toques on the Frontier; Chasing the Umbra Through Time and Space): the earlier, major expeditions resulting in more science and less adventure; the latter, more idiosyncratic expeditions, less science and definitely more adventure.

Readers interested in applications of the eclipse (and there were such targets of opportunity), might go to Chapters 13–15 ("A Darkness That Can Be Felt"; Standing on the Edge Looking Up; Vulcan). For the sociological aspects of the eclipse—and there are many—read Chapters 19–20 (Americans in Totality; In the Shadow of Benjamin Banneker and "Even Thoughtless Women and Children Hush their Restless Tongues …"; Fire Cloud). Chapters (What Did It All Mean?; …And What Happened After That?) are a denouement.

History does not predict, but it may foreshadow. As you read, keep in mind the following: Coincidently, a total solar eclipse, mimicking the one that is the subject of this book's tale, will occur in 2024.

A few details for the bibliophile: Quotations herein are indicated by double inverted commas or are in indented block style. They are *verbatim*, in words and punctuation, with the caveat that the American form of spelling is used throughout for consistency. Capitalization is inserted so as to highlight proper nouns.

Single inverted commas indicate a quotation within a quotation, marked using the double-inverted comma convention. They also are used to signify anachronistic metaphors.

This convention is used for the parenthetical: braces within brackets within parentheses.

Citations tend to interrupt the flow of reading; I insert them in the back of the book. Here, you also will find extended illustration credits, a placement that keeps captions from becoming too lengthy.

Units are metric unless something—*e.g.*, a telescope objective—was constructed to a particular size in English units (one with a small number of significant digits). Units are most typically those of length: meters and kilometers (inches and miles). Time is local unless otherwise noted. Dates are Gregorian.

Locations should be assumed to be in the United States [USA]. Notice is made of exceptions.

Titles are those of the period. Gender is included if it adds meaning.

No product endorsement is intended. I am not big into drugs and firearms, but failing to mention them would be anachronistic.

I tried to include illustrations that one would not normally encounter through casual web browsing. Images used as illustrations are of the highest resolution and picture quality that I am able to obtain. I come from a family in which the issue of vision comes up frequently: If you find it difficult to see the text, please let me know *via* the publisher.

This is the first book that I wrote completely paperless, before submission to the publisher. No electrons were harmed in the making of this work.

Cedar Falls
October 2022

Abbreviations

APS	American Philosophical Society
HCO	Harvard College Observatory
MIT	Massachusetts Institute of Technology
NAO	Nautical Almanac Office
NAS	National Academy of Science
NLT	New Living Translation
RAS	Royal Astronomical Society
USA	United States of America/United States Army
USCS	United States Coast Survey
USN	United States Navy
USNA	United States Naval Academy
USNO	United States Naval Observatory
VMI	Virginia Military Institute

Foreword

I am so enjoying Tom Hockey's nostalgic romp through the observations and the shenanigans associated with the 1869 passage of the totality path across his residential state of Iowa. He takes us back to that innocent time, some fifty years or so before the Music Man Henry Hill came on his con to Iowa, as celebrated in the musical theatre in Meredith Willson's 1957 masterpiece on Broadway and subsequent film. Now a long-term professor of astronomy at the University of Northern Iowa in Cedar Falls, Hockey puckishly investigated perhaps the major astronomical event in which his state starred: this first major solar eclipse that crossed the United States. So it isn't necessary to go to high mountains like Palomar, Maunakea, or Cerro Tololo to have important astronomical observing sites. Hockey starts in Iowa, though the second half of the book extends to other states, including Kentucky and Tennessee, and even the not-yet-state of Alaska.

The circumstances of the eclipse path, well-illustrated with a wide variety of maps and charts, brought the nation's leading astronomers there. Key astronomical royalty of the time included especially Simon Newcomb of the U.S. Naval Observatory from far-off Washington, DC. Hockey describes the science and the motivation of the U.S. Navy for understanding time signals and more, and in individual chapters describes varying sets of observations and the visiting and local personnel.

Photography in 1869 was still primitive, with the single, positive images known as daguerreotypes being superseded by "wet plates," which gave negatives that could be later reproduced. But the question of exposure time was key. Nowadays, we bracket widely in exposure time, which we are able to do because of the sensitivity of our computer chips, since we are free of the slow chemical reactions of film, two orders of magnitude slower.

The story of assistants preparing plate after plate to fit within the minutes of totality, and sometimes running out of time or plates, reminds me today of my own

21st-century eclipse experience of the partial eclipse two weeks ago that my wife and I viewed from Viña del Mar, Chile, with my colleague from the University of Chile ferrying camera chips from our viewing balcony into the room where I was verifying that the data were being adequately recorded, simultaneously checking on focus—universal tasks for astronomers.

It's fun to learn about some contemporary eclipse music of the time, such as Solar Eclipse Gallop and a march by John Phillip Souza. See also https://eclipse2017.nasa.gov/eclipses-and-music and lists by Andy Fraknoi at http://bit.ly/astronomymusic.

We learn about the astronomical dignitaries of the day, and how they were arranged along the path of totality, as it extended past Iowa into Tennessee, Kentucky, and elsewhere. For many of us, we have a special interest in the expedition of Maria Mitchell, who was by then professor of astronomy at Vassar College and one of the first female professional astronomers. Her students, encumbered by their long dresses, hoops, et cetera, got good views of the eclipse—and show dramatically in photos that Hockey has selected.

Hockey's readable narrative takes us through space and time, and out into the solar system. We learn of individual personalities and of their actions and interactions in this wonderful book. Hockey's admirable thoroughness describes and eventually summarizes an important era in the history of astronomy, both from the scientific side and from the national history.

Jay M. Pasachoff
Field Memorial Professor of Astronomy, Williams College, Williamstown, MA, USA
Chair, International Astronomical Union Working Group on Solar Eclipses, Paris, France
Veteran of 75 solar eclipses as of 2022

Prolog

I was channeling a total eclipse of the Sun. On the afternoon of 7 August 2019, I stood across the street from what was once the location of J. H. Stanley's retail store in downtown Cedar Falls, Iowa, my hometown. In a community where storefronts often exhibit plaques claiming provenance of their antiquity and ownership over the years, Stanley's shop is no longer among them. It did not survive, as brick-and-mortar replaced wood.

There were not many stores in 1869 Cedar Falls. Even though through the Civil War it was the terminus of the Illinois Central Railroad, the mill town only recently had surpassed 3,000 citizens. It had not been long since so little mail arrived here that the postmaster carried letters under his hat and delivered them to addressees when he saw them on the street.

Stanley was a jeweler. That season, the item in great demand was a filter with which to peep at the Sun. Stanley was, of course, exceptional in that he was a city-dweller in a time when almost everybody lived on the farm.

Exactly one-hundred fifty years before I took my station at the northeast corner of Main and Second Streets, Stanley took a break from selling his wares to step outside and cross the street in order to get a better view of something unusual. He looked up. Perhaps the facing lot was vacant, or he positioned himself to look between buildings. Perhaps the wooden storefronts of the day were not tall enough to obscure the view of many objects in the sky. It helped that it was early afternoon in the summer; the Sun was high. Here he waited to observe a total solar eclipse.

While no settler from the Eest had seen such a thing in the American West before, Stanley had learned that astronomers wished for the local duration of the event. This was especially true for towns like Cedar Falls, where the latitude and longitude could easily be measured, and the total part of the eclipse was to be brief. My neighbor (in space, if not time) may not have understood the particulars of why such a measurement was coveted. However, we can guess that—for the

Fig. 1 Contemporary view of Main Street, Cedar Falls, Iowa. {Cedar Falls Historical Society}

moment—he wished to be part of the emerging culture of science in the latter nineteenth-century US.

The weather was uncharacteristically cool for August, but Stanley probably still wore his cotton frock coat. He had in hand a white-faced stopwatch, accurate to half a second. In my mind, it was pulled from a vest pocket. Earlier, the watch was set by use of telegraph signals transmitted by an operator consulting a master clock in the bigger city to the east, Dubuque. Stanley also held a piece of colored glass. At his feet was a lantern: While still long before summer's late sunset, it was about to get dark.

Stanley managed to memorize the two solar-eclipse "contact times" corresponding to the onset of total obscuration of the Sun by the Moon and its end. He would record these times for posterity shortly. In between, he beheld the marvel taking place above.

In an era when more public life took place on the streets, passersby undoubtedly stopped to take a look, too. It likely was quiet, except, perhaps, for the noise associated with the construction of the new stone railroad depot nearby (the only Cedar Falls building Stanley would recognize today). When the rare spectacle was over, all may have wondered how such a show—one that easily might become a life-long memory—could have taken place so quickly. Had they held their breaths?

Fig. 2 The only extant artifact I have found of J. H. Stanley's jewelry store. {Sold at estate auction}

1

Introduction: "The Rush of this Black Wing of Night"

Prior to 2017, the last total eclipse of the Sun to pass over the United States took place in 1991. The state was Hawai'i. I flew to the Big Island to see it, thinking that if I was clouded out, the worst-case scenario had me stuck—in *Hawai'i*. No downside here. The result was an unexpected large dose of endorphins. I was hooked. Later I traveled to Antigua, Argentina, Chile, Mongolia, Romania, Turkey, (and Missouri!) just for the chance of seeing another total eclipse. Usually, I dodged clouds and was successful. I cannot put into words *why* I made these peregrinations. I suspect that the answer is different for every 'eclipse junkie.' If you saw, for instance, the 2017 solar eclipse, you know what I am talking about. 200 million people (80 percent of all American adults) did so, in person or *via* electronic media. The rest of you will not have to wait long: By great coincidence, another transcontinental total solar eclipse will occur in 2024.

In 2017, historian Steven Ruskin deemed the total solar eclipse of 29 July 1878 "America's First Great Eclipse."[1] In his 2017 book by that title[2], he does mention the total eclipse of 1869. But if 1878 was "Great," than 1869 was surely, at least by definition, "First." For the first time in the United States, scientists and tourists in large numbers made scrupulously organized pilgrimages to see this awesome, preternatural pageant in the sky.

---

[1] Like Ruskin, I use "American" as shorthand for "of the United States of America." I use "North America" for the continent. In reality, citizens of Honduras have as much claim to being Americans as those of Houston or Detroit.

[2] *America's First Great Eclipse: How Scientists, Tourists, and the Rocky Mountain Eclipse of 1878 Changed Astronomy*. Staffordshire (UK): Alpine Alchemy Press. This is a terrific book, by the way.

© The Author(s), under exclusive license to Springer Nature   1
Switzerland AG 2023
T. Hockey, *America's First Eclipse Chasers*, Springer Praxis Books,
https://doi.org/10.1007/978-3-031-24124-6_1

Indeed, it was the earliest total eclipse of the Sun where the zone of totality—see the next chapter–spanned the central North American continent[3] (of which there is textual record). The 1878 total-eclipse path left land at the Gulf of Mexico and never crossed the Appalachians. It missed the most populated portion of the USA altogether.

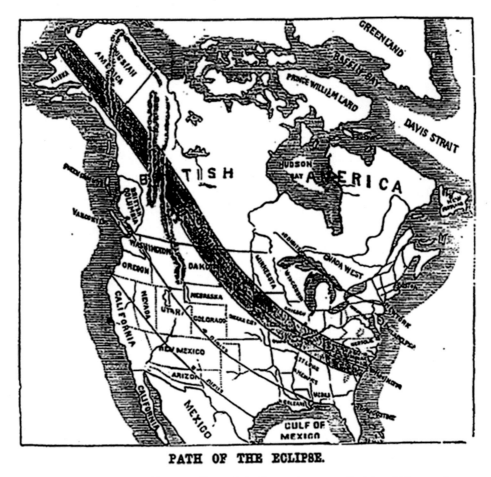

PATH OF THE ECLIPSE.

Fig. 1.1 Path of the 1869 total eclipse of the Sun as it was predicted and illustrated at the time. {Stella Cottam}

More significantly, the 1869 solar-eclipse milieu had all the hallmarks of a contemporary eclipse. It involved professional expeditions to the zone of totality, scientific results, public enthusiasm, and hucksters ready to exploit the event.

---

[3] between 31° and 49° North latitude

The total solar eclipse of 7 August 1869 is a vivid example of the manner in which events, at the time of great importance, fade in memory with the decades. They meld into a faint and obscure background, replaced by interest in more recent affairs...

Who remembers when *War and Peace* was published? (1869.) The fanfare and excitement about the total eclipse that year were enormous at the time. (Someone even wrote a song about it!) Yet ask Americans today what they know about *anything* that took place in the 1860s. I have tried it; the result is a short list.

There was the Civil War, the 13th Amendment to the Constitution, and Reconstruction, of course. Known of, but less often spoken about, is the murderous rampage of the Ku Klux Klan. The Presidencies of Abraham Lincoln and Ulysses Grant. Perhaps the impeachment of Andrew Johnson? The pony express? Stanley and Livingston? A 'dark horse' candidate: the first rodeo (literally)? But the percentage who can give any other answer falls into the single-digit category.

The 1869 American total eclipse of the Sun is, in many ways, itself 'eclipsed,' by attention given the similar, but better remembered, continent-crossing total eclipse of 29 July 1878. I hope to cure this historical solar-eclipse amnesia. (Coincidentally, 2019, the year that I started this project in earnest, was the 'forgotten' total-eclipse's 150th anniversary.)

Total eclipses of the Sun have passed over North America since plate tectonics formed the continent. Emigrants from Asia doubtlessly saw them. Unfortunately, we have no documentation of these sightings.

The fledgling United States experienced several total eclipses of the Sun, beginning while its Revolutionary War still waged. In 1780, Founding Father John Hancock wrote to the opposing commander on behalf of Reverend Samuel Williams (1743–1817; Harvard College), the first astronomer in the United States. Might the academic be given Right of Passage across battle lines, to observe the solar eclipse? It was to be visible from Maine, then occupied by the British.

This was a more genteel time; permission was given. It is too bad that Williams missed the central event: "totality." He experienced only a partial solar eclipse from where he camped on Penobscot Bay, as it turned out, just outside the path of totality.

Williams said that the current theory of lunar motion, based upon which he predicted the path of the Moon's shadow, was wrong or his map was off. Alternately, he just screwed up his math. It must have irked Williams that the total eclipse was observed successfully by Harvard graduate and Tory doctor John Clarke, from uncontested British territory.

The total eclipses of the Sun in 1803 and 1825 both crossed Florida. Notwithstanding, that peninsula did not belong to the United States at the time (1803), or the eclipse was not total over the territory (1825). The total eclipse of 1860 began near Washington Territory but then headed north.

The shadow of the Moon moved across the Louisiana Purchase all the way to Massachusetts; it was 6 June 1806. Meriwether Lewis and John Clark would have seen a partial solar eclipse as they returned home from their historic expedition to the West Coast. Of this total eclipse, the famous American author James Cooper (*The Deerslayer*, *The Last of the Mohicans*) wrote in his *circa* 1831 essay, *The Eclipse*: "I shall only say that I have passed a varied and eventful life, that it has been my fortune to see Earth, heavens, oceans, and man in most of these aspects; but never have I beheld any spectacle which so plainly manifested the majesty of the Creator, or so forcible taught the lesson of humanity to man, as a total eclipse of the Sun."

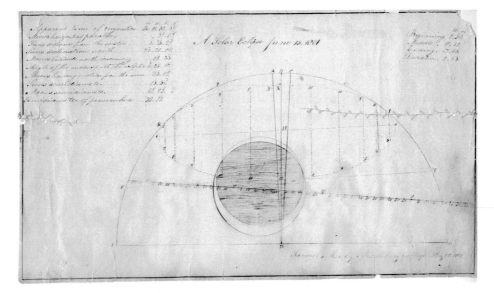

Fig. 1.2   Drawing from the 1806 total solar eclipse. {Middlebury College}

The total solar eclipse of 30 November 1834 began in what is now the Yukon (now Canada), entered the United States proper at Missouri, and departed land in South Carolina. However, those were the times of another United States: a parochial country with the majority of its population hugging the eastern seaboard. This solar eclipse was observed by anyone remotely an astronomer only at its terminus. (Total eclipses always move from west to east.)

The first total eclipse of the Sun over the United States as we know it, a now continent-spanning actor on the international stage, took place on 7 August 1869.

1869. Folk were singing *Little Brown Jug*. 'Medicinal' root beer was concocted. Uncle Sam appeared in art. H. J. Heinz bottled a delicious red sauce. A new game

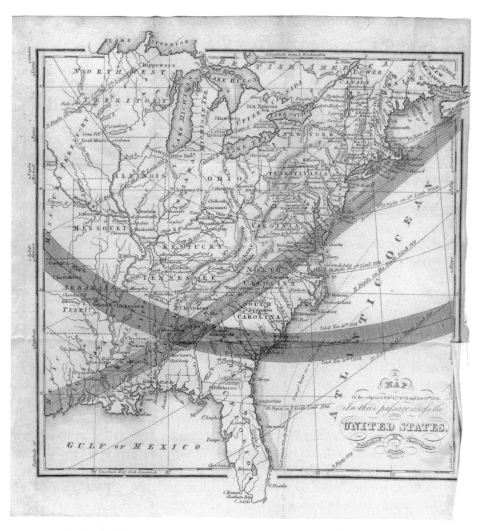

Fig. 1.3 Paths of the 1831 and 1834 total eclipses. {Michael Zeiler}

called football first was played between two college teams. (Rutgers beat Princeton[4] 6 to 4.) And there was a total eclipse of the Sun.

7 August was *the* 1869 total eclipse. The solar eclipse of 11 February 1869 was not total and passed over only ocean (excepting the tips of South Africa and Madagascar); it has no particular historical significance.

The most remarkable thing about the geometry of the 1869 total eclipse was the route its path took, and the width of that path (Chapter 3.)

---

[4] then College of New Jersey

8:45 PM Universal Time. The path of our total solar eclipse began 600 kilometers east of Lake Baikal, in Russia, not far from the Mongolian border. It dipped into China briefly before returning to Siberia and swept across the Sea of Okhotsk, the northern Kamchatka Peninsula, and Gulf of Anadyr. It passed over the Bering Sea and then into the newly American-purchased Department of Alaska, at Norton Sound. (Nome would not exist for another twenty years.) The Moon's shadow continued through western British America. It met the USA's contiguous border in the newly organized Montana Territory–and encountered the Dakota Territory– before covering a small part of the state of Minnesota (the sparsely inhabited southwest corner); an insignificant portion of the newest state, Nebraska; almost all of the state of Iowa; the state of Missouri (the northeast corner); more than half of the state of Illinois; less than half of the state of Indiana; most of the state of Kentucky; the state of Virginia (the mountainous, far western tip); the state of Tennessee (the far eastern tip); an Appalachian sliver belonging to the state of West Virginia; the state of South Carolina (just barely); and nearly all of the state of North Carolina. (Through the capriciousness of political borders, Ohio was ever-so-closely missed.) It ended 600 kilometers into the Atlantic Ocean, near Bermuda.

11:15 PM Universal Time. The passage of the Moon's shadow over the face of the Earth took two and a half hours.

Let us consider the census of 1870. The population of the United States at that time was likely even a bit greater than it was on 7 August 1869, plus not everybody responds to the census. Nevertheless, it is a reasonable place to start. Census counts are done by county. (The few American citizens and visitors who traveled to or from a county on the day of the total solar eclipse, especially during the era in which we are interested, can be dispensed with as negligible.) If we include the official population of every county through which the umbral shadow passed completely within, we get 5,163,576. However, nature does not respect county borders. There were people in adjacent counties who experienced totality as well. Our number, so far, is an undercount. We also must consider counties through which only portions did the umbral shadow did pass. If we include the populace of these counties, we arrive at 6,765,816. This time, we get an overcount. Let us take the number of Americans in total eclipse to be the mean of the two preceding numbers and make the difference between it and both the over- and undercount to be a conservative level of uncertainty. The result is 5,964,696± 801,120. Some folks were infants, some were too ill, some chose to sleep through the event, and others may have hidden in a root cellar until it was over. Still, a number not far from 6 million could have experienced the total eclipse. At the time, our final figure was 15 percent of the United States' population!

Flashing forward, prior to the 2017 eclipse of the Sun, there was debate over how many American states would experience totality. The result would be a

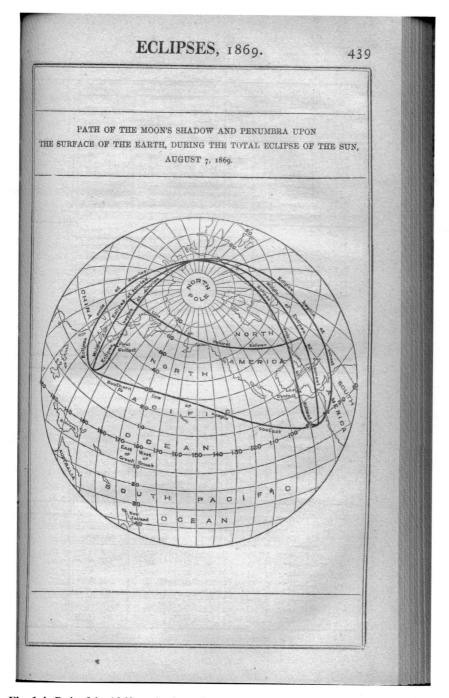

Fig. 1.4 Path of the 1869 total solar eclipse from a global perspective. {Michael Zeiler}

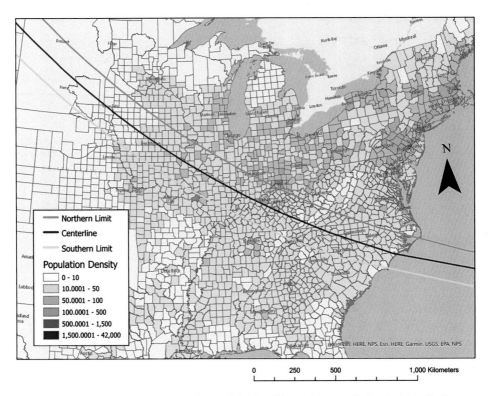

Fig. 1.5 Path of the 1869 total eclipse of the Sun illustrating population density. Dark = comparatively high population; light = comparatively low population. {GeoTREE Center, University of Northern Iowa Geography Department}

record. The number ranged from 12 to 14. Two states were problematic. The total-eclipse path only bit one corner of Montana, and that territory was rugged, requiring back-pack hiking up a mountain to reach. Less than 2 square kilometers of Iowa's southwest Fremont County experienced 33 seconds of totality. (At least the site was accessible by automobile.) Depending upon whether you counted Montana, or Montana and Iowa, varied the sum of states by 2. Iowa most often was ignored.

Yet this was to be that state's only shot for the foreseeable future: The total solar-eclipse paths of 8 April 2024, 23 August 2044, 12 August 2045, and 14 September 2099 will all intersect nearby states. However, they will trace a frustrating quadrilateral of eclipse exclusion surrounding Iowa.

Ironically, in 1869, the total-eclipse path made a nearly perfect diagonal across Iowa, and here it was to be at its widest compared to other states. This assured that proportionately more people in this state were under the shadow of the Moon than in any other. Iowa, too, was the prime target for most 1869 solar-eclipse wayfarers.

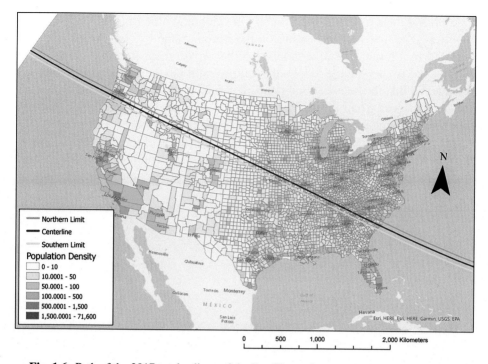

Fig. 1.6 Path of the 2017 total eclipse of the Sun illustrating population density. Dark = comparatively high population; light = comparatively low population. {GeoTREE Center, University of Northern Iowa Geography Department}

Not 'Just' History: 2024

Those who read this book early in its publication may have an opportunity to witness a total eclipse of the Sun for themselves. On 8 April 2024, the path of a total eclipse will begin in the Pacific Ocean. It will travel from Mexico, enter the United States at Texas, and cross the well-populated states of Oklahoma, Arkansas, Missouri, Illinois, Indiana, Ohio, Pennsylvania, New York, and those of northern New England. It will then visit Quebec, Canada, and a bit of the Maritime Provinces before heading out into the Atlantic after exiting Newfoundland. (*Everyone* in the continental US will experience a partial solar eclipse; but such a sight is much less spectacular than a total—See Chapter 3.) Those for whom the eclipse is total will experience as much as four minutes under the Moon's shadow. The amazing thing about 2024 is that those sitting under the darkness created by the Moon will include citizens of major cities such as Mazatlán, Durango, Austin, Waco, Dallas, Fort Worth, Hot Springs, Little Rock, Carbondale, Terre Haute, Bloomington, Indianapolis, Dayton, Lima, Toledo, Cleveland, Akron, Erie, Buffalo, Rochester, Syracuse, Montréal, Montpellier, and Presque Isle. It is all but guaranteed that the 2024 total solar eclipse will be experienced by more North Americans than that of 1869, more of us than during any other total eclipse, and perhaps more people than any single such eclipse ever.

Several years ago, I joined the country in celebrating the fiftieth anniversary of Apollo 11: the first human landing on the Moon, made by Neil Armstrong and Buzz Aldrin. I found myself wondering. What would the astronomers and others who flocked to the August 1869 solar-eclipse path have thought had, somehow, they learned that the celestial body casting the lunar shadow would be visited by men a century and a half forward in time? Total-eclipse expeditions, the subject of this book, were the nineteenth century's missions to the Moon.

There were 525,600 minutes in 1869. This is the story of three of them. Let me take you back four generations: to when real people; with real lives, problems, and aspirations for the future; dropped everything to see the rare, the beautiful, and the frightening—a total eclipse of the Sun.

2

A Trans-American Eclipse

"Arrive[d] at Des Moines at 10 o'clock." (diary entry)

Simon Newcomb stepped off the train onto the platform adjoining the depot and presumably looked about his surroundings. It had been a three-day trip from Washington, DC. It was now a week before the total eclipse of the Sun.

Newcomb was an imposing man, about 77 kilograms, though of only 1.7 meters in stature. He had a "fair, ruddy skin, dark brown hair and beard, and full eyes far apart and of such depth of blue that in the shadow they seemed black."

The traveler and his party likely wore three-piece attire. Such costume was expected when one journeyed by the state-of-the-art (and most luxurious) form of in-land transportation available: long-distance railroad—first class. (While he could remove his hat, the stock around his neck must have been uncomfortable, riding in the enclosed car during the West's hot and humid season.)

Newcomb knew he would be standing now, at this station, since the previous May. All this time, travel, and 'boots on the ground' was devoted to a singular event in the sky.

One can imagine that Newcomb tipped a porter to load a large steamer trunk onto a wagon for transport to the Savery Hotel. Clothing was bulky and laundry facilities uncertain. His personal items also might have included a shaving kit with which to touch up his beard, a folded or medium-sized, fixed-blade knife, a sewing kit, and a candle with matches. Like other travelers of the time, it is possible that he discreetly carried a small revolver.

An astronomer, Newcomb no doubt instinctively looked upward at the sky. Was it cloudy or clear?

Newcomb was destined to become the best-known scientist in the United States—a household name in educated families. He has been described as

© The Author(s), under exclusive license to Springer Nature Switzerland AG 2023
T. Hockey, *America's First Eclipse Chasers*, Springer Praxis Books, https://doi.org/10.1007/978-3-031-24124-6_2

Fig. 2.1 Nineteenth-century railway station in Des Moines, Iowa. {Zachary De Vries}

everything from Walt Whitman's poetic prototype, the introspective *Learn'd Astronomer*, to the model for the fictional Sherlock Holmes' nemesis, Professor Moriarty. Most likely something in between, he was comfortable hobnobbing with the political and scientific intelligentsia of Insiders' Washington, though less patient with those of a lower social stratum from which he had himself once emerged.

But that Newcomb lived in the future. At age 34, his notoriety outside the world of academia was limited to possessing a physically large head. On this day the younger Newcomb had reason to be uncharacteristically worried.

While he dwelt for most of his life in the cosmopolitan-by-comparison city of Washington, Newcomb was a recently naturalized US citizen who had grown up in rural Canada. Thus, even though he never previously visited this new 'city' in a new state, Des Moines, Iowa, it should have held no special trepidation for him in itself. Nevertheless, his seemingly parochial past did cause him to have a life-long need to prove himself.

Fig. 2.2 Simon Newcomb at about the right age for the 1869 total eclipse of the Sun. {United States Naval Observatory}

Newcomb loved to travel, but this was no holiday. He was the appointed head of an expedition to observe a total eclipse of the Sun. Professional astronomers (mostly in Europe) had junketed to see total eclipses since 1851. To out-of-the-way locales? Only since 1860.

Moreover, this trip was different. Newcomb undertook his adventure at the bequest of his employer, the United States Naval Observatory [USNO]. Though a civilian, he held the titular rank of lieutenant, commissioned by Abraham Lincoln himself.

Newcomb may have earned this particular job because he was a veteran of the 1860 total eclipse of the Sun. It could have been due to his reputation as a multi-tasker and 'workaholic.' He might have successfully lobbied for it; he was good at that sort of thing.

Established forty years before as a Bureau of Charts and Instruments, the USNO originally was charged with setting maritime clocks accurately by timing the apparent movement of the stars. The Observatory grew into the paramount source of time for the nation. (For instance, its authoritative time ball was dropped from the building daily at noon.) The goal of all its activities was officially the same, to aid navigation.

What was the United States Navy [USN], in the person of its civilian employees led by Newcomb, doing in the land-locked West?

Fig. 2.3 United States Naval Observatory *circa* 1869. {DC Preservation League}

Before radio, the singular reference a ship beyond view of shore had, in order to establish its location, was a featureless seascape and the lights in the sky. Imagine a shell of the heavens, called the Celestial Sphere, surrounding the Earth and only interrupted in our view by the horizon. Every night—or in the case of the Sun, day–a ballet performs there. This dance is staged by celestial components (the Sun, Moon, and planets) perambulating against a background of stars, which seemingly rotate together like a barrel around us.

Fortunately for the navigator, all these objects move in predictable fashions, on repeating paths. These motions are relatively simple compared to those found in most of nature, if only one knew their itinerary. Much of this apparent motion is due to the movement of the Earth, but no matter. It looks like the *sky* advances, and *we* sit beneath. Ignorance of the azure could result in ships sunk by the enemy, wrecked upon the shore, or, most grimly, lost to starvation and the elements.

While latitude is comparably easy to fix, longitude is a challenge. This is because the Celestial Sphere always turns, causing objects projected upon it to rise and set, all due to the real diurnal motion of the Earth. Knowledge of both the position of celestial objects on the Celestial Sphere and *time* were necessary to establish longitude. Indeed, a census of ships foundering on the rocks might show that more tragedies occurred because a vessel's helmsman sailed too far east or west, rather than too far north or south.

Thus, occurrences in the sky were more than useful. An example: Eclipses were standard observing fare for nautical astronomers, who wished to keep track of and time the exact location of the Moon for their ocean-going colleagues. An eclipse places the Moon precisely with respect to the location of the Sun. The instant at which the event occurs according to a ship's chronometer, compared to that listed in an ephemeris for locations on land, yields a difference in longitude—how far east or west the ship is from a reference meridian (perhaps that of a port).

The time of eclipse onset (sometimes measured to a fraction of a second on shore, though not on the pitching deck of a vessel) and the exact location from which the event is observed (also precisely measurable on land with the right, stabilized instruments) in turn refine the ability to predict future eclipses, and ultimately other sky events involving the Sun and the Moon. This is how an eclipse in the Midcontinent could ultimately benefit a ship's master in the North Atlantic, who it is said guarded an accurate ephemeris and clock even above the shipboard stock of rum.

Christopher Columbus

The ephemeris occasionally was consulted for more than navigation: It benefited the art of intimidation. In 1504, Cristóbal Colón was marooned in Jamaica, awaiting resupply. He told the islanders that if they did not provide him and his crew with food, he would take away the Moon. Columbus was able to demonstrate the power of his threat since he knew that a *lunar* eclipse was to occur on 29 February. He timed his pronouncement accordingly. He let the natives squirm until his book said the eclipse would end. The ruse worked. Regrettably, this was one of Columbus's more *benign* interactions with the original residents of the New World.

Does this trick work twice? See Chapter 18.

Eclipse measurements continued to be made over and over again. If the Sun or Moon seemed to slip from their appointed routes, their predicted tracks would be updated (using eclipses or other methods).

However, by 1869, a competing method for establishing, at least, terrestrial longitude had come into existence. Telegraph signals—often called the nineteenth-century internet—were used to establish geographical coordinates. Once clocks were set by observation of celestial denizens, the sky was removed from the calculation. Comparing two clocks by telegraph message yielded longitude: regardless of whether it was day or night, clear or cloudy. This could be done whether the operators knew anything about astronomy at all.

By that of the 1869 total eclipse of the Sun, it was possible to see a time when the navigational role of naval astronomers would wane. Indeed, finding the exact *locations* of eclipse-observing sites had defaulted to the United States Coast Survey [USCS]. Newcomb and others at the USNO saw the sea change.

On the other hand, there was another vision for the United States Naval Observatory, going back all the way to that of President John Quincy Adams. It was the function of *de facto* National Observatory, potentially in every way as competitive scientifically as the great state observatories of Europe. And in the 1860s, that began to mean not just astrometry but also physical astronomy.

In the 1860s, the USNO staff saw for themselves also this latter new, additional role, concerning itself with the *constitution* of the Sun and Moon—luminaries, heretofore regarded as mere bright disks moving against the firmament and acting like the hands of a timepiece. Physical astronomers provided data with which physicists learned about the 'clock hands' themselves.

The whole idea of USNO *cum* National Observatory was politically controversial. However, as long as the Navy's astronomers continued to provide their deliverables, the government was willing to look the other way as the USNO swayed from its mission statement: Its staff had worked tireless to keep the vital chronometers calibrated aboard warships fighting the Civil War. A little payback was in order. Apparently, it was thought that there was no harm in allowing the USNO astronomers a bit of 'mission creep.' The result was that the immediate *post bellum* became what historian Steven Dick calls the USNO's "golden era."

Reverend Samuel Haughton's 124-page *Manual of Astronomy*[1] from the 1850s spends only two paragraphs on the physical constitution of the Sun. And this passage includes such eclectic information as, "Each point of his surface is estimated to give out 146 times the light of a Drummond lime-light, and each square yard of his surface produces a heat equal to that of the complete combustion of six tons of Swansea coals per hour."

Anything more became possible only later in that decade with developments outside of astronomy: for instance, the perfection of photography by chemists and spectroscopy by physicists and opticians. With these acts what eventually was called *astrophysics* was born, distinct from classical astronomy. More about how these new tools worked later.

It seems like a big jump for practitioners of the old astronomy. That is not necessarily so. Is mensuration of a precise wavelength of light that much different from that of a precise angle in the sky? Professional astronomers of the day were measurers. Each of the quantified outputs from devices that registered intrinsic properties of astronomical objects (the afore-referenced spectroscope, as well as the polarimeter, and photometer or actinometer) consisted of just another measurement. Exactitude was the benchmark just as it was for more-traditional reading of angular orientation of a telescope. Much of the skill set required was the same.

---

[1] London: Cassell, Petter, and Galpin

The USNO did not presume to interpret these measurements; it is unclear that they were interested in doing so. Obtaining the numerical quantity itself was *raison d'etre*.

With this back story, what was it that now concerned—indeed, compelled—Newcomb, on this fine summer's day at the limits of the United States? The familiar apparatuses of positional astronomy were fairly uncomplicated and had not substantially changed for hundreds of years. They were devices with which to gauge small angles and, later, chronometers (clocks capable of indicating precise time). On the other hand, the truck associated with physical astronomy was specialized, complex, and unwieldy. It all had to be imported on Newcomb's train. It required unloading, reassembly, and testing. It must work, the first (and only) time. For this total eclipse of the Sun, Newcomb had traversed 1,400 kilometers to study what would repeat but few times in his lifespan. He must be standing on the right spot, at the right minute—and (against his predilection) praying for good weather. Everything had to go right: Capital, both political and monetary, already had been spent. There were no 'redos.'

Fig. 2.4 Contemporary railroad train. 1969 or 1870. {Ebay print}

Newcomb's normal science involved, for instance, waiting for stars to pass the Central Meridian (an imaginary, half circle dividing the sky east-west) in the cross hairs of a task-specific telescope called a transit instrument. Referencing a chronometer established that star's right ascension, or celestial longitude. A similar device, called a meridian circle, measured the declination (celestial latitude) of the star. Together, these two coordinates pinpointed a locus on the Celestial Sphere.

This activity was undertaken from the comfort of a familiar observatory not far from home. Later requirements for astronomers seeking out the Universe's most distant and faint objects, those of remote dark skies and difficult-to-climb elevations above the thickest of the Earth's obscuring atmosphere, did not yet apply.

While the transit telescope of which Newcomb was in charge back in Washington happened to be the largest such instrument in the United States, using it was a comparatively leisurely affair. Missed a star? Wait for the next. Or repeat the attempted measurement the next evening, or the one after that. Had enough? Declare the remaining night 'cloudy,' and retire with one's observing assistant to a nearby restaurant for oysters.

What about an eclipse? A total eclipse of the Sun lay outside the boundaries of the familiar. This easily could have agitated Newcomb's mind. For a man who was used to setting his own agenda, so much about a solar eclipse was out of human hands. For a politically figure like Newcomb, it is understandable how human nature easily could produce an unusual feeling of nervousness—the pressure to succeed.

* * * * *

The total solar eclipse of 1869 entered the United States proper at the location of the unincorporated mill-town of Beloit, Iowa (latitude +43° 17′, longitude −96° 34′[2]). It left at the fishing village of Snead's Ferry, North Carolina (latitude +34° 33′, longitude −77° 23′).

The last total eclipse of the Sun to pass over (what would become) an extensive portion of the United States occurred nearly thirty years before: 5 September 1840. However, at the time, this territory belonged to Mexico.

Even if the term "eclipse expedition" had meaning then, a United States effort to place astronomers on the path would have been nearly impossible. Novelist Max Byrd envisions how it might—just might—have been done: first by train and then,

> The expedition would travel down the Ohio River and up the Mississippi to Saint Louis, over the famous Santa Fe Trail southwest toward the Rio Grande and Mexico, and then due west into the vast, fabulous, and unexplored deserts beyond the borders of the Independent Republic of Texas.

---

[2] North latitude is given as a positive number, south as a negative number. East longitude is given as a positive number, west longitudes as a negative.

There were a dozen reasons why nobody thought they would succeed.

In the first place, the southwestern deserts of America, notoriously and literally *terra incognita*, lay a good five hundred miles from the last outposts of civilization—no more than a handful of white men had ever ventured into them, Indian tribes were violently hostile, and such maps that existed at all were works of imagination and fiction rather than science. For all anybody knew, latitude 37, longitude 103, was at the bottom of an immense inaccessible canyon or the top of an unscalable mountain peak.[3]

In the above quoted fiction, from *Shooting the Sun* (2004), the protagonists set about doing just this: They travel most of the way by horse-drawn wagon over non-existent roads. In reality, there is no record of astronomers observing the 1840 total eclipse. It does not even matter that, in the novel, the path of the eclipse is shifted slightly, for dramatic effect.

In contrast, the state of Iowa was in many ways optimally placed for watching a total solar eclipse in the summer of 1869. Until one reached the highest longitude regions of North America, the further west one was located on the centerline (Chapter 3), the longer the total eclipse. Iowa was the westernmost state to experience totality. (At greater longitudes lay sparsely populated United States' Territories, where battles between the Army and the people they inaccurately called the Sioux still flared.) The 1869 total-eclipse path, 260-kilometers wide, cut Iowa in half as the umbra traveled from the state's northwest corner to its southeast corner. The spectacle happened midafternoon; the Sun was near the zenith. The weather was supposed to be good, and this expectation was largely met. (This was *not* the case for European total-eclipse expeditions staged in the year immediately prior to and the year immediately after 1869's American eclipse.) Iowa was to be host to the first large-scale, solar-eclipse expeditions in the USA.

Why? The first North American eclipse expedition of any kind took place only a decade earlier. It had required a minimum of equipment.

Between this one and the next, astronomy underwent its transformation. Starting with the eclipse of 1868, the equipment crated to observe the event increased dramatically. By 1869, solar-eclipse expeditions carried many delicate, weighty, and costly astronomical instruments. Table 2.1 is one manifest[4].

New technology was popping up everywhere in 1869. The motion-picture projector, typewriter, vacuum cleaner, and—lately 'infamous'—electric voting machine (of Thomas Edison) were all patented in 1869. Plastic was invented. Construction of New York City's Brooklyn Bridge began. August also saw the first fatal automobile accident.

---

[3] Note that Byrd is writing in the vernacular of the day.

[4] Recuring astronomical terms appearing in this table are defined in subsequent chapters.

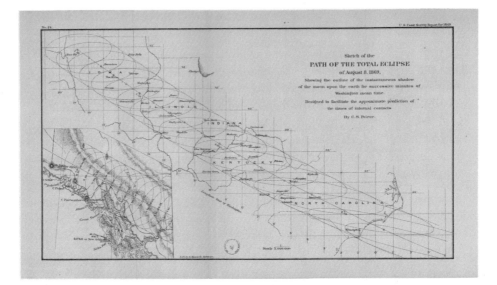

Fig. 2.5 Contemporary United States Coast Survey map of the 1869 total-eclipse path through the Midwest and Eastern Seaboard. {University of Navarra}

Table 2.1 Example List of Equipment Brought to Observe the 1869 Total Eclipse of the Sun

polariscopes (2X)
1-prism spectroscope
3-½-inch-aperture telescope
3-inch-aperture achromatic telescope
4-inch-aperture comet seeker
50-foot metallic tape measure
6-inch-aperture refraction-circle objective
7-½-inch-aperture achromatic telescope
8-½-inch-aperture transit-circle objective
actinometer
aneroid barometer
binoculars
black-glass artificial horizon
blank forms for latitude & time
blanks for spectroscopic observations
books (9X)
break-circuit telegraph key
camera
chronometers (3X)
colored glasses (3X)
mercurial artificial horizon
photographic supplies
photometer
pocket achromatic telescope
pocket compass
prismatic compass
rain gauge
reflecting level
scale of tints with which to compare colors
set of drawing instruments
sextants (3X)
thermometers (3X)

This was true in science, too. In the view of famed British astronomer Norman Lockyer (1836–1920; eventually Royal College of Science[5]), "Certainly, never before was an eclipsed sun so thoroughly tortured with all the instruments of Science."

A similar list to Table 2.1, for the total eclipse of the Sun just nine years earlier ("a telescope, a sextant, a chronometer, a spy glass, and a polariscope") would include little that one could not carry on one's back! No more.

Hauling the utensils for the 1869 total solar eclipse *depended* upon the use of the railroad. *The optimum eclipse-monitoring station was the intersection of totality's path with the westernmost depot of the current railroad network.* This favored the bigger cities in the state of Iowa.

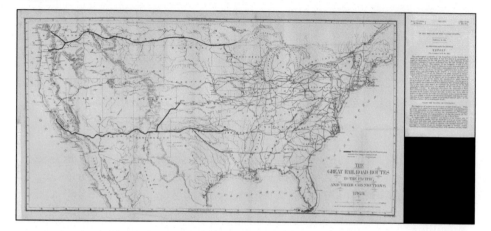

Fig. 2.6 United States railroad network. 1869. {American Photolithographic Company}

Heavy baggage was unavoidable. Transport by rail was the safest way to get astronomers and gear to the observing site. Indeed, the Des Moines eclipse expedition put together their observing base not far from the local train station, minimizing the handling of luggage.

A newspaper elucidated this new fact of eclipse life to its readership in a nearby community:

At no point were the people more highly favored than in Iowa, and at no point in Iowa more than Oskaloosa. The party who were here for the purpose of taking observations informed us, that the city of Oskaloosa is located within two miles of the center of the shadow; and, consequently was the most favorable point upon the continent from which to have viewed the whole matter. There was only one reason why this was not made the head

---

[5] now part of Imperial College

Fig. 2.7 Des Moines railroad depot today. 5th and Court Street. {Ron Reiring}

quarters [*sic*] of the entire company who came west to take observations, and
that was, the want of a [suitable] railroad to the city. Some of the instruments
used at Des Moines, Ottumwa and other points, were very large and as a
consequence, very expensive; so that those connected with them, were
unwilling to risk bringing them over the road from the depot to the city.

Perhaps this conversation was designed to enthuse locals about an invasion of
astronomers. Regardless, it was also true.

What were all these appurtenances for? A telescope magnifies. This was especially useful to the nineteenth-century photographer. (Think of a photograph as a
two-dimensional measurement.) A larger image meant that more chemical crys-
tals on the photographic plate would be activated, resulting in a sharper picture.

A refracting telescope (refractor) at its simplest is built out of two lenses: Light
from a distant object first encounters the objective; behind it is the eye piece
through which the observer looks. The larger the objective (light-gathering) lens
of the telescope, the higher resolution it provides.

An objective convex lens brings light from a distant source to a focal plane,
where an image can be examined by the eye piece (or captured by a camera). It is
much as one might use a hand magnifier to examine rows of print.

The distance from a lens at which focus occurs is called the focal length of that
lens. Magnification is the ratio of the two focal lengths, that of the objective and
that of the eye piece. Both photography and the visual observing of details on the
Sun favored long telescopes.

The larger the objective's diameter, the greater its light-gathering ability, too. Making images brighter is useful for observing faint astronomical objects in the nighttime sky. However, in the case of the Sun, it can be a hindrance.

The filters used in the 1860s failed to eliminate the Sun's telescopically concentrated infrared radiation. It is a wonder that more astronomers were not blinded by using a telescope to observe the partial phases of a solar eclipse; see Chapter 3. (Casual naked-eye spectators mostly were spared this risk—they likely could not find or afford a telescope.) Nonetheless, brighter images meant shorter photographic exposures.

Fig. 2.8 Few telescopes used to observe the 1869 total eclipse are still to be found in their original homes. Here is an exception: the objective of John Eastman's (Chapter 13) portable refractor, in storage at the United States Naval Observatory. It could be used today. {Photograph by the author}

(A particular style of telescope, called a comet seeker, is utilized for exactly what its name implies. A comet seeker has a large-aperture objective but also a large field of view. It proved useful for observing the extended corona of the Sun.)

The refractors in the hands of professionals almost always were multiple-lens achromats. An achromat lessens chromatic aberration, the effect that occurs when different wavelengths come to different foci. This phenomenon results in objectional colored fringes around the object in view. In an achromat, a compound objective is constructed of different transparent materials, each of which brings a particular wavelength to focus at a constant focal length.

A telescope is made easier to use by a smaller 'scope, called a finder or sighter, mounted on the larger instrument's side and pointed in a parallel direction. Normally such an ancillary piece of optics has a smaller aperture and provides less magnification but exhibits a wider field of view than the principal. This auxiliary makes it simpler to find an object in the sky. Once an object is spotted in the sighter, presumably it is within the much narrower field of view in the main telescope, too.

Other helpful support features of a telescope include an equatorial mount, in which one of the two axes about which the telescope is allowed to swivel is pointed toward the Celestial Pole (the projection of the Earth's axis of rotation onto the Celestial Sphere). With such a cradle, the telescope need be swung only about a single axis to track objects as they appear to move due to the rotation of the Earth. Even better is an attached clock drive, which automatically turns the telescope at the speed of the Earth's rotation, significantly reducing the number of times the operator must cease observing and recenter the instrument.

While the telescope had been known for centuries, the spectroscope is a product of the 1850s. It is attached to a telescope and uses a prism[6] to split (in the case of an eclipse) the Sun's light into its component wavelengths. (Different wavelengths are refracted through different angles.) The result of this dispersion ordinarily is a 'rainbow' of light.

The rainbow analogy is literal; we perceive different wavelengths of light as different colors. Multiple prisms spread the light out further and can provide better resolution of information embedded in the spectrum. For instance, in the mid-1800s, the spectroscope yielded the chemical recipe for the Sun (or, at least, the bright photosphere of the Sun; see Chapter 3).

Less well known are other instruments in the astronomer's kit. A polariscope is used to measure the polarization of light—in the case of an eclipse, the degree to which the Sun's light waves are oriented in a given plane of vibration. In the middle of the nineteenth century, at the new polariscope's heart was a transparent material (often Icelandic spar) that admitted only one polarization. Polarization indicates the presence in the light's path of particulate matter.

A photometer compares the brightness of an unknown object to that of a known light source, thereby quantifying it. An actinometer measures the temperature of radiated heat—a sort of pointable thermometer.

In addition, there were instruments (with similar purposes of extracting information from light) specifically engineered for the eclipse. Long gone were the days of Galileo, when the astronomer could place his instrumentation in his coat pocket!

*   *   *   *   *

The state of Iowa is blessed with many rivers. Withal, except for the mighty Mississippi and Missouri (which establish its east and west borders), these rivers are not navigable very far from their confluences with the above. This is true even for the flat boats and shallow-draft steamboats designed for transporting

---

[6] a diffraction grating produces the same effect.

agricultural goods. Moreover, Iowa soil does not lend itself easily to road building; Pre-Civil War attempts to link Iowa centers of commerce by grading paths
were abandoned afterward.

Already in the 1850s, Iowa municipalities were petitioning the railway companies for service. A steam locomotive reached the state as early as 1855 (Davenport
to Iowa City by 1857). However, during the War, rail-building westward was put
on hiatus. It was revived when President Lincoln decided that Council Bluffs,
Iowa, should be the starting point for completion of the Transcontinental Railroad.

With tracks eventually crossing the state, Iowans now could travel year-round.
More importantly, it often was cheaper to freight Iowa commodities to Chicago,
Illinois, by rail than it was from the interior to the waters of the Mississippi River
for transshipment.

Track made possible a change in the Iowa economy. Thereafter, it would be no
longer solely agricultural, but would include the grain-processing and meat-
packing industries.

Commodities flowed east, and cash flowed west. When the train finally arrived
at Adair, in western Iowa, Jesse James immediately robbed it.

The following table refers to railroad towns mentioned in this book and at
which organized expeditions studied the 1869 total eclipse of the Sun. In most
cases, these locations only recently had received train service. *It is interesting to
think that our knowledge about the Sun would have been delayed, perhaps by
years, if it were not for coincidental railway expansion in the American West.*

Table 2.2 Years in Which the Railroad Reached Select Cities

Springfield, Illinois	1852
Kewanee, Illinois	1854
Mattoon, Illinois	1855
Davenport, Iowa[7]	1855
Iowa City, Iowa	1856
Mount Pleasant, Iowa	1856
Monroe, Missouri	1859
Ottumwa, Iowa	1859
Cedar Falls, Iowa	1861
Oskaloosa, Iowa	1864
Boone, Iowa	1865
Des Moines, Iowa	1866
Jefferson, Iowa	1866
Burlington, Iowa[8]	1868
Le Mars, Iowa	1868
Cherokee, Iowa	1870
Sioux Falls, South Dakota	1872

[7]See note below

[8]This was the first iron railroad bridge across the Mississippi River; the wooden bridge at
Davenport was destroyed by a colliding steamboat shortly after it was constructed and had to
be rebuilt

I end this chapter as I began it. In my vision, Newcomb is standing at the railway station with "a look full of strength—steady, direct, penetrating." But his feelings are mixed. Simply put, he soon was to be a fish trying life out of water.

Yet recall that this was not to be Newcomb's first total eclipse of the Sun. As a young man in 1860, he accompanied a group of three slogging its way through the wilds of Manitoba, Canada, by birchbark canoe. It was a three-month voyage, only to be foiled during the time of totality by a view of—not the total eclipse—but the underside of clouds.

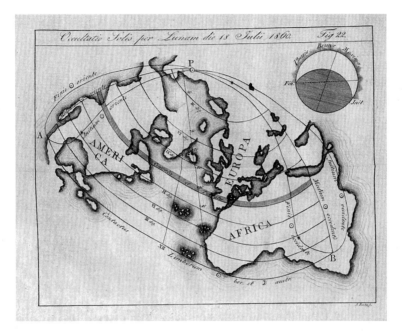

Fig. 2.9 Path of the 1860 total solar eclipse. {Michael Zeiler}

Now, Newcomb was in charge. Regardless, despite being an organizer nonparallel, Newcomb himself would do little to 'illuminate' the Sun. His autobiography mentions surprisingly little about the summer of 1869. It would be his colleagues and others working alongside of him who would make this contribution.

Each of the government-sponsored solar-eclipse expeditions would spend incredible time, effort, and resources—ahead of time, on site—in order to facilitate the classical total-eclipse practice of timing the event. However, during the eclipse itself, the actual mechanics of clock work was mostly left to junior members of the expeditions.

During totality (Chapter 3), Newcomb, as well as the majority of senior astronomers in attendance, concentrated on the time-honored praxis for observing the

eclipse: mentally recording its morphology, color, and brightness; and then trans-
ferring that impression to the written word or artwork (in one of several media
available at the time). In other words, they intended to 'do' astronomy the way
Galileo had. *Unlike* the master, they would have less than three minutes available
to them.

What follows is a brief tutorial on the geometry of total solar eclipses and
eclipse phenomena. Many such summaries are readily available. Mine differs in
that it is not a modern one but rather one that was meant to be understood in the
century before last, our period of historical interest.

3

"Some Light upon This Dark Subject"

Nothing there is beyond hope,
Nothing that can be sworn impossible,
Nothing wonderful,
Since Zeus,
father of the Olympians,
Made night from mid-day,
Hiding the light of the shining sun,
And sore fear came upon men
(classical Greek poem)

On 6 August 1869, with a total eclipse of the Sun to occur in the local sky the next day, a local newspaper printed, "All of our readers, of course, know that an eclipse is an obscuration of one of the heavenly bodies by the interposition of another . . ." Did they? This is politic, but as unlikely to have been true then as it is today.

To say what a solar eclipse is, I must first say what it is not. Do not confuse phases of the Moon with eclipses. Eclipses are rare compared to phases, which occur every month.

Often there are 'facts' that get stuck in our head, perhaps from as early as elementary school, that are not quite right. Yet pseudo-facts can lodge in the brain for so long that it is difficult to extract them. A common myth is that the phases of the Moon have something to do with shadows. They do not. That is eclipses.

The Moon orbits the Earth in a period of time called the month. With patience, we can see the Moon move by its own diameter with respect to background stars

T. Hockey, *America's First Eclipse Chasers*, Springer Praxis Books,
https://doi.org/10.1007/978-3-031-24124-6_3

in about an hour. Its rising and setting due to the spinning Earth is delayed night-to-night as the Moon makes its way eastward, opposite the direction of our planet's rotation.

Technically, it is the *sidereal* month to which I refer; the adjective distinguishes it from the *synodic* month. The sidereal month is straightforward: The Moon appears at one location on the Celestial Sphere, say, in front of a particular constellation of stars. When it returns, one sidereal month has passed.

In the case of the synodic month, the interval is measured not with respect to the stars, but with respect to the Sun. The Sun appears to change its place by about one twelfth of the Celestial Sphere's circumference each month, as the Earth and Moon revolve together about it during the course of the year. The Moon and Sun look as if they move in the same direction on our sky, so the Moon must catch up with the Sun every month. When the angular relationship between the Moon and Sun repeats, one synodic month has passed. This is why the synodic month is slightly longer than the sidereal month, 29.5 days as opposed to 27.3 days. Practically speaking, of course, our calendar cannot have a month with half a day in it. So, we round the month to 30 days, which sounds more like the 'hanging-on-the-wall' month with which we are familiar.

The phases of the Moon cycle during the synodic month. They are more obvious than the Moon's changing location amongst the stars. For this reason, we use the synodic month and not the sidereal month to establish the western calendar. Moreover, the sidereal month has little to do with eclipses. Therefore, when I refer in the future to "the month," without adjective, I will mean the synodic month.

The older Simon Newcomb was both the most famous and biggest popularizer of physical science in the English-speaking world: Stephen Hawking and Bill Nye. Let me allow him to explain, in order to maintain a feel of the nineteenth century. After all, in Newcomb's words, "Most persons will desire to know something . . . of eclipses . . ."

Every one [*sic*] knows that the Moon makes a revolution in the celestial sphere in about a month, and that during its revolution it presents a number of different phases, known as "New Moon," "Full Moon," and so on, depending on its position relative to the Sun. A study of these phases during a single revolution will make it clear that the Moon is a globular dark body, illuminated by the light of the Sun, a fact which has been evident to careful observers from the remotest antiquity . . .

In my experience, the phrase "Everyone knows" usually is followed by something most people do not know. In words other than Newcomb's, fifty percent of the Moon is illuminated by the Sun at all times, just like any sphere. How much of the lit half of the Moon is visible from our vantage point here on the Earth determines the phase that we witness. Of importance to us, in considering solar eclipses, is the New Moon.

Twice each month, the three bodies, Earth, Sun, and Moon, are lined up in one of two ways: Earth-Moon-Sun, which we call New Moon, and Moon-Earth-Sun, which we see as Full Moon. The former is required for a total eclipse of the Sun to occur. (There is such a thing as a total eclipse of the Moon/lunar eclipse, when the Full Moon enters the Earth's shadow—it is the kind of eclipse that Columbus exploited—but the phenomenon lies outside our scope of interest in this book.)

Why does a total eclipse of the Sun not occur every month? New Moon is not all that is required for such an eclipse.

The Moon's orbit about the Earth inclines to the plane of the Earth's orbit about the Sun; that latter plane is named the ecliptic. ("Ecliptic" ... "eclipse": Get it?) We know that the Sun does not really travel around the Earth—the Earth (and Moon along with it) revolves around the Sun—but it looks the other way around from our point of view on the seemingly motionless Earth.

The places at which the inclined celestial paths of our two celestial orbs cross are called nodes. ("Node" comes from the Latin word for "knot," as in tied together for weaving.) There is nothing at a node most of the time. Nodes are just points in space favorable for eclipses.

New Moon does not have to be a perfect line-up. In fact, most of the time, the Moon is a little above or below the line running from the center of the Sun through the center of the Earth.

What happens when these bodies *are* lined up perfectly, though? For instance: Earth-Moon-Sun (a perfect New Moon)? Imagine shooting an arrow and striking all three spheres through the middle. Then, and only then, does the Moon physically *block* light emitted by the Sun from reaching the Earth. It is a total solar eclipse.

Newcomb picks up the narration, using quant gendered pronouns that today sound a bit odd to the ear, but did not when he wrote them. Notice that astronomers are heroes of his story.

The early inhabitants of the world were, no doubt, terrified by the occasional recurrence of eclipses many ages before there were astronomers to explain their causes. But the motions of the Sun and Moon could not be observed very long without the causes being seen. It was evident that if the Moon should ever chance to pass between the Earth and the Sun, she must cut off some or all of his light. If the two bodies followed the same track in the heavens, there would be an eclipse of the Sun every New Moon; but, owing to the inclination of the two orbits, the Moon will generally pass above or below the Sun, and there will be no eclipse. If, however, the Sun happens to be in the neighborhood of the Moon's node when the Moon passes, then there will be an eclipse. . . As he crosses both nodes in the course of the year, there must be at least two solar eclipses every year on some points of the Earth's surface.

The Moon occludes the Sun. So what? Objects pass in front of other things in the sky all the time. The heavens are a busy place. The Moon covers up a star, but nobody notices one star's worth of light missing. A planet, such as Venus, transits the solar disk. But it is a barely detectable dark spot; a filtered telescope is necessary to see it. These are unspectacular events because the bodies involved are of such disparate apparent (angular) sizes.

However, by a wonderful coincidence, the Sun and the Moon appear to be the same size in our sky. The Moon is about the extent of an aspirin tablet held at arm's length: half a degree. We tend to think that the Sun is bigger (because it is so much brighter?). Yet, in reality, you can blot it out with that same pill.

This concomitance happens nowhere else in the Solar System. The Moon is 3,476 kilometers in diameter. If it were only 700 kilometers smaller, we could not see a total eclipse of the Sun: The Moon never would appear large enough to cover the whole solar disk at one time. Other planets have satellites (moons), but none are the right diameter or distance from those worlds to see on them a total-eclipse footprint of any significant angular size at all. It is convenient that such a solar eclipse happens on the one planet where there are people looking up to see it.

For a total eclipse of the Sun to occur: It has to be New Moon, *and* at the same time the Moon must happen to occupy one of the nodes of its orbit. As the time it takes the Moon to go from node to node to node is different and unrelated to the time it takes to go from a phase to that same phase again, for these two events to happen simultaneously requires a coincidence. If these two conditions are not met, there is no total eclipse. If the Moon is just a little off from a node point, there only is a partial solar eclipse: The disk of the Moon 'bites out' part of the Sun's disk, not all of it.

Using the male pronoun was a convention of the time that most writers, even female writers, sometimes have followed. Newcomb again:

> [The form of the Moon's shadow] is that of a cone, with its base on the Moon, and its point extending towards the Earth. Now, it happens that the diameters of the Sun and Moon are very nearly proportional to their respective mean distances, so that the point of this shadow almost exactly reaches the surface of the Earth. Indeed, so near is the adjustment, that the dark shadow sometimes reaches the earth, and sometimes does not, owing to the small changes in the distance of the Sun and Moon. When the shadow reaches the Earth, it is comparatively very narrow, owing to its being so near its sharp point; but if an observer can station himself within it, he will see a total eclipse of the Sun during the short time the shadow is passing over him. ... [This is] why a total eclipse of the Sun is so rare at any one place on the Earth. The shadow, when it reaches the earth, is so near down to a point that its diameter is not generally more than a hundred miles; consequently, each total eclipse is visible only along a belt which may not average more than a hundred miles across.

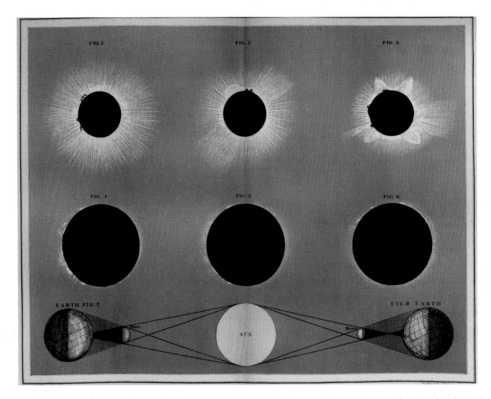

Fig. 3.1 This (then) new graphical illustration of how eclipses occur was popular in 1869. {by Alexander Johnston}

The shadow print on the Earth under the Moon—for that is what a total eclipse of the Sun really is—is stretched into a path by the rotation of the Earth during the total eclipse. At its greatest, this path may be 267 kilometers in width. Its length may be thousands of kilometers. A partial solar eclipse proceeds and follows a total.

A partial eclipse also is seen to either side of the total-eclipse path, in a band up to 3,200 kilometers wide. Thus, many more people are apt to view a partial eclipse than a total.

A total eclipse of the Sun begins with first contact. During the subsequent partial stages, the Moon's shadow makes the Sun look like Pacman. You know those little circles that appear underneath a leafy tree when the Sun is shining? Those actually are pinhole-projected *images* of the Sun. During a partial solar eclipse, they appear as crescents.

Shadows have two parts: a darker, inner part—the umbra, and a lighter, outer part—the penumbra. After first contact, we (the solar-eclipse observers) are in the Moon's penumbra. Then, if the eclipse is to be total, there is a second contact at

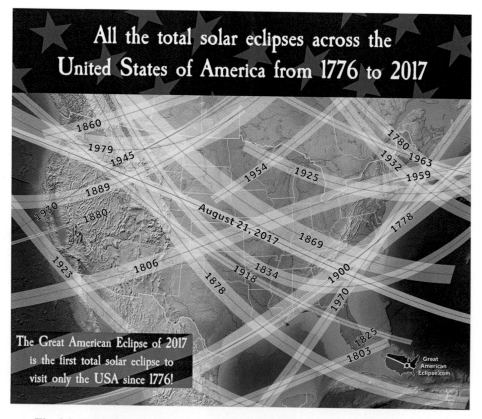

All the total solar eclipses across the United States of America from 1776 to 2017

1860
1979
1945
1889
1930
1880
1923
1806
1878
1918
1834
August 21, 2017
1954
1925
1869
1900
1970
1780
1932
1963
1959
1778
1825
1803

The Great American Eclipse of 2017 is the first total solar eclipse to visit only the USA since 1776!

Great American Eclipse.com

Fig. 3.2 Total-eclipse paths across the United States, 1778–2017. {Michael Zeiler}

which time we are now in the Moon's umbra. The solar disk cannot be seen; it is "totality." Again, if a total eclipse, at third contact we return to the penumbra. At fourth contact we are out of the Moon's shadow altogether.

Only people within the path of totality, the width of the Moon's narrow umbra, experience a second and third contact. Everyone else sees a partial solar eclipse at most. Maximum totality duration is to be found in the middle of the eclipse path, on the "centerline."

Solar eclipse magnitude refers to the fraction of the Sun's (normally seen) apparent radius that is eclipsed: 1.0 or more is totality, less is a partial eclipse. Few people notice a solar eclipse of eclipse magnitude 0.7 at a minimum, unless it is pointed out to them.

A solar eclipse magnitude greater than 1.0 indicates how much larger the apparent size of the Moon at the time is than that of the Sun. Greater eclipse magnitude means longer total eclipse. The bigger the solar eclipse magnitude, the more likely that a total eclipse will be seen. (Total eclipse magnitude will vary according to where you are on the eclipse path and how close you are to the centerline.)

As you can understand, if you stay in one place, it is unlikely that it will be the right piece of real-estate from which to see any given total eclipse of the Sun. Your odds are about one in four-hundred years. Total eclipses are for voyagers.

Newcomb further explains:

The duration of a solar eclipse depends upon the time required for the Moon to pass over the distance from where she first comes into apparent contact with the Sun's disk, until she separates from it again; and this, in the case of eclipses which are pretty large, may range between two and three hours. In a total eclipse, however, the apparent disk of the Moon exceeds that of the Sun by so small an amount, that it takes her but a short time to pass far enough to uncover some part of the Sun's disk; the time is rarely more than five or six minutes, and sometimes only a few seconds.

Fig. 3.3 Mid-eclipse (totality) during a recent solar eclipse. {Jay Pasachoff, Bryce Babcock, Stephan Martin, *et al.*}

The Moon steals the Sun. That was a big deal to peoples in the past. (It still is to us.) A reason for past concern/superstition about solar eclipses was their seeming unpredictability. What governs the circumstances of an eclipse? How are eclipses predicted?

This has not changed since the time of Newcomb. Newcomb was, above all, a man about arithmetic: the more, the better. In fact, he was more comfortable 'doing astronomy' at a desk with pencil and paper than he was at the eye piece of a telescope. So, if the following is not your proverbial cup of tea, you might wish to move along. Take a deep breath. Newcomb:

> Returning, now, to the apparent motions of the Sun and Moon around the Celestial Sphere, we see that since the Moon's orbit has two opposite nodes in which it crosses the ecliptic, and the Sun passes through the entire course of the ecliptic in the course of the year, it follows that there are two periods in the course of the year during which the Sun is near a node, and eclipses may occur. Roughly speaking, these periods are each about a month in duration, and we may call them seasons of eclipses.

> Owing to the constant motion of the Moon's node[s] . . . the season of eclipses will not be the same from year to year, but will occur, on the average, about 20 days earlier each year.

Let me unpack this. Remember the nodes? When the Moon appears to travel north, 'up' through the ecliptic, it is at ascending node. When it appears to travel south, 'down' through the ecliptic, it is at descending node. The Line of Nodes is an imaginary one connecting these two points. This line slowly shifts around the Earth. It would not do so if the Earth and Moon existed in isolation; it is the gravitational effect of the Sun that causes this westward regression.

It takes 18.6 years to complete one twist. So, the time from the Moon's presence at ascending node to its return again, or passage from descending node to descending node, is not exactly one month: It is 27.22 days, the *nodical* month (or "draconic" month, from the myth that an eclipse happens when a dragon consumes the Sun).

Reviewing, solar eclipses happen when the Moon is at a node (so that the Earth, Moon, and Sun are all in the same plane). It must, at the same time, be New Moon, a specific time of the synodic month. This is a rare occurrence—but fairly predictable if the nodes were always in the same place on the Celestial Sphere. Then, if the New Moon approached one of these special celestial spots, an eclipse easily could be prognosticated.

Making prediction more difficult is the fact that the nodes appear to drift through the ecliptic, so it is more difficult to keep track of their locations on the Celestial Sphere and to foretell eclipses of the Sun. Or where the path of a total eclipse will be (where an imaginary line going through the center of the Sun and Moon will intersect the Earth).

By turning the eclipse 'rules' into equations, the constants verified by observation, Newcomb knew to be standing in downtown Des Moines, Iowa, looking in the direction of the river, on 7 August 1869—awaiting a total eclipse of the Sun.

It is harder than it would appear to be, at first glance, to predict the *type* of solar eclipse, because the apparent size of the Moon is always changing. The line (called an "apside") connecting the Moon's apogee point (that which it occupies when furthest from us in its elliptical orbit about the Earth) and perigee point (that which it occupies when closest to us in its elliptical orbit about the Earth) slowly precesses eastward around the Earth with a period of 8.85 years. Therefore, the time between apogee and apogee, or perigee and perigee, is slightly different from the sidereal, synodic, or nodical month. It is 27.55 days—the *anomalistic* month.

The Moon appears bigger when near perigee and smaller when near apogee. This difference in apparent size could mean the difference between an annular eclipse (when the Moon's shadow cone does not quite reach the Earth's surface) and a total solar eclipse (when it does). In an annular eclipse of the Sun, a particular kind of partial solar eclipse, the center of the Moon's disk even may pass through the center of the Sun's. Yet a ring (or annulus) of light that is the outer edge of the solar disk still can be seen. The eclipse is not total.

(Occasionally, a hybrid solar eclipse will be annular over part of its path and total over another. The previously mentioned eclipse of 1825 is an example.)

Moreover, the apparent size of the Moon affects the width of the total eclipse path as well. (If the Moon is closer to us, it casts a wider shadow onto the Earth.) The time of the anomalistic month may determine whether you are on the total eclipse path at all. These different intervals of time also affect the duration of solar eclipses.

Even given all these variables, Newcomb, or any 'card carrying' astronomer of his day could predict a total eclipse of the Sun. The right to map its official path was left to the then-most famous American mathematician/astronomer, Superintendent Benjamin Peirce (1809-1880; United States Coast Survey). (It can be argued that in 1869, Newcomb 'eclipsed' Peirce in this public role.)

Most astronomers seeking out the path of the 1869 total solar eclipse targeted Iowa. I follow the events of 7 August 1869 everywhere on its intercontinental swath, from the umbra's appearance on the Earth in the morning sky of Asia to its disappearance in the evening sky of the Atlantic Ocean.

When Nearly Everyone Had *Heard* of a Solar Eclipse—Maybe

More than any other book, the *Bible* was required reading in the nineteenth-century USA. It may have been the only book in a household.

The *Bible* might possibly record ancient solar eclipses. "And when the Sun was going down . . . great darkness fell upon him" (Genesis 15:12) may refer to the eclipse at dusk on 9 May 1533 BCE. Does Joshua's stopping of the Sun refer to an eclipse that happened on 30 September 1131 BCE (Joshua 10:13)? (The eclipse, not the "stopping.") As for, "And on that day, says the Lord God, I will make the Sun go down at noon, and darken the Earth in broad daylight." (Amos 8:9), some historians speculate that this passage is about the nearly total eclipse of 15 June 763 BCE.

All these eclipses of the Sun were total in Old Testament lands. Yet trying to date such events as historical may miss the authors' point: that they were miraculous and the work of a supreme deity.

Indeed, so out of the ordinary are total eclipses of the Sun that they have long been considered important omens, for good or ill. Venerated or feared, they rarely have been ignored.

Elsewhere in the *Bible*, what sounds like a solar eclipse accompanies the crucifixion of Jesus. However, no literal eclipse circumstance matches fully with other associated events in the New Testament story. If an 'eclipse' occurred, it was transcendental.

It is impossible to say what people first thought of eclipses, of course. That report is lost to prehistory. Despite this, we do know that *homo sapiens* have a long history with eclipses. The circumstances of 13,200 lunar and solar eclipses all over the globe have been calculated, 'just' between the years 1207 BCE and 2161. We note that the Chinese wrote about something that sounds like the corona (coming up) of the Sun, only seen during total solar eclipses, as early as the eighth century BCE. There might be a Mesopotamian record from even earlier.

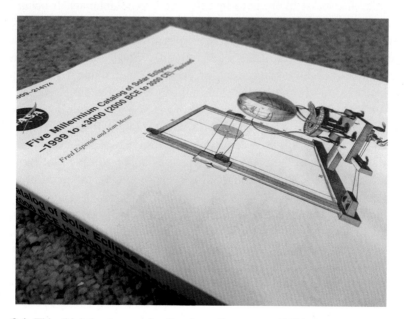

Fig. 3.4 This thick book records all solar eclipses over 5,000 years (in tiny print). (Espenak, F and Meeus, J. "Five Millennium Catalog of Solar Eclipses." *NASA/ TP-2009-214174*. (2009)). {photo by the author}

Simon Newcomb may have been able to relate to the famous legend from these earlier times of two Chinese court astronomers, Hsi and Ho, who supposedly were executed for failing to predict an eclipse. This tale, though, almost certainly is apocryphal.

Considering the consternation that total eclipses of the Sun brought, when did they first become predictable?

We can group together the related, overlapping, and (sometimes) conflicting societies that dwelt at the confluence of the Tigris and Euphrates Rivers, during the first several millennia BCE, as the Mesopotamians. Here was the so-called Cradle of Civilization, home to the first great city-builders. And it was here, fittingly, that people noticed and recorded that there is a pattern to eclipses.

The Mesopotamians found a cycle of eclipse activity that repeated every 223 months. This is the period of time at which the synodic month, nodical month, and anomalistic month are nearly commensurate with each other. It is called the Saros cycle. There are approximately 71 eclipses of the Sun each Saros (an interval of 6,585.3 days). Similar solar-eclipse geometry happens at the same time each Saros. For instance, 18 years, 11 days after one eclipse, there is almost always another.

What if, for instance, a total eclipse of the Sun occurs? The next solar eclipse, a Saros later, also most likely is to be total (or maybe annular). It occupies the same place on the Celestial Sphere. Only that fraction of one-third of a day (one-third Earth rotation) guarantees that each member of these total-eclipse series will transpire for those one-third of the way around the Earth from the previous one. The second solar eclipse is pushed out of view if you stay put. You have to wait three Saros cycles to experience the same eclipse at the same longitude.

And even then: That "same" total eclipse of the Sun will occur up to thousands or so kilometers, north or south, of the last "same" solar eclipse. Each successive eclipse in a given Saros will be deeper (meaning less partial and more likely total) during the first portion of the Saros. The opposite happens during the second half. After about 1,300 years, an individual Saros cycle (now seen to represent a family of similar eclipses) 'dies.'

What the Mesopotamians *circa* the eighth century BCE actually did was this: Apparently, they figured out that eclipses tend to be separated by five or six New Moons. An eclipse was not by any means *guaranteed* at that interval, at least witnessed in Mesopotamia.

The Saros is a good model of eclipse behavior, but not exact. Predictions based upon it alone eventually will fail. Still, it is a tribute to ancient record keeping that the Saros was discovered as early as it was.

Ultimately, the answer to the question, 'who first predicted an eclipse?' depends upon our criterion for success. Societies the world over have made the attempt. More than one have succeeded in establishing eclipse *probabilities*: when an eclipse was more or less likely to occur.

For instance, there is the legend of Thales. Nineteen-century scientists like Newcomb would have been expected to be well-versed in the history of their discipline. Among the famous names in astronomy, that of Thales (a pre-Socratic philosopher) would have been as familiar as that of Galileo. Thales, largely on the word of Aristotle, was considered to be the wisest of the Seven Sages of Ancient Greece. He traditionally is the father of western science.

The historian Herodotus claims that, during a war between the Medes and Lydians, Thales predicted a solar eclipse. It sounds more likely that he warned of the *possibility* of an eclipse. Anyway, both sides wanted peace, and this was as good an excuse as any to call off the war. The story can be associated with the eclipse of 28 May 585 BCE.

In the Victorian era, the Anglophile world fell in love with all things Ancient Greek. As archeologists extracted more and more artifacts, curiosity grew about the roots of western civilization.

For instance, we have the verse by Archilochus, quoted at the beginning of the chapter. Archilochus lived during the seventh century BCE. A total eclipse of the Sun was visible from Greece on 6 April 648 BCE.

Astronomers in Newcomb's time thought that ancient-eclipse observations could service modern theory. Such events established data points on the mathematical curves that were the orbits of the Earth and Moon. They were neither as accurate nor precise as modern data, but their antiquity gave them great weight.

Most astronomers were classically trained. Newcomb was an autodidact. He was skeptical. Did Helicon of Cyzicus predict the 12 May 361 BCE solar eclipse? Or did he simply observe it? One scenario suggests greater timing accuracy than the other. Newcomb used such data sparingly.

Newcomb eventually *would* apply historical data to modeling the exact motion of the Moon, but that from records of more modern (and more precise) eclipse timings. He would eventually employ mathematics of greater and greater finesse. His algorithm for predicting an eclipse; as well as for locating in the sky at any time the Sun, Moon, and planets; would eventually supersede all that had gone before.

Newcomb's final equations—still in the future in 1869—describe the orbits of any two bodies involved: for instance, the Moon about the Earth or the Earth about the Sun. They also include the shapes (very nearly spherical), diameters, and gravitational mass distributions of the Sun and Moon. (Of course, for any of this to be of practical use, one had to know from where one was observing: the shape of the Earth and one's latitude and longitude on it.)

This was the good news.

On the other hand, in classical astronomy of time and position, the Sun might as well be a giant motorcycle headlamp. Its nature is irrelevant. It is just *there*.

What of physical solar astronomy? What was understood about the unique Sun itself in 1869? The answer is, very little.

The approximate size and mass of the Sun were pretty much axiomatic. The combination yielded its average density. That was about all.

Most significantly, in 1869, the nature of the Sun's vast interior was a complete mystery. Earth-like analogies were popular. A few held out that the inside of the Sun was solid (and maybe you could walk around on it!). However, its average density, a little greater than that of water, suggested that something like an (incompressible) liquid was more likely. (Wrong, too.) No one had ever seen a *gas* with such a high density. Air, for instance, has a density a thousand times lower.

Of course, there obviously was a very unearth-like outer layer to the Sun of great luminosity. It was named the photosphere and is identical to the bright, yellow-white disk we think of when conjuring our star. A good telescope and eye just resolve detail in the photosphere: a global pattern of slightly lighter and darker granules.

There are much larger, discrete dark spots on the photosphere. The number of sunspots seen at any given time follows an approximate 11-year cycle from minimum to maximum to minimum again. In the 1869s, sunspots were thought to be some sort of depression or hole in a porous photosphere. (The opposite—discrete bright regions near sunspots—are called faculae.)

1860s investigations of the Sun during eclipse can be thought of as starting at the photosphere and working upward. During a total solar eclipse, a fainter layer of the Sun above the photosphere immediately becomes visible. Norman Lockyer named it the chromosphere, just in our year of interest, 1869. The word comes from its distinctive red color. Rising above the chromosphere seemingly are flame-like appendages originally called "protuberances" (though I will use the modern term, prominences, too).

Also during a total solar eclipse, a still fainter corona is seen to surround the Sun, extending much further from the photosphere than the chromosphere. As it happens, this solar feature was named during a total eclipse visible from the United States (1806): It was so called by the Spanish/Basque astronomer José Joaquín de Ferrer (1763–1818; Observatorio de San Fernando) who observed it from Kinderhook, New York. (It also was he who mentioned that the irregularity of the Moon's profile is very apparent during a solar eclipse, a fact of great interest to selenologists.) Yet as late as a 1856 textbook, the corona still was described as a sometimes thing.

Most observers agreed that the corona has two components: inner and outer. But they did not agree always on what the distinction between the two was. In 1868, Old World attention was directed toward the chromosphere and prominences at the expense of the mysterious corona. Right after that eclipse, it was realized that the spectroscope it was possible to study the prominences without a total eclipse (though it was still easer to do so during one and awareness of this trick may not have reached the USA by the time of its own eclipse). The corona got scant attention in 1868 and little at all was learned about it.

Using the total eclipse as a tool, the lack of knowledge about the Sun was about to change. The physical astronomy of our star was to begin its surge from behind moribund classical astronomy having to do with solar position in the sky. It was the great eclipse of 1869.

4

Navy Astronomers 2,000 Kilometers Ashore

Are the gods angry? Is this their frowning you see?
Not at all! They're as jolly as jolly can be!
The Goddess of Night hides her cheek in eclipse,
While the God of the Morning his arm 'round her slips
And yields to the temper, and touches her lips!
Young Eros has stormed the dominions of Jove,
And, false to their nature, the gods are in love!
– Stanza 2 of (admittedly inferior) poetry written on the occasion of the 1869 total solar eclipse

After the Civil War began in 1861, Simon Newcomb was appointed to the United States Naval Observatory in Washington, replacing staff who had departed to go on active duty—or join the Confederacy. Though a 'book learned' theoretical astronomer, he eventually mastered the practical side of astronomy at the USNO, standardizing procedures, and tracking down systematic errors in stellar position catalogues (though he exhibited little patience with this kind of observational work). He was now Professor Newcomb (for that is what Navy astronomers were called).

T. Hockey, *America's First Eclipse Chasers*, Springer Praxis Books, https://doi.org/10.1007/978-3-031-24124-6_4

Simon Newcomb and Fiction

"The first condition of our success is absolute loyalty on the part of each member of our organization ..." Newcomb did not really say this, as far as we know. It is extracted from the one (bad) science-fiction novel that he wrote, and these are the words of his protagonist.[1] Nonetheless, it appears that Newcomb patterned the character after himself. And the passage reminds us that total-eclipse observing requires teamwork.

As early as the 1842 total eclipse of the Sun, part-time astronomer Englishman Francis Baily (1774–1844; businessman) recommended that different scientists be given different tasks to accomplish during the brevity of totality. (This division of labor often broke down in the field.) Far more individuals contributed to the USNO and other 1869 total-eclipse expeditions than are named in this book.

Newcomb urged the USNO to place its astronomers under the totally eclipsed Sun. The USNO Superintendent, the Commodore Benjamin Sands (1811–1883; United States Navy), agreed. Sands had no particular scientific qualifications, but he knew how to get things done. Soon, the USNO solar-eclipse expedition was on. While the venture was modest compared to some expeditions, it is still to Sands's credit that he was able to launch it within his budget and without any additional funding from Congress. Moreover, he did it almost completely 'in house,' with USNO personnel.

Fig. 4.1 Benjamin Sands. 1861–1865. {United States Naval History and Heritage Command}

---

[1] Newcomb, S. *His Wisdom, the Defender*. New York: Harper & Brothers. (1890)

Two other, principal USNO scientists accompanied Newcomb, as well as a small staff to help with the auxiliary observations and technical aspects of the expedition. Sands authorized the Navy astronomers to set up anywhere within one-hundred miles of Des Moines, Iowa, as they saw fit.

In choosing Des Moines itself as their final destination, the Washingtonians disappointed the citizens of Mitchellville, Iowa, (founded 1856). They had *moved their entire city* in order to become a train stop on the Chicago, Rock Island, and Pacific railroad.

Mitchellville sat on the total-eclipse centerline. However, the duration of totality was only two seconds less at Des Moines than at Mitchellville. And Des Moines provided more amenities than that more modest whistlestop. E. P. Austin, of the United States Coast Survey [USCS] (more *anon*), calculated the latitude and longitude of Des Moines and helped convince the USNO expedition to settle in there.

Des Moines evolved from Fort Des Moines along the same-named river. It had been incorporated in 1857 and was now the state capitol. Local industry in 1869 included the Central Oil Works (linseed oil), the Iowa School Furniture Company, and A. W. & F. Voodry's Carriage Factory. The city's population had more than doubled since the Civil War and the beginning of passenger train service. In an 1867 census, the number of citizens stood at 10,511.

So it was that Newcomb and his party were greeted at journey's end by local dignitaries such as the mayor of Des Moines and judge of the county court. The date was 31 July 1869.

The plan was to erect a temporary observatory building at the observing site. Newcomb selected his locus the day after arrival.[2] His local contractor (a Mr. F. T. Nelson) then hastily threw up a shelter in the northeast yard of the Polk County courthouse. The courthouse was a new edifice in the town center, only ten years old. Its precise geographical coordinates were measured by the USCS. "The occupation of a station near my lodgings and near workshops, stores, &c., saved much trouble and expense." A stone used to mark the site lasted there until the courthouse was demolished at the turn-of-the-century to make way for a new one.

The Italianate building must have served as an elegant backdrop for observing a total eclipse of the Sun insofar as it was constructed for the then-staggering sum of $79,151. It sat across from the post office, where later, another commemorative stone used to mark the elevation established by the 1869 astronomers.

Volunteer carpenters, curious to see what all the fuss was about, helped Newcomb erect four rough board walls supported by the courthouse itself. The Western Union Telegraph Company wired the facility *gratis*. Its president and

---

[2] Newcomb was used to working on two important tasks in tandem. We know from his diary that the high-reaching Newcomb was at this very same time secretly trying to obtain employment at the University of California.

Fig. 4.2 Polk County Courthouse. 1858. {Jim Dove}

general manager provided to the USNO astronomers free use of their company's service, starting several days before the total eclipse of the Sun."

Newcomb assembled telescopes from the objectives taken from the transit and refraction circles[3] at the home Naval Observatory. That of the biggest, from the transit circle, had an aperture of 8–1½ inches. To do the job that the USNO astronomers had them made, these telescopes, particularly the transit circle, required precise positioning. Yet Newcomb's USNO team members were willing to upset this arrangement by ripping these instruments' most precious and delicate parts from them. They then transported the fragile lenses afar, without certain knowledge of what might befall the glass. Such was the seriousness with which this group of observers undertook its first total-eclipse expedition.[4]

The telescopes that housed the precious optics were really just long wooden boxes with their insides darkened and lenses mounted at each end—a far cry from the stately brass or polished mahogany tubes professional astronomers were used

---

[3] A transit circle is used to make observations of right ascension by observing and noting the times at which stars pass through the Central Meridian. The USNO's "refraction circle" was tasked with measuring the distortion of star positions due to atmospheric refraction as a function of altitude.

[4] It was Newcomb's second.

to. The crude-looking instruments then were placed on makeshift wooden mounts that tipped and swiveled so that they could be pointed at the Sun.

Surprisingly, we do not know who used the converted transit circle telescope during the total eclipse of the Sun. This telescope was new in 1865, and though he had completed his four-year project of fundamental stellar positional measurements with it back in the DC, it still was the instrument with which Newcomb was most familiar—and responsible. (This remained true even though it was not best suited to his self-assigned tasks on Eclipse Day.) Yet nowhere in Sand's definitive *"Reports on Observations of the Total Eclipse of the Sun ..."* this major instrument, the reconfigured transit telescope, mentioned again.

As for who used it, there are two suspects. After all, to every party there are those who simply 'must' be invited.

Others on hand with the USNO expedition included Professor Truman Safford (1836–1901; University of Chicago). Safford would seem to been—literally— born for this expedition: He was predicting eclipses before he was age 10.

Sanford's Dearborn Observatory served as a geodetic reference for the USNO expedition. (In only two years, the Great Chicago fire would force Safford to seek other employment.) Yet Harkness (below) mentions in passing that Safford was with him elsewhere, early during the eclipse. Did Safford then run to the courthouse for totality? While he published prolifically during the 1860s, no report from him on the 1869 total eclipse seems to exist.

A dark horse is Lord Sackville Cecil (1848–1898; British railway executive), on tour and possibly taking a seat of privilege (or, rather, a stand) as a representative of the railroad industry. He was the personification of a major benefactor to the expedition. If it was Cecil who used the transit-circle objective fashioned into an eclipse-viewing telescope, it was as a gratuity. We are told by his industrial colleagues that he was an "odd and peculiar man" but "very popular."

The Naval Observatory's other contingent was headed by Scottish American astronomer Professor William Harkness (1837–1903; USNO). (At the time of the eclipse, he was temporarily assigned to the United States Hydrographic Office.) While Newcomb ran the USNO's overall total-eclipse operation, we already know that he did not intend to make physical measurements himself. That was a bridge too far. Newcomb studied the enigmatic corona by eye: The corona became the principal focus of the 1869 total solar eclipse among all expeditions.

It would be Harkness, known as an instrument virtuoso, who was assigned command of the expedition's station dedicated to new kinds of measurements. Harkness was the expedition's guiding spirit for physical astronomy. He does not get credit always for co-discovering one of the 1869 total solar eclipse's major revelations.

Harkness's biography gives us an inkling of his interest in such matters. After medical school and war service as a surgeon, Harkness sailed on the *Monitor*-style warship *Monadnock* from 1865 to 1866. Aboard her he made an exhaustive study

Fig. 4.3 William Harkness. {National Library of Medicine}

of terrestrial magnetism and the influence of iron armor on the behavior of the compass. He spent the rest of his career with the USNO. While Newcomb thrived on it, Harkness eventually suffered a breakdown from overwork. (His living situation was unusual: He spent much of his adult life in rooms within Washington's exclusive Cosmos Club; since 1878 "a private social club for [those] distinguished in science, literature and the arts, a learned profession or public service."[5])

For three days after arrival in the Iowa capital, Harkness reconnoitered. He finally eyed a spot for a temporary observatory even closer to the centerline than Newcomb's. It was to be situated on the north side of Short Street, between Second and Third. Slightly east of the courthouse, and one-and a-half kilometers north of the senior astronomer, at the time it was on the north edge of town. The real estate was a high, vacant lot on the bank of the Des Moines, free from obstructions. Its coordinates were not measured by the USCS; Harkness would have to determine those himself. But, "… as the city has not built up so far north, there is no travel, and it is perfectly free from that great enemy of photography, dust." The spot is now in the middle of Interstate 235, behind the Wells Fargo Arena.

Harkness saw his mission as being all about the camera. He was attracted by the presence of a nearby well to meet the needs of the darkroom.

Harkness paid his respects at the address of the agent for the property. There, he received permission of the widowed owner to use it as he would like. Harkness contracted another local builder to erect another temporary observatory. His

---

[5] Once upon a time, I was a slightly impoverished college student. An aged great aunt, married into the Washington illuminati, invited me to lunch at the Cosmos Club. There, I shocked the *maître d'* with the fact that I did not own a dining jacket.

photographers were pleased; they had thought that they might have to build the required facility themselves.

With an alacrity that I have never experienced when it comes to home remodeling, a construction crew started work right away and completed the project in less than a week. This was even though there was a five-day series of intermittent thunderstorms in-between. The astronomers immediately collected their instruments out of storage. The building was a cramped 5 X 10 meters of "rough pine," with a roofline parallel to north-south such that the long sides of the edifice faced east-west. The roof slanted, so the west wall was only one-and-a-half meters above the ground, no doubt requiring a great deal of stooping. A 2- by 15-meter darkroom was partitioned off. Part of the roof was covered only in canvas, supplied by the United States Army Medical Museum in Washington. Rolling it up exposed the telescopes to the sky.

Fig. 4.4 Harkness/Curtis temporary eclipse observatory. I am not able to identify the figure in the hat. {Trudy Bell}

The clay surface on which the observatory was built turned out to be a source of dust, which all photographers feared. Installation of a floor over the pernicious ground was required; its irregular pieces resulted in a tripping hazard—especially during nighttime observing. No architectural awards were to be handed out here. Still, that this facility was so much more elaborate than that of Newcomb's reflected the physical and chemical demands of photography.

A Clark 7-¾-inch-aperture achromat, on loan from the United States Naval Academy, was put to use for photography. It was operated by Brevet Major Edward

Curtis, M. D., (1838–1912; United States Army [USA] Medical Museum) a surgeon who had marched alongside General William Sherman. He was the one soldier amongst a group of nominally naval officers. (Curtis is most remembered as one of the pathologists who performed the autopsy on Abraham Lincoln.)

Fig. 4.5   Edward Curtis. {Historical Collections of the National Museum of Medicine}

Curtis was present because he was also an amateur—soon to be professional—photographer. Photographic skills were sadly lacking at USNO.

The refractor had all the trimmings: an equatorial mount, and clock drive. Prior to departure, another small observatory had been erected on USNO grounds to temporarily house Curtis and the telescope, so that he could practice under conditions simulating what he would encounter in Des Moines.

The other optical instruments were Harkness's own. The first company to manufacture large high-quality achromats was Dollond & Sons in England. They re-popularized the portable telescope. In the nineteenth century, Dollond was joined in manufacturing telescopes by Merz & Mahler (later Merz & Söhne) in Germany. Fitz was the first telescope factory in the United States.

However, Harkness's refractor was produced by the newer Alan Clark & Sons firm of Cambridge, Massachusetts. His 3-inch-aperture Clark was mounted equatorially on a portable tripod. (Clark made many of these telescopes for the USNO; the retail price at the time was $160.00[6]) According to Harkness, it was "a remarkably fine one."

---

[6] $1,975 in today's dollars

A brass box contained a spectroscope. This new purchase had been manufactured in Germany for use in a chemical laboratory, but Harkness had made alterations so that it suited astronomical work. The spectroscope attached to the 3-inch achromat.

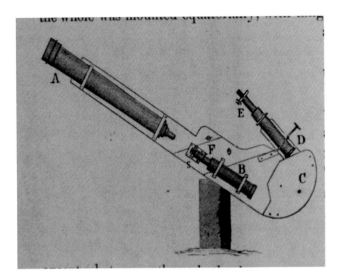

Fig. 4.6 Contemporary spectroscope attached to a telescope. {by Friedrich Westermann}

Harkness also had with him two polariscopes, two sextants (one of his own design), two artificial horizons, two compasses, and three chronometers (made by Thomas and John Negus. (The latter were the best such clocks of their day). Redundancy clearly was Harkness's word to live by.

Most of the other paraphernalia listed on the manifest in Chapter 2 was Harkness's, too (even the rain gauge). Just the books must have been weighty: Those that he could not be without included The *American Nautical Almanac, and Supplement, for 1869*; William Chauvenet's *Astronomy*; Elias Loomis's competing *Practical Astronomy* (1855); Chauvenet's *Trigonometry*; plus, tables of logarithms, multiplication tables, and star maps.[7] (The books still are in print.) Harkness was invested heavily in the 1869 total eclipse of the Sun.

After deploying the instruments, the next task was to establish better where the 'observatory' *was*. Eclipse timings were of no use if the point at which the observations were made was unknown. Latitude and longitude were determined astronomically, but the longitude calculation required accurate time.

---

[7]The information provided by all these references can now be accessed with a single smart phone.

A sextant and artificial horizon were used to check the observatory's latitude using the stars by night and its local time using the Sun by day. The reduction of both sets of data takes more than seven pages in Harkness's final report.

Time was compared to the master clock, kept back east in the USNO's Foggy Bottom-district building. On several occasions before the eclipse, a chronometer was carried to the Des Moines telegraph office, where time signals from Washington were received. Harkness then sent his time back to Washington by breaking a continuous telegraph signal on the second. (According to Harkness, a break in a telegraph signal is easier to accomplish at a given instant than the creation of a signal.)

Fig. 4.7 Contemporary chronometer. {National Museum of American History}

There was no direct telegraph line from Des Moines to anywhere in the District of Columbia. The route the information took was literally circuitous: for instance, Washington to Philadelphia; Philadelphia to Pittsburg, Pennsylvania; Pittsburg to Crestline, Ohio; Crestline to Chicago; Chicago to Keokuk, Iowa; and Keokuk to Des Moines. At each intermediate location a repeater forwarded on the signal. In this way, the astronomers in the field reconciled their time with the definitive source back home. Another nine report pages.

Each step risked the introduction of errors. Still, after many repetitions of the process, the astronomers were confident that they had determined their latitude to within an arc second and longitude to a tenth of a second with respect to the Observatory in the nation's capital.

Finally, the different stations in Des Moines, two of the USNO and that of the Litchfield Observatory (Chapter 12), raised flags high into the air. Surveying by use of these markers established the geographic relationship (horizontal *and* vertical) between them. Eight pages.

Harkness's 'after action' report ends with tables summarizing all preparatory measurements—in 29 pages. (In contrast, Harkness's summary of the total eclipse of the Sun itself is a comparatively succinct eight pages!) There was no shortage of things to do for an eclipse astronomer in Des Moines.

Harkness: "By Thursday, July 22, the large telescope was mounted, and on the following day a photograph of the Sun was made … and from that time till [*sic*] August 9[,] photographs were taken and observations [were] made every day when the weather permitted." Both Newcomb's and Harkness's observatories were ready. As a final touch, a rope was stretched around the latter to keep on-lookers at a distance; Des Moines policemen guarded the observatories on the afternoon of the total solar eclipse.

By the solar eclipse of 1715, predictions of total- eclipse paths were good enough that astronomers and others could plan on the place and time to see one. But recent eclipses had been plagued by clouds. Might history repeat itself? The two days before the 1869 total eclipse of the Sun were overcast! Mist caused the floor of Newcomb's observing shack to become covered with mud. The day of syzygy began cloudy but cleared to a mere haze, which dissipated by onset of the solar eclipse.

It was time.

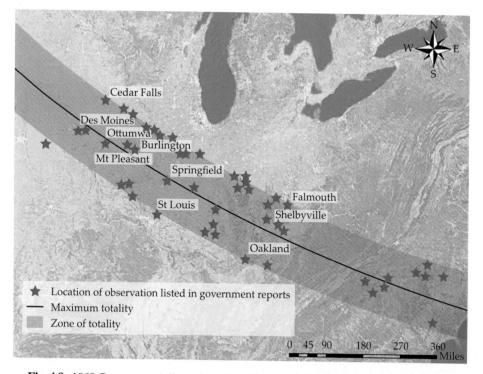

Fig. 4.8 1869 Government eclipse-observing stations in the United States. {Stella Cottam}

5

"The Vast Black Orb"

As when the Sun, new risen,
Looks through the horizontal misty air,
Shorn of his beams, or from behind the Moon,
In dim eclipse, disastrous twilight sheds
On half the nations and with fear of change
Perplexes monarchs

(from John Milton's *Paradise Lost* (1667), quoted on the occasion of the 1869 total solar eclipse)

A note to the reader: Descriptions and quotations regarding historical observations of the Sun include practices that are now known to be hazardous to the eye.

First Contact: 03:57:34 PM CST
Altitude = 37° Azimuth = 258°[1]

Day began to become, if not night, deep twilight. With no electric lights, buildings were illuminated from within by lamp light.

Simon Newcomb:

… before the advent of darkness a crowd of people who had taken their position on the roof of the court-house [*sic*] greeted the progress of the phenomenon with loud and prolonged cheering, rendering me quite uneasy for our time observations. But I

---

[1] Altitude refers to the angle of the eclipsed Sun above the horizon. 0° equals the horizon, 90° the zenith. The greater the altitude, the less likely clouds will appear along the line of sight.

Azimuth refers to cardinal direction and is measured clockwise as viewed from above. 0° = due North.

© The Author(s), under exclusive license to Springer Nature Switzerland AG 2023
T. Hockey, *America's First Eclipse Chasers*, Springer Praxis Books,
https://doi.org/10.1007/978-3-031-24124-6_5

think they grew quite silent before the Sun had entirely disappeared, and did not renew their cheers on its reappearance.

Quiet was important. The only other sound was vital. It was that made by a college student, named Armstrong (University of Chicago[2]), drafted to call out the seconds as they passed by on the chronometer. We do not know what he was doing in Des Moines on his summer break. He ended up steadfastly keeping his gaze upon the chronometer. During totality he never looked up. Even Newcomb was impressed.

Newcomb had ruled the eyepiece of his equatorially mounted 4-inch-aperture Clark comet seeker with a grid. Rotating this grid by the known angle at which the limb of the Moon would approach that of the Sun, he knew just where to look for first contact. This worked, though it took Newcomb two or three seconds to realize that what he had anticipated for so long was really happening. When the silhouette of a lunar mountain appeared, he knew for sure.

The astronomer stopped his telescope down to 1-inch in aperture and placed a yellow filter—often called a "shade glass" at the time—into the optical path so as not to allow in too much light. He then checked his pointing one last time. The following thing that Newcomb did might seem a little odd.

Newcomb climbed into a box. It sounds strange to hide away from the solar eclipse just as it started, but there was reason to this madness. Newcomb wanted his eyes adapted to darkness. He intended for his pupils to open to their maximum extent, a process that takes a good twenty minutes for most people. This is why at first we cannot see when awakened at night.

It must have been nerve wracking for Newcomb to know that something extremely unusual was going on (literally, 'outside the box'), hear others' reaction to it, but not peek. He trusted his companion, one professor Frasier (University of Kansas), to warn him when totality approached.

"I emerged from the box a few seconds before total darkness, but did not look up until Professor Frasier had given me the word." Note the self-control!

The Sun's photosphere dissipated until it was no more. Frasier gave the word.

Second Contact: 05:00:02 PM CST
Altitude = 26° Azimuth = 270°

Before leaving Washington, Newcomb had what was now supposed to happen all planned. He had attached to the courthouse screens of different sizes, their angular sizes measured with a sextant. He did so with the view that one of them *occult* the eclipsed Sun. That is right—he intended to intentionally block most of the apparition in order to better see various total-eclipse phenomena far from the center of the Sun. (See Chapter 14.)

Just the corona 'should' not take too long to examine. Everyone had been told by those who had seen it before that this phenomenon consisted of a fuzzy circle around the eclipsed Sun and nothing more. There would not be much to see.

---

[2] not the present University of Chicago

During the real total solar eclipse, the details of this plan went out the proverbial window.

> I then took a single glance at the corona through two thicknesses of green glass. I
> then attempted to hide the corona behind the outer and the larger of the screens, and
> thought I had done so, but after the total phase had passed I was convinced that I had
> mistaken the Moon itself for the screen. Though I knew theoretically that the sky in
> the direction of the Moon ought to seem darker than that outside of the corona, I was
> wholly unprepared for so strong an illusion of a black globe hanging in mid air [*sic*].
> The corona itself was far less bright than I had anticipated.

While everyone understands that the Moon is a sphere, it appears in the sky to us
as a disk. Except during a total eclipse of the Sun, that is. More than one observer
in 1869 used the three-dimensional word "globe" to describe the appearance of
the dark Moon suspended in front of the Sun.

The only illumination now was from the thin, red chromosphere and then the
faint, amorphous corona. (Both are normally drowned out in the glare and blue
sky produced by the much brighter, yellow photosphere now absent.) Long creamy
streamers radiated from the corona. This had not been reported thusly before. The
Sun was no longer a circle.

Newcomb was soon singularly transfixed by the corona. The language used by
this man of numbers is remarkable:

> … I first took a general view of the phenomenon through the comet seeker, with the
> full aperture and no screen. Nearly or quite the entire corona was then visible at a
> single view. As thus seen, I regret that I can describe it with no more accuracy than
> as an effulgence which, while not at all dazzling to the eye, was yet glorious beyond,
> description.

Newcomb also spent time with the prominences. His observing plan was in tatters
if not shreds.

> An immense protuberance on the upper side of the inverted corona attracted my
> attention. During the hurried inspection I made of it, the following points were
> noticed:

1. It did not seem materially brighter than the corona; it could be viewed
   directly without the eye being at all dazzled.
2. The large predominance of red in its color was so evident and so strongly
   marked that I could not entertain doubt of its reality. The tint was a most
   beautiful pink.
3. Its structure was not uniform, nor did the protuberance bear the slightest
   resemblance to a flame. It looked like an immense pile of cumulus clouds,
   illuminated by a white and red sun, and thus exhibiting different shades
   of color as the light from the one sun or the other chanced to predominate
   at different points.

Maximum eclipse: 05:01:43 PM CST
Altitude = 25° Azimuth = 270°
M = 1.0214
Velocity of the Moon's shadow = 1.42 km/s
Path width = 256 kilometers

Newcomb then *abandoned* his telescopes, which he had chaperoned all those many kilometers, and examined the corona with his naked eye.

> ... it seemed to be of a jagged outline, extending out into four sharp points, nearly in the horizontal and vertical direction, while midway between these points the serrated edge hardly seemed to extend beyond the body of the Moon. The greatest distance to which the extreme points seemed to extend did not exceed a semi-diameter of the Moon, and there was nothing like long rays of light extending out in any direction whatever.

Nobody writes "semi-diameter" today. But just as a semi-annual white sale occurs every six months, a semi-diameter of the Moon is half of a lunar diameter. In other words, Newcomb is referring to the Moon's radius.

Maybe eschewing the telescope made everything all the more real. Such an unfamiliar sight was this that Newcomb moved his head about in order to convince himself that nothing he was seeing was an optical illusion. He finally proved to himself that this was not the case by observing the exact same visage through his green glass filter.

Newcomb summarized the corona, "A fish tail gas-light, seen against a dark ground at a distance of fifty yards or more, presents to my eye a similar appearance." If only anybody alive knew that looked like! The most common description of the corona's shape by the perambulating astronomers observing the total solar eclipse was that of a trapezoid, one rotated with respect to the Sun's equator.

The story of carefully laid observing plans being hijacked by a compulsion to gawk with the naked eye at the corona, without anything between the witness and the spectacle, was repeated over and over during the 1869 total eclipse of the Sun. As Samuel Langley (Chapter 11) explained, "The special observations of precision in which I engaged would not interest the reader; but while trying to give my undivided attention to these, a mental photograph of the whole spectacle seemed to be taking without my volition." Spectroscopist Charles Young (Chapter 7) also seems to admit the same when he writes:

> I cannot describe the sensation of surprise and mortification, of personal imbecility and wasted opportunity, that overwhelmed me when the sunlight flashed out. I think it was shared by other observers to a greater or less degree.

The 1869 corona was not what astronomers had been led to expect. It always had been depicted as a more symmetrical ring without further morphology. This mental picture had been shared with them by those who had observed previous total eclipses of the Sun. What else had they been 'sold'?

Third Contact: 05:03:09 PM CST
Altitude = 25° Azimuth = 270°

Totality was over.

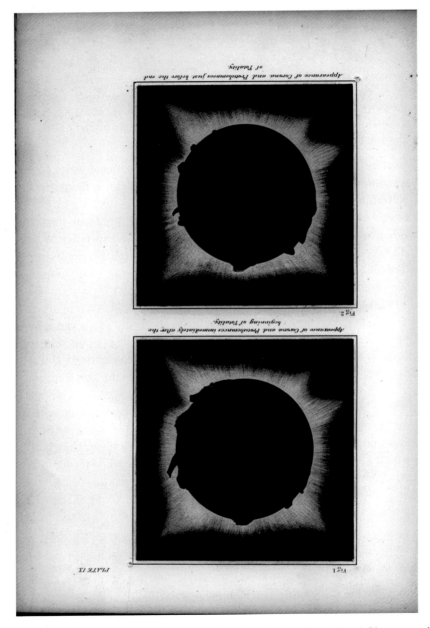

Fig. 5.1 John Eastman's color renditions of totality. {United States Naval Observatory}

Newcomb's report eventually appeared in a government document. "Beautiful"? "Dazzling"? "Glorious"? It certainly does not sound like one. His forthrightness about mental pratfalls he took during the total eclipse of the Sun contrasts with that revealed in modern documents of the same ilk.

Fourth Contact: 06:00:12 PM CST
Altitude = 14° Azimuth = 279°

The alternate reality of a total solar eclipse ceased to be. For Newcomb, it was back to the familiar world of numbers. He never would observe regularly again. In his report, less ink was spent describing the total eclipse than on a table showing comparison of his chronometer values with time signals transmitted by USNO/Washington over the telegraph. A mind-numbing appendix records his chronometer errors as a function of time by reference to both exchanged telegraph signals from, and astronomical measurements made in, Des Moines with the sextant.

And then, of course, there were the contact times. Why? It is unclear how all this data, which by Newcomb's time was already of only specialized interest, might ever be put to use. Perhaps he simply could not help but collect it.

Meanwhile, Harkness did not even bother making his team take contact times. Instead, he reported, "Presently the sunlight went out like the snuffing of a candle, and simultaneously a suppressed, but very audible 'Oh, oh,' arose from the spectators in the neighborhood."

Before the eclipse, Curtis, who was with Harkness, had worried. He was to take photographs through the Naval Academy refractor. This led to a complication: The instrument's objective lens was designed for visual observation—not photography.

A visual achromat brings to a common focus wavelengths optimal for the sensitivity of the human eye. A photographic plate is most sensitive at a different wavelength. So, the photographic focus of the telescope is different from its visual focus. (Moreover, exposure time was not much more than a guess.) Values for many variables must be chosen wisely for the result to be an acceptable photograph.

Photography from Des Moines started on 24 July and continuing every clear afternoon thereafter until just before first contact. Curtis and his assistants practiced their photographic technique on the uneclipsed Sun. Sunspots served as test objects. Curtis finally became convinced that he could make the telescope perform well with a camera attached.

But then, things do not always work smoothly when it is no longer just practice. Test plates on total-eclipse morning began to appear with black spots. It was only hours until totality. The immediate developing required of Curtis's wet-plate photographic method made use of lots of *aqua pura*. Ironically, the well that had attracted them in the first place began to run foul. Curtis:

Fig. 5.2 Naval Academy telescope with Edward Curtis's camera attached. I take the figure as intended to represent Curtis. {Trudy Bell}

Fortunately, in order to provide against every contingency, I had previously had hauled to the observatory a large cask of cistern water; so we rapidly emptied our barrel of the well-water and filled it from the cask, while Mr. Brennan started off at a run to engage a man to bring more water from the nearest cistern in the neighborhood during the progress of the eclipse, so that there might be no danger of our running short.

Just before first contact and after second, plates were sliding into and out of the camera so fast that movement of the telescope could not be avoided. Harkness made the re-pointing adjustments, while Curtis photographed. Photography was paramount. Some of Harkness's own instruments went untouched.

After second contact but before third, exposures were longer, the necessary mechanical adjustments became less frequent, and it was assumed that Curtis could handle the telescope himself. So, Harkness finally was able to escape to his spectroscope. He probably wished that he, too, had somebody to help point his telescope.

Together, the astronomer and photographer managed to take 122 photographs, each 4-inches in diameter as projected through an eye piece. 120 were of the partial phases of the solar eclipse (though three of these "were spoiled.") But of the total phase? Satisfactory images amounted to just two.

Fig. 5.3 Edward Curtis's camera. {United States Naval Observatory}

Wearying travel. Many man-hours day and night. Exotic equipment. Noxious developing chemicals. All for two photographs of a total solar eclipse. What happened?

Harkness repositioned the telescope one more time and left. Pointing was accomplished by using the sighting telescope to project a solar image on a white card. (Projection protected the eye from the brilliance, even in the finder, of the uneclipsed Sun.) Curtis had been informed that the corona was as bright as the Full Moon. Indeed, tests with the Moon back in Washington showed that he could see that orb projected through the sighter without difficulty. So, he was horrified when, during the total eclipse of the Sun, he could not see a thing on his card! Curtis was using the full aperture of his photographic telescope; it was not stopping down. But that did not affect the sighter. He knew that the telescope eventually would need to be repointed and refocused by someone skilled. How would he do so after Harkness was gone, when he could not even trust the sighter? Recall that Curtis was not a professional astronomer, and he no longer had USNO personnel to help him. Thinking that his first might be his only shot at capturing the corona and prominences, Curtis held off activating the shutter until well into totality.

Why was the corona so dim? Rather than blame his predecessors for overestimating the coronal brightness, Curtis chose without proof to accuse atmospheric "haze."

With little time to spare, Curtis's eyes adjusted to the darkness. He was able to make out a faint eclipsed Sun on his incommode card. He finally had the opportunity to swing the telescope back into position and go for another photograph not long before third contact.

How was the first photograph? How was the exposure? Curtis called out for guidance from the darkroom. The attendants still were processing the first totality plate. After a pause, he heard shouted, "double the time!" However, in the interim, Curtis decided that totality was near its end—unbeknownst to him, he had almost ten seconds still—and already was preparing to re-cover his second corona plate. To his dismay, he made an exposure even shorter than the first.

(Curtis attributed an overly short prediction of totality's duration to Truman Safford. Perhaps it had not sunk into the photographer's mind that astronomers could not yet predict the time between contacts to the second. Rectifying this fact was a stated *objective* of the solar-eclipse expedition.)

Curtis's normal stance so as to operate the camera put his back to the solar eclipse. Between all the drama he allowed himself a (furtive?) direct glance at the real eclipsed Sun. Sure enough, the corona was not as he had imagined. What else he saw, though, filled him with astonishment. Curtis was near-sighted and never expected to be able to see naked-eye prominences. "…they looked to me of a pure, rich carmine tint, and seemed to glow and sparkle as if the Moon were …studded on its eastern side with rubies or garnets flashing in the Sun." Most eclipse-observing astronomers preferred to use the less artistic color description "pink"

In retrospect, Curtis made photographic compromises to exposures supposedly of the corona, in order to include images of the prominences with which he was much intrigued. ("Antelope horn", "the head of … long-eared varieties of owls" …he goes on for pages in his official report.) But with help from his darkroom crew, who (recognizing that the coronal exposures were short) pulled every suggestion of light from the developing collodion, his result was a success, in one post-second-contact, pre-eclipse-maximum image and in one post-eclipse-maximum, pre-third-contact image. Nine prominences and something of the trapezoidal corona appear. Yet, like all photographers, Curtis experienced the irony that his best view by far was the one recorded 'live' in his mind.

Curtis's images were self-assessed as "… of a beauty and delicacy of detail never approached before." This is something of an exaggeration. Of course, "never before" was a low bar. Des Moines was one of the westernmost encampments attempting serious total-eclipse photography. Curtis's was a more-or-less true statement—for at least the few minutes that it took the Moon's shadow to cross the country. In fact, it is difficult to make out even the brighter inner corona in reproductions (which most people saw, instead of the original plates) of either of Curtis's photographs.

a

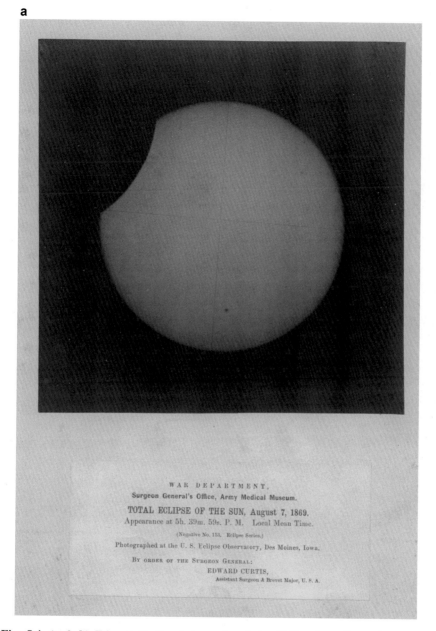

WAR DEPARTMENT,
Surgeon General's Office, Army Medical Museum.

TOTAL ECLIPSE OF THE SUN, August 7, 1869.
Appearance at 5h. 39m. 59s. P. M.   Local Mean Time.

(Negative No. 113.   Eclipse Series.)

Photographed at the U. S. Eclipse Observatory, Des Moines, Iowa.

BY ORDER OF THE SURGEON GENERAL:
EDWARD CURTIS,
Assistant Surgeon & Brevet Major, U. S. A.

Fig. 5.4 (**a** & **b**) Edward Curtis photographs of the 1869 solar eclipse. {Historical Collections of the National Museum of Medicine}

b

Fig. 5.4 (continued)

Solar photography was continued for a few days after the eclipse. Then non-expended materials and equipment were packed and shipped. The observatory was disassembled. As soon as he could, Curtis turned his plates over to Harkness. They eventually were stored at the Army Medical Museum in Washington. The Naval Academy telescope was returned; it remains in use for public 'star parties' today.

Fig. 5.5 The Naval Academy telescope has been refurbished and remains in use. {James Dire}

The pet project of Harkness himself was, of course, spectroscopy. His spectroscope was inferior to Charles Young's (Chapter 7); it was difficult to isolate particular parts of the eclipsed Sun with the telescope/spectroscope combination available to him. In spite of this, Harkness's innate artistry with any kind of instrumentation made it work.

A less colorful figure, Harkness's fellow Naval Observatory Professor, John Eastman (1836–1919; USNO) was in attendance, too. His intention was to stand in the nearby woods in order to conduct meteorological studies. Eastman wanted little to do with any 'new' astronomy.

(Perhaps Eastman had had enough of excitement. He had participated in the Valley Campaigns of 1864.)

Instead, Eastman was recalled from the trees and sought out prominences for Harkness to observe, through the telescope's sighter. He placed a needle point that Harkness had installed in the sighter upon a prominence, and Harkness recorded the featured' spectrum. One prominence. A second. A third. He asked then for the corona. Eastman obliged, but with a disappointment that mirrored Curtis's, Harkness no longer saw anything through the spectroscope. Eastman tried another spot. The result was the same. Harkness had the sudden idea for an on-the-spot adjustment to the spectroscope. He dexterously made it in record time. Meanwhile, Eastman aimed at a particular bright spot on the inner corona. Success. A continuous spectrum appeared. A perfect rainbow. Wait. Harkness saw something peculiar: a single line, a shade of green, much brighter than the rest of the spectrum. It should not have been there. His discovery, while others similarly equipped failed, speaks to the work of an outstanding observer.

Yet even Harkness allowed himself to become distracted. Nobody had expected there to be so many prominences! Harkness drifted from thinking about the corona. He asked Eastman to point the telescope/spectroscope once again toward the most conspicuous prominence.

"You have had it already."

"Never mind, give it to me again ..."

Harkness now had observed spectra of both prominences and the corona. The telescope itself otherwise attendant upon the spectroscope, he stole a peek through its finder. "My view of it may have lasted five or ten seconds, certainly not more, yet the magnificence of the spectacle is so indelibly impressed upon my memory that it will be years ere I forget it." Harkness switched places again with Eastman for one more prominence spectrum, but the flash of light accompanying third contact caused Eastman to pull away and state the obvious: "All over."

Harkness had drawn the spectra he observed in the pages of a notebook. After totality, Eastman sat down and drew the corona and prominences, as seen through the finder, from memory. The two astronomers matched which prominence belonged to which spectrum. It turned out that, except for some variations do to differing brightness, the spectra of all the prominences were the same.

What do astronomers do the day *after* a total eclipse of the Sun? Sleep in? Relax taut muscles? Have a good meal? A celebratory drink?

The next morning, Harkness compared the spectrum of the uneclipsed photosphere with that of salt (sodium) dropped into a flame. From the resulting fiducial lines, he could better place the spectra he had drawn during the previous day's total eclipse of the Sun.

Harkness realized that small differences in the prominence spectra correlated with when during the eclipse the prominence was observed, and with how far it was from the center of the Moon. In other words, if the same part of each prominence was observed—its base, or its top—each resulting spectrum would be identical.

Finally, the best overall description of the total solar eclipse, as seen from the USNO physical-astronomy station, comes from Eastman:

The corona seemed to be composed of two portions [inner and outer], both visible to the naked eye, in which, with the small instrument which I used, I was unable to trace any similarity of structure."

The portion nearest the Sun appeared to be 1'[3] high, forming a nearly continuous band about the Sun, and appeared to be a mass of nebulous light, resembling in structure the most brilliant, irresolvable portions of the Milky Way. Its color was silvery white and, like its density, appeared the same throughout its whole extent. The outer portion consisted of rays of light arranged in two different ways. In five places they were arranged into groups resembling star points composed of slightly convergent and radial rays, but elsewhere were disposed as radial lines. The color of the bases of the star points and of the radial lines was the same as that of the inner portion, while the outer portion of the points had a very faint greenish-violet tint. The radial lines were the most prominent.

Four of the star points project[ed] farther from the Sun than the ordinary radial lines, and give the contour of the corona the form of a trapezoid. Between the protuberances Nos. 4 and 5 scarcely any corona was observed.

A Minority Viewpoint

John Eastman was nearly alone in finding the total eclipse of the Sun to be less than the overwhelming spectacle that he had wished for. He also wrote this:

> I was considerably disappointed with the appearance of the color and brilliancy as well as with the extreme contour of the corona. Most observers have described the color as "pure" or "clear" white and the light as very brilliant, while nearly all the published sketches represent the contour as nearly circular and regular and the coronal rays as radial and equally distributed about the body of the Sun.
>
> The color of the corona, as I observed it both with the telescope and without, was a silvery white, slightly modified in the outer portions by an extremely faint tinge of greenish-violet, and I could not detect the least change in the color or position of the rays during totality. The light of the corona was not brilliant—perhaps from the effect of haze—but appeared more like the pale light from the train of a meteor than like anything else that I could recall at the time.

Eastman's aesthetic appears to have favored the classical Greek view of the heavens as a place for absolute symmetry and uniformity. One man's "magnificence" is another's disappointment.

---

[3] one arcminute, or 1/60th of one degree

Des Moines was abandoned by the astronomers. Newcomb, Harkness, Eastman, and Hall—see below—prepared for a trip to Europe, in order to observe the total solar eclipse of 1870, just sixteen-and-and-a-half months in the future.

Harkness summarized the 1869 experience, but it could have been Newcomb: "The event, for the proper observation of which we had traveled so far, and spent so much time, and thought, and money, had come and gone, and now it remains for the scientific world to judge whether or not we made the most of our opportunity."

Local lore eventually forgot about the USNO astronomers; it told of a visit from scientists belonging to (for some reason) the Smithsonian Institution. Anyhow, the event itself lived on in the name of the Eclipse Coal Company.

$$* \quad * \quad * \quad * \quad *$$

Far from Des Moines, we now come to Sand's and Newcomb's most audacious plan: to station an observer as far up the eclipse path as it seemed was humanly possible. The point of maximum eclipse could not have been more inconveniently placed: It was in the Saint Elias Mountain Range on the Alaska-Yukon border (duration of totality = 3 minutes, 48 Seconds). Mount Saint Elias (Yas'éit'aa Shaa in local Tlingit) is the second highest mountain in the United States and rises to 5,489 meters elevation in just 14 horizontal kilometers. (There are extensive alpine glaciers; even today geologists call the vicinity the Eclipse Icefield.) The range is inaccessible for practical purposes—First Nation people crossed it, but only in winter with snowshoes. Yet a USNO man would get as near as he could.

The astronomer chosen to play the role of Odysseus was Professor Asaph Hall III[4] (1829–1907; USNO). Why? This was to be not only the most distant eclipsed station established by anybody, but required a much, much longer time to reach than any other. It demanded a sea voyage that was to be arduous and disproportionately lengthy. This was to be also the arm of the USNO total-eclipse expedition least likely to yield results (due to typical weather at the locale). How did this journey fall to Hall? It was not that Hall was the most junior USNO astronomer; he was a near contemporary of Newcomb. According to Sands, Hall volunteered. According to Hall, he was ordered. We do know that Newcomb and Hall never really got along. It may have seemed prudent to separate them by a continent.

Regardless, on 21 May 1869 Hall and his traveling companion found themselves on a steamship and bound for San Francisco, California. The second man was not a stranger; he was Joseph Rogers (1840–1917; temporarily USNO) of the United States Hydrographic Office. The two crossed the Isthmus of Panama by rail, caught another commercial ship, and arrived at their destination on 12 June.

---

[4] It was Hall who was on duty alone, according to a famous story in which President Lincoln supposedly knocked on the Observatory door and asked to take a look through the telescope.

They then boarded the Navy's sloop-of-war, USS *Mohican*. The steamer put into port on Vancouver Island and then stopped at Unalaska in the Aleutian Islands.

Their final target? It was not the USA's newly acquired Alaska. It was Providence/Plover Bay, *Russia*. [Бу́хта Провиде́ния].

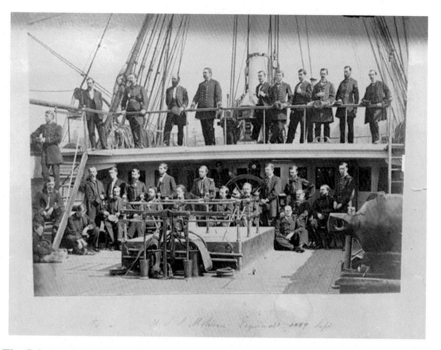

Fig. 5.6 Asaph Hall aboard ship on his way to the total-eclipse path in Russia. 1869. I believe that Hall is the civilian standing at the extreme left upon the upper deck. {United States Naval Observatory}

The following reasons led to the selection of this place for observing the eclipse: the computed line of central eclipse passed only six or eight miles off the coast; an excellent harbor, ease of access, furnishes a secure anchorage for all kinds of vessels; and, from all the information I could obtain, the chance of having good weather is better there than on the American coast, and as good as at any point in those regions, near the line of central eclipse, that the short time remaining for the voyage would permit us to reach.

The Russo-American Treaty of 1824 allowed American shipping to call at Russian ports. Hall does not communicate whether permission was solicited from the Russian government to moor an American *warship* off that nation's coast.

Siberian locals pointed out anchorage at Emma/Komsomolskaya [Комсомольской] Harbor. And so, two Washingtonians, desk-bound sailors at best, found themselves deposited a quarter of the way around the world, at a

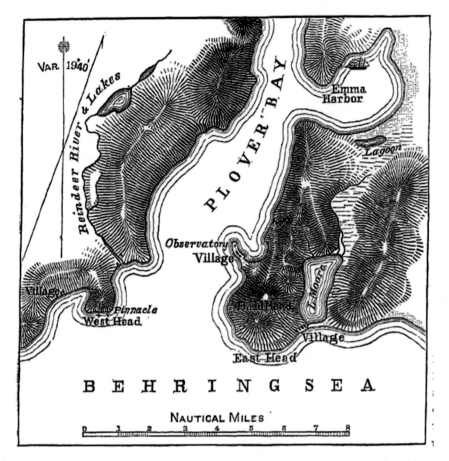

Fig. 5.7 Section of the United States Coast Survey Bering Sea chart showing Providence Bay. 1928. {National Oceanic and Atmospheric Administration Historical Map and Chart Project}

whaling station, nearly on the Arctic Circle. It was one week before the total eclipse of the Sun.

The day after arrival, Hall and Rogers mountaineered about the surrounding range. They soon realized that lugging all their instruments up (and down) the steep trails to these peaks was asking for trouble. The two decided to split up, making their primary stand on the shore upon Napkum/Plover Spit.

Sergei Maslikov (President, Novosibirsk Astronomical Society [Новосибирское астрономическое общество]) researches solar eclipses visible in Siberia. Thanks to him, I have learned that (certainly unbeknownst to Hall) another astronomer was preparing to observe the total eclipse of the Sun several hundred kilometers to Hall's West. He was Estonian Karl Neiman [or Neumann] (1830s-1887; Russian Geographical Society [Русское географическое общество]), who was a member of the Chukotka [Чукóтка] Expedition to explore far eastern Siberia. Neiman knew when the total eclipse was to occur. His problem was that he did not know exactly where he was in relation to the path. Neiman described his location as, "...in the upper reaches of the Nerpichya [now Kanchalan = Канчалан] River ...flowing into the Anadyr [Анадырский] Bay." He just missed his mark. Neiman was ready with his 3-inch-aperture Fraunhofer refractor. Unfortunately, he only saw a partial eclipse of magnitude 0.996. So close! The difference, though, was literally that between night and day. Alas, it appears that Neiman would have been the first astronomer to see the 1869 total eclipse of the Sun. His is the farthest solar-eclipse report West or North of which Maslikov or I know.

Back on the beach, The *Mohican*'s carpenter threw up an observatory for the American astronomers. It housed a telescope on loan from the University of Pennsylvania and one much smaller (2–3/4 inch) used by their ship's navigator. Miraculously, nothing had been lost or damaged on the voyage.

> Almost the only vegetation to be seen in this desert were mosses; besides these there were a few small, wild flowers [*sic*], and a little stunted grass. There seems actually to be no soil. The roots of these plants were matted together, forming a sort of crust; upon the removal of which, no soil is found beneath. The rock of which the hills are formed is crumbled up into small pieces—produced, I presume, by the continuous action of the frost.

Surrounded by this tableau, Rogers manned the observatory. Hall climbed up the easier-to-reach peak of Bald Mountain with a 3-1/4-inch refractor. Hall's earlier assertion about climbing, *vis à vis* doing so would be despite the inherent difficulty and danger, loses some impact when he later refers to the ten locals who guided him and helped with his baggage.

Fig. 5.8 Sketch of the lower portion of Providence Bay, including Emma Harbor, showing the location of the two temporary 1869 solar-eclipse observatories and the anchorage of the *Mohican*. {Alamy}

First Contact: 08:14:36 AM ANAT
Altitude = 31° Azimuth = 120°

Second Contact: 09:18:19 AM ANAT
Altitude = 36° Azimuth =137°

Maximum Eclipse: 09:19:56 AM ANAT
Altitude = 36° Azimuth = 137°
M = 1.0236
Velocity of the Moon's shadow = 1.02 km/s
Path width = 244 kilometers

Third Contact: 09:21:34 AM ANAT
Altitude = 36° Azimuth = 138°

Fourth Contact: 10:27:27 AM ANAT
Altitude = 40° Azimuth = 157°

Both men were in place plenty of time before the early morning (8 August) total eclipse of the Sun. The *Mohican*'s Captain, officers (and—presumably—crew) watched the total eclipse from offshore.

Clouds. "Light, ashy" clouds. The fear of every eclipse enthusiast came true. There was a peek of eclipsed Sun, and then more clouds. And for those who had ventured farthest to see it. The remainder of the day was warm and sunny.

A naval officer passed his verdict: "…for scientific purposes, I do not think [Hall's and Roger's] observations were of much avail." Nevertheless, the two observers kept their chins up. They had made useful geographic measurements of latitude, longitude, and magnetic bearing along the way; they would continue to do so on the way back.

The long journey to the Bering Sea in order to observe a total eclipse of the Sun was a climatological long shot from the start. Hall and Rogers departed Plover Bay hours later. Hall returned to the USNO no worse for wear.

Ironically, it was Newcomb who was felled. Only in his diary do we read that he suffered an unspecified illness after the total solar eclipse, for which he was treated with quinine. (I 'diagnose' it as a recurrence of malaria, which he had contracted as a younger man.) Newcomb writes privately,

Sick at the Savery House, Des Moines, Ia., 1200 miles from home. A dismal prospect.

When Newcomb finally was able to take his leave of Iowa, it was almost three weeks after his arrival, nearly half of which time he spent in bed. On the train connections back, he was treated to the best of accommodation: double seats for laying prone, then a sleeping car, and finally his own suite-like coach. Still, he "fe[lt] more the worse for journeyed ." It did not help that the proud Newcomb lost his pocketbook *enroute* and had to borrow $5 just to get home.

Newcomb was unable to walk for the rest of the month. That Newcomb, upon return to Washington, seemed to care little about the volume of data collected on his expedition might find a partial explanation in the bad memories of this disagreeable coda.

6

New Astronomy in the Old West

... Great Pan is dead."
ancient, but quoted at the time of the 1869 eclipse

The necessity of bringing a physical astronomy armamentarium to a total eclipse of the Sun was not recognized uniquely by the Naval Observatory. The USNO expedition had home-town competition. Simon Newcomb tells us it was from "a rather dilapidated old dwelling-house, about half a mile or less from the Observatory, in one of those doubtful regions on the borderline between a slum and the lower order of respectability."[1]

The organization was the United States Nautical Almanac Office [NAO], also on hand for the total solar eclipse. The NAO was charged by Congress with preparing publications of use to navigators. While the USNO set time, the NAO packaged everything else of need to the navigator. The two were separate institutions in 1869. The NAO's work product was the *American Ephemeris and Nautical Almanac*. Preeminent American astronomers, foremost Benjamin Peirce, had argued successfully for a distinctly American almanac for use by the Navy as opposed to those printed in Europe.

The NAO was staffed by scientists even more number-obsessed than those of the USNO. Starting the very year of the total solar eclipse, Simon Newcomb had begun to petition for transfer to NAO. It did not happen until later in life, when he had risen through the ranks such that he could have nearly any job that he wanted. In fact, he would be named Superintendent of the NAO, sparing himself without question any further evenings occupied by routine telescope work.

---

[1] at Washington's Nineteenth and Pennsylvania Avenues

© The Author(s), under exclusive license to Springer Nature Switzerland AG 2023
T. Hockey, *America's First Eclipse Chasers*, Springer Praxis Books,
https://doi.org/10.1007/978-3-031-24124-6_6

This was to be the NAO's second eclipse expedition: It had sponsored Newcomb and company's ill-fated voyage to Hudson Bay for the total solar eclipse of 1860.

At first glance, the NAO mathematicians had less reason to be in the path of a total solar eclipse than did the USNO astronomers. Nevertheless, starting by at least Autumn 1868, an eclipse expedition coalesced. It was the brainchild of Superintendent John Coffin (1815–1890; NAO), a charter member of the National Academy of Sciences [NAS]. Coffin seems to have considered himself foremost a Naval officer; he literally wrote the textbook on navigation at sea.[2] Yet, Coffin, too, saw a role for government astronomers within the umbra of the Moon.

Fig. 6.1 John Coffin. {Special Collections, University Library, University of California at Santa Cruz}

While still primarily an administrator, Coffin was more hands-on than Benjamin Sands regarding his total-eclipse endeavor. Unlike his counter-part Sands, Coffin had experience as a telescopic observer, which ended only after he contracted a disease of the eye.

---

[2] Not long before the total eclipse he published a new edition. Coffin, J. H. C. *Navigation and nautical astronomy.* 4th edition. New York: D. Van Nostrand. (1868)

In the social structure of the NAO Expedition, there was no one dominant *astronomer* the likes of Simon Newcomb at USNO. The NAO expedition model was quite different from that of the USNO. It solicited help and accompaniment from a slew of 'free agents' outside the Almanac Office. The greatest success of the NAO Expedition was not the result of work done by NAO staff, it was that done by these 'civilian' participants largely from academia. NAO provided assistance, logistics, and bankroll.

Coffin was a good salesman. He obtained the endorsements of the American Academy of Arts and Scientists, the American Philosophic Society, and Franklin Institute. He then went to the United States House of Representatives with open palm. Whether or not it understood that Coffin was just as interested in physical observations of the Sun during eclipse as he was measurements of time and position, nonetheless, Congress appropriated $5,000 (over $100,000 today). It was "for observation of the eclipse of the Sun in August, 1869, under the direction of the Superintendent of the Nautical Almanac." The budget, authorized by the year's Naval Appropriations Act, was "for defraying expenses of observers, of moving instruments of sufficient power to the line of the central eclipse, and of preparing for abundant photographic records and spectroscopic observations." (In it unknowable what the effect on solar-eclipse expeditions would have been if Black Friday, the financial panic of 1869, had occurred two months earlier.) While government money for exploration dates to Lewis and Clark, this was the first allocation of such special funding for an astronomical expedition.

(I can find only one example of a Congressman actually traveling to the path of totality in order to see the eclipse of the Sun for himself. This was Senator Garrett Davis, who watched it from his home state of Kentucky.)

By Spring 1869, Coffin already had published a pamphlet, as a Supplement to the *Almanac*, with general advice for the public interested in observing a total solar eclipse. That this addendum was well read is indicated by the repetition in newspapers and magazines of a phrase that is unique to it: In throwaway hyperbole, Coffin writes that 1869 will be the last total solar eclipse to cross the United States during the century. This statement was false, and that it was so would have been known to anyone at the NAO who might have proofread the text.

Within government, there is always a committee, of course. In this case it was a subgroup of the fledging National Academy of Science [NAS], a body established by President Lincoln. The eclipse committee was chaired by another charter member of the NAS, total-eclipse veteran Professor Stephen Alexander[3] (1806–1883; Princeton University). Coffin, after checking where others were going and with Alexander's advice, chose Burlington, Iowa, as the Expedition's main destination. Establishing where in the world Burlington was again became the job of the United States Coast Survey. (The USCS did this for most of the NAO stations; often the employee surveyed before or after taking his own position for observing the solar eclipse.)

---

[3] also an ordained minister

Burlington sits on bluffs overlooking the Mississippi River, about 2,500 kilometers from its mouth. The city was established in 1833, not long after Native

Fig. 6.2 Burlington, Iowa. 1889 perspective map. {Library of Congress Geography and Map Division}

Americans had been driven from the area in the Black Hawk War. Settlers streamed in *via* ferry across the river. Burlington eventually became a steamboat port.

The city was the territorial capital of Iowa and eventually a major national railroad nexus. (The first-ever iron bridge across the Mississippi River, at Burlington, was completed less than a year before the total eclipse of the Sun.) Its population, according to the 1870 census, was 14,930—greater than that of Des Moines. In 1869, Burlington also was another source of support for the ever-resourceful Coffin: "Previous to my arrival, the city council of Burlington appointed … a committee 'with full power to extend all needed facilities to those who should visit the city to take observations of the eclipse.'" After Coffin's arrival:

> Besides other considerate attentions, they provided a special police officer, to guard the observatory at night; and their presence, with several city officers, at the time of the eclipse, was sufficient to prevent the pressure of a crowd, and to secure for us the quiet and freedom from interruption or distraction which were essential to the success of our observations.[4]

---

[4] During a total eclipse of the Sun, I have had a crowd surround me while observing through only an extremely modest two-inch-aperture telescope.

Further largess came from the Pennsylvania Central; Chicago, Burlington, and Quincy (Great Burlington Route); and Burlington and Missouri railroads. They not only ponied up free transportation, but outfitted a special car for the expedition participants (and most of their many assistants) with certain seats removed for equipment. An employee of, and guide to, the railroads accompanied them, also at company expense. The eclipse car patiently waited through the event to take the men and instruments home. The estimated expenditure spared on transportation was $1,500.00.

The NAO had no telescopes! Philadelphia Central High School[5], Pennsylvania College[6], and the University of Pennsylvania granted Coffin the use of theirs. (See "Philadelphia axis" in Chapter 9). Coffin even talked Sands into lending him chronometers and other equipment, which one might have thought the USNO superintendent would have horded for his own eclipse purposes.

NEW BOYS' HIGH SCHOOL.
BROAD STREET.

Fig. 6.3 Philadelphia Central High School, with its unusual feature of an observatory dome. Corner of Broad and Green Streets, Philadelphia, Pennsylvania. {Philadelphia Architecture and Buildings}

---

[5] The refractor was destroyed by fire in 1947.
[6] now Gettysburg College

Western Union gave Coffin free use of the telegraph. Finally, a civic-minded landowner lent a vacant lot between 7th and 8th streets and Elm and Maple streets for a temporary NAO observatory.

Today we would call Coffin a master of either grantsmanship or the handout. After all, the United States was just coming out of a resource-depleting war; the national coffers were not completely restocked. Yet Coffin had been around the proverbial block for a while. He had the right words, the right connections, and the right favors to be called in. With such an abundance of cash and materiel, he was able to outfit, not just Burlington, but two other Iowa stations. This was a smart move in case one or more experienced clouds under the region's generally clear but unpredictable skies of August.

Dr. Benjamin Gould (1824–1896; formerly of the Dudley Observatory) was the first American to receive a PhD. in astronomy. Though prominent in his field (*e.g.*, a charter member of the NAS), Gould was forced to resign from USCS when Benjamin Peirce was named Superintendent. The two did not get along. So, he was between jobs (for instance, his at the United States Sanitary Commission) and had plenty of time on his hands to devote to the total eclipse of the Sun. Gould acted as a sort of attaché to Coffin. He brought several cronies with him to Burlington.

Fig. 6.4 Benjamin Gould. {Dudley Observatory}

Also accompanying Coffin to Burlington was a traditional Professor, Charles Young (1834–1908; Dartmouth College).[7] As was Young's father. As was his grandfather.

---

[7] Early in his career, Young founded the Loomis observatory at Western Reserve Academy. It is today the oldest observatory in the United States still in its original location. In 1988, I was fortunate in having the opportunity to attend the Observatory's *sesquicentennial* celebration.

The youngest Young agreed to join the NAO eclipse team. His intend was not just to observe the eclipsed Sun, but to try to ascertain its nature. While not as rare in Europe, Young was among the few scientists in the USA who could practice observational astronomy but also had a physics background that prepared him to interpret physical data. (In the American model, most astronomers' preparation was in mathematics only.)

Fig. 6.5 Charles Young. {Department of Special Collections, Princeton University Library}

Young led the kind of life that defeats popular biographers. It was filled with neither heroic highs nor scandalous lows. Young, "Twinkle" to his students, was both liked and admired by those who met him. A youthful female astronomy devotee present at the 1869 total solar eclipse wrote of Young that, "We were all charmed with Prof. Young, whose modesty, tho' he was beginning to be famous, was in striking contrast to the "sirs" of some of the other masculine scientists".

Young brought physical instrumentation 'cred' to the NAO party. He was an expert in astronomical spectroscopy, particularly as it applied to the Sun. The laws that govern the break-through analytical technique of spectroscopy were discovered only a little more than a decade previously. Spectroscopy was a watershed gateway to physical understanding of astronomical bodies and was as 'cutting edge' as it got in 1869. Congress's request for "spectroscopic observations"? Check.

Recall that Coffin and the rest of the NAO could more accurately be called mathematicians than physical astronomers; they knew little about spectroscopy and the only slightly older tool of physical astronomy, photography. This may seem strange today as we often *equate* astronomy with photography. Astronomers now rarely look through the telescopes they use. Today's imaging techniques are far more sensitive than the human eye.

However, in 1869, photography was still a specialized trade. Many astronomers would no more think of mastering the array of mechanical contraptions and bottles of chemicals necessary to produce a photograph than they would the tools of the blacksmith to shoe their horse.

So, the NAO eclipse expedition contracted with the Franklin Institute's *ad hoc* Philadelphia Photographic "Corps," camera enthusiasts from the City of Brotherly Love, who would accompany the NAO astronomers. (Today, we call the result a government/NGO consortium.)

The Franklin Institute side was organized by newly minted Chemistry Professor, Henry Morton (1836–1902; University of Pennsylvania), who was also the Secretary of the Institute. He was known for his organizational skills and not a little showmanship. Morton staffed his half of the expedition with "… those whose position or engagements would allow them to volunteer without other compensation than the moral contingent [up]on success." Together, the entire operation was the biggest expedition to observe the 1869 total eclipse of the Sun.

Professor Alfred Mayer (1836–1897; Lehigh University) was in charge of photography *in situ* at Burlington. The astronomers plus the photographers set up their equipment in what is now Burlington's South Hill Park (where a bronze plaque still marks the spot at which their stone pier once stood). "… abundant photographic records"? Check.

The NAO Burlington station sat a safe "… 156 feet above [the] high-water mark of the Mississippi River … and about 630 feet above the ocean." It was built on clover grass, surrounded by a hedge. Most importantly for the astronomers' precious optics, the site was comparatively dust-free compared to the streets of the average western town.

The observatory consisted of a single building, five meters square, with a one-and-a-half-meter-square annex at its corner. The complete structure sheltered the Philadelphia Central High School telescope and was equipped with two photographic darkrooms. There would have been room for little else in this glorified chicken coop. The roof was paper and gave the astronomers concern during a local storm.

Within the station, a cross of eight-inch beams was erected with which to stabilize the large refractor. The USCS was not content with measuring the position of the lot or even this building. They were happy to report the exact location of the post. The observers would be twenty kilometers from the total-eclipse's centerline.

Fig. 6.6 Temporary NAO solar-eclipse observatory at Burlington and its staff. The visiting woman seated in black is Maria Mitchell (Chapter 12). She is dressed in mourning for her recently deceased father. Charles Young also is seated, near the center. {United States Naval Observatory}

The expedition arrived in Burlington on a sunny 4 August, leaving little time to spare. The next day they mounted their telescope; the Philadelphia Central was placed on an equatorially aligned mount with clock drive. The significance of this was that it could track the Sun (for short intervals) on its own, with only a little help. The length of the refractor would have been such that manipulating it to fit within the confines of the building must have constituted hot, inglorious labor, in closed "thick and smoky" air. Astronomers' normal, day-to-day work rarely produced a sweat.

Then the clouds rolled in. And the wind, which rattled the uncovered telescope.

After their lengthy trip from the East Coast, what if the instruments had become maladjusted or, worse, damaged? Attempts to test them by observation of the stars were delayed by the weather. Coffin, Gould, and Mayer completed this exercise between the clearing of the sky after midnight on 7 August and just before sunrise. As if scripted for drama, the telescopes checked out in the proverbial 'nick of time.'

First Contact: 04:00:45 PM CST
Altitude = 35° Azimuth = 261°

7 August was completely clear after 10:00 AM. Preparations for the eclipse were made, though—as always is the case with the fastidious—not to the complete satisfaction of the astronomers.

As totality approached, Young, Mayer, and the rest of the latter's photographers worked inside the observatory. Coffin and Gould stayed outside about six meters west, presumably so that the chatter between the two groups with differing tasks would not interfere with each other's.

Young had brought along a 2-¾-inch-aperture refractor on a stand set up for group viewing and contact timing: It had a card mounted on it onto which the Sun's projected image, 15 centimeters in diameter, was shown to Gould's consociation.

Meanwhile, another team member stood just outside the door. He shouted out the time, second by second, so that all could hear.

Rank has privileges. Coffin studied the total eclipse of the Sun through a telescope specially made by Alvan Clark & Sons for solar observing. It was an equatorially mounted refractor, 3-inches in aperture. A flat mirror inside it directed light to the eye piece at a 45° angle to the optical axis for comfortable observing. Instead of being painted with a patina of shiny silver, Alvan Clark had left this mirror uncoated. Light hit the surface of the glass itself. That way, only about half of the Sun's rays reached Coffin's eye.

Even so, it was still necessary to filter the telescope. Coffin had three neutral-shade filters that he used during different parts of the eclipse: the optically thickest around first and fourth contact, the optically thinnest near second and third. The result was a safer (though not actually safe) view of a perfectly defined solar disk. Coffin reported that the corona through his special refractor was the color and brightness "of the Nebula of Orion, as seen through a large[r] telescope."

Gould was trained on the great permanently housed telescopes abroad. That did not translate necessarily into experience with smaller instruments. He intended to use for visual observing a specially engineered five-inch-aperture comet seeker of his own design. A special accompanying feature was a set of eye pieces blacked out to different radii from the center of the lens; these could be used to blot out brightness close to the Sun so as to see fainter structure or objects farther from it. (Gould's telescope is now in the Smithsonian Institution.) Unfortunately, it did not work as intended.

Thus, instead of using his own telescope, Gould borrowed a look through the four-inch-aperture Clark refractor of Professor/Reverend[8] James Clarke (1810–1888; Harvard University), out from Boston, Massachusetts, for the event. Half a minute before first contact, he broke it—the filter, at any rate.

In a spiral of decreasing aperture, Gould then hopped over for a peek through Young's 2-¼-inch Clark. Luckily, Young came to the solar eclipse 'telescope heavy.' Nevertheless, considering Gould's bad luck with telescopes so far that afternoon, would you lend him yours? Gould ended up by watching the total part of the eclipse with opera glasses. At one point he was stuck using a peculiarly tinted violet filter.

---

[8] of Natural Religion and Christian Doctrine

Fig. 6.7 Benjamin Gould's custom-built comet seeker. {Division of Medicine & Science, National Museum of American History}

While we do not know the fate of most small refractors used during the 1869 total solar eclipse, we do know the present location of one of them. It is safely in the hands of Clark telescope expert Kenneth Launie. He tells me that James Clarke was

> … the owner of an early (unsigned) telescope made by Alvan Clark. The two shared a common ancestor, although I do not know if they were aware of it at the time of purchase. I think that it was paid for by Clarke's wealthy father-in-law. Clarke wrote a letter to his daughter describing seeing the 1869 eclipse with his telescope in Burlington.

> The telescope survives, though not its original mount (or filter remnants, not surprisingly). The tube consists of tin wrapped with papier-mâché that is painted to look like wood, similar in construction to the 1852 Williams College Clark.

Retracing our steps to one sunny day in Iowa.

Second Contact: 05:02:09 PM CST
Altitude = 23° Azimuth = 272°

Gould was very honest in his subsequent recollection. He admitted that he became confused between Baily's Beads[9] and the emergence of the chromosphere (and by the premature reaction of the surrounding crowd) and that he had a hard time judging second contact.[10] Despite this, in a moment, the total eclipse was obvious to all the observers.

---

[9] The last (or first) bits of photosphere seen are broken up by the irregular limb of the Moon, as sunlight glints through lunar valleys and is blocked by lunar mountains.

[10] I feel that Baily's Beads are misnamed. "Beads" implies a constant phenomenon. In reality, their brightness is always in flux.

Totality! Coffin saw what must have been the solar chromosphere as a "… band of rose-colored light, closely skirting the Moon's limb … [standing] in bold relief from the corona."

And "eagles wings":

… a large prominence appeared suddenly to shoot up from the eastern limb of the Moon, pyramidal in shape, but rounded at the top, and its outline broken and irregular. It appeared to me of a beautiful rose-color, very soft and delicate, more deeply tinted at the top than at the base, and to stand out in bold relief from the white corona, which apparently formed the background.

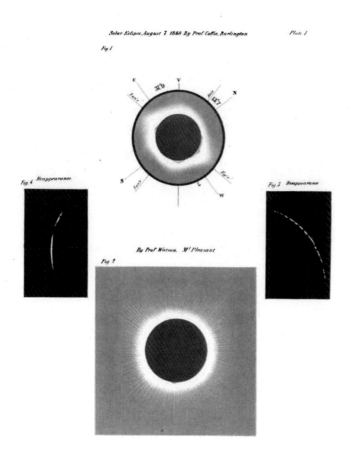

Fig. 6.8   Top: John Coffin drawings of first ("fig 3") and second ("fig 4") contact and of prominences, including the one visible to the naked eye ("fig 1"). Bottom: artwork ("fig 2") by James Watson (Chapter 7) showing the corona. {United States Naval Observatory}

Echoing his Naval Observatory colleagues, Coffin noted the bright, rhombic corona. To him, it seemed elongated at the azimuth of the prominences. Its light reminded him of the streamers of the aurora borealis ('northern lights') or the cloudy patch in a winter's evening sky (again), the Orion Nebula.

Maximum Eclipse: 05:03:36 CST
Altitude = 23° Azimuth = 272°
M = 1.0202
Velocity of the Moon's shadow = 1.54 km/s
Path width = 255 kilometers

An Ancient Instrument Put to Modern Use

Conversely, Gould had better-than-might-be-expected luck with 'low tech.' He brought with him to Iowa one of the oldest astronomical instruments known, the Jacob's Staff (though he did not call it that).

In Gould's version, a disk slid up and down a graduated sighting bar. When the disk obstructed the naked-eye corona, simple trigonometry yielded its apparent angular size. But as Gould's eyes adjusted to the darkness, the appearance of the corona changed, giving the impression of movement within it. He had to pull the disk in farther and farther to incorporate what he could now see.

Gould's experiment was predicated on a *round* corona; how to account for the appendages, which, in his words, were, "… like streaming remnants of some pale, tattered banner"? He finally decided that the whole corona fit into a circle a little less than a degree in diameter. He thought that if it were possible to block out just the bright inner part, he would be able to see the corona extend farther still.

Gould made hurried sketches of the totally eclipsed Sun that focused upon prominences, including the large one that was easily visible to the naked eye. It even was visible to him slightly before second contact or a bit outside the zone of totality by some. Gould was in friendly artistic competition with his companion Brigadier General Edward Wild (1825–1891; formerly USA)—Gould liked to travel with a posse—who made a more careful drawing. He did so even though he was single-armed due to one of his several Civil War injuries. (Wild is a problematic figure in history: He is respected for his abolitionist views and leadership of African American troops from 1863–1865, but he also committed what today would be regarded as unpunished war crimes.)

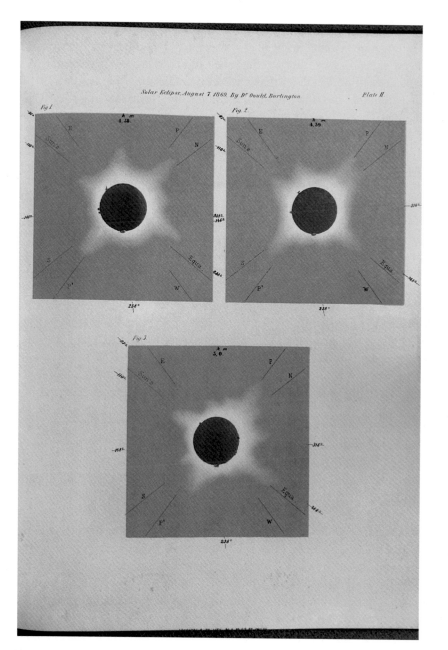

Fig. 6.9 Benjamin Gould's drawings of totality. {United States Naval Observatory}

Meanwhile …

Young's colleagues regarded the total solar eclipse as variations of brightness and color in the normal two dimensions across their field of vision. Through his spectroscope, Young observed a single-dimension version of the solar eclipse—quite different from our usual perception of space.

Most hot, glowing objects produce a continuous spectrum: All wavelengths within a range are present, and one color seems to blend into another.

In a few special cases, though, the result of spectroscopy is different. When a continuous spectrum (such as that produced by the interior of the Sun) passes through a cool, thin gas, the chemical elements present in that gas absorb specific wavelengths. This pattern of dark absorption lines superimposed over the continuum provides a fingerprint identifying the elements present.

Similarly, a hot, thin gas itself will glow, but only at those *same* wavelengths characteristic of each element that exists within it. The result is a pattern of discrete, bright, emission lines. Once again, the elements identify themselves.

Spectroscopy is physical measurement because it is a means of determining the state of matter, temperature, and composition of a distant object without having to travel the impossible distance in order to sample it. Even more so than photography, spectroscopy is synonymous with the physical side of astronomy.

Young purchased his spectroscope for $350.00 in 1866 ($7,000 today!) from Alvan Clark & Sons.[11] It incorporated no less than five prims. Each prism spread out the spectrum more, so the resulting view was one of high resolution. Situated between the 4-inch-aperture objective of a Merz & Söhne comet seeker and its eyepiece, Young's instrument was unlike any spectroscope that had been used to examine the Sun before.

The telescope was mounted equatorially, and was aimed by "… my assistant, Mr. Emerson, with an accuracy of hand which answered almost every purpose of a clock-work [*sic*] …" Young oriented his spectroscope with respect to the vector on which he knew the Moon would encroach onto the Sun. This way, he was able to observe a spectrum beginning at first contact.

Looking through his spectroscope during a total eclipse of the Sun, Young's reality became brightness and color only—a succession of photosphere, chromosphere, and corona without morphology. No longer merely a white-saturated yellow, in truth an amalgam of wavelengths combined by our eye to appear nearly colorless, Young's Sun was broken down into its component hues. In the time it takes to boil an egg, he lived in a completely alien worldview without geometry, without shape.

Did Young's mind crave the normalcy of a naked-eye view? Time was precious searching out dark or light lines interrupting the continuous rainbow. Still, people had been looking at the Universe through a spectroscope for only two decades. Did Young raise his head from the eyepiece to peek at the, unusual, but nonetheless more familiar, sight? One made up of apparent width and height for which untold generations of genes had evolved his sense of vision? Regardless, using the meaning that you and I would give the word, Young acquired all those layers of traveler's dust only to–quite intentionally—miss the *sight* of a total solar eclipse.

---

[11] This instrument was in use until 1963; it is now in the Dartmouth College Museum.

First, Young saw the expected spectrum of the photosphere, continuous but interrupted by many dark absorption lines produced by its cooler, more tenuous, outer layer. Then, as he aimed at a selected prominence after totality commenced, the continuum grew faint and a few bright emission lines replaced those of absorption. Finally, Young targeted the corona and saw a still fainter continuum merged with the pattern of a few, hard-to-see emission lines: The particular one that attracted his notice was *green*. All this repeated in backward order as totality ended. A frenzied two minutes, fifty-two seconds of action, observe, remember, record … action, observe, remember, record …

Young had a tremendous advantage over William Harkness inasmuch as his total-eclipse obligation constituted just spectroscopy. Spectroscopy. And spectroscopy.

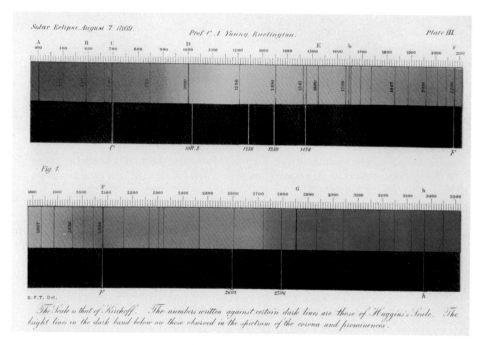

Fig. 6.10 Coronal spectrum observed by Charles Young during the 1869 total solar eclipse. {United States Naval Observatory}

Third Contact: 05:05:01 PM CST
Altitude = 23° Azimuth = 272°

Gould was fooled again. The total-eclipse phase was over.

During the expedition, the Philadelphia Central telescope was damaged. The government did not reimburse the high school until 1872.

Fourth Contact: 06:01:08 PM CST
Altitude = 12° Azimuth = 281°

Fig. 6.11 The marker at the spot of the NAO's temporary eclipse observatory in Burlington, Iowa. The location is now South Hill Park. The stone and plate were placed by the historic-preservation organization, Daughters of the American Revolution. {Photograph by the author}

We cannot leave Burlington without mentioning poor Professor Darwin Eaton (1822/23–1895; Packer Collegiate Institute), who tagged along with the NAO expedition. He had borrowed a two-inch-aperture refractor from (of course) Young. Unfortunately,

1.   Eaton's loaned telescope came with no filters.
2.   Eaton had to project the image of the Sun onto a white disk attached by wires. He was required to be content with looking at this much dimmer reflection instead of the direct view through the refractor.
3.   This was done necessarily to protect his eyesight during the solar eclipse's partial phases. Notwithstanding, it was clearly a 'Rube Goldberg' arrangement, done on the cheap.
4.   While trying to adjust the resulting contraption to view the eclipse, Eaton threw the Sun far out of field.
5.   There was no time for the inexperienced Eaton to try to repoint the refractor and no one not enjoined with their own eclipse-related task to help him.

6.   Eaton ended up observing the eclipse with his naked eye.

7.   "I attempted, however, to catch a view of the shadow [Chapter 13] in its flight eastward across the prairies on the opposite side of the river; but, unfortunately, several gentlemen happening to stand so as to intercept my view in that direction, I was compelled to change my position some ten feet or more, and before this could be done the shadow was gone."

Could it be worse? You decide. John Stockwell's (1832–1920; NAO), unmounted, 2-inch-aperture achromat with a red filter was adequate for solar-eclipse viewing. All the same, he was so enamored of the prominences that he seems to have *forgotten* to make any careful examination of the corona at all! While some were moved by the total eclipse of the Sun, and others were haunted, Stockwell appears to have been wholly confused.

7

Observing in Style

"… Eclipse Saloon, corner of Main and Eighth"

In 1869, the very word "expedition" conjured up great adventure in exploring. And sacrifice. There had been the Sir John Franklin expedition of 1845, in search of a northwest passage. It ended in entrapment within Arctic ice, starvation, and cannibalism. The Burton-Speke expedition, to locate the source of the River Nile (from 1857 to 1859), faced desertion, illness, wild-animal attack, hostile locals, and mind-numbing duration. (Neither example achieved its goal.)

In comparison, a total-eclipse expedition involved no terrific risk, deprivation, absence, nor solitude. While the train trip west was not an opulent by today's standards of travel, neither was it arduous. Seldom was food, lodging, or alcohol out of reach. After all, it was the era of the Grand Hotel.

The Canadians of Chapter 12 approached peril but could joke about it in the end. The major casualty of the 1869 total solar eclipse appears to be an astronomer who developed a case of eczema.

These were never excursions into the totally unknown. The sky was a familiar place. Astronomers knew more-or-less what to expect to see and when to expect it.

Perhaps in an unconscious attempt to elevate what they were doing to the popular connotations associated with "expedition," the junketing eclipse astronomers seemed to linger on description of any minor hardship that they encountered. However, as we shall see, asceticism is not a requirement for observing a total eclipse of the Sun.

Forty-five kilometers west of Burlington is Mount Pleasant, Iowa (incorporated 1856), a town situated on a bluff near the Skunk River. It had used to be the literal 'end of the road' insofar as it was connected to Burlington by a wood thoroughfare that later became the railroad right-of-way.

T. Hockey, *America's First Eclipse Chasers*, Springer Praxis Books, https://doi.org/10.1007/978-3-031-24124-6_7

Fig. 7.1 Mount Pleasant, Iowa. 1870. {B. J. MacDonald}

According to the 1870 census, Mount Pleasant was home to 4,245 people. The location of multiple educational institutions, it was the "Athens of Iowa." It boasted a new 1,000-seat opera house. Compared to many alternative eclipse-viewing locations, it was indeed "pleasant." It is telling that Mount Pleasant was one of the few expedition total-eclipse stations for which we know much about the hotel at which professional visitors stayed.

One cannot underestimate the positive effect of having a town marshal:

> The marshal's tasks also included rounding up stray livestock and warning the residents to clean up woodpiles and "filth and manure" in alleys because it was as menace to public health. The livestock were impounded until the fines imposed by the 1867 and 1869 ordinances were paid.

Mount Pleasant was an official substation of the United States Nautical Almanac eclipse expedition. In charge was Professor James Watson (1838–1880; University of Michigan). (Some sources erroneously place Watson at Burlington.) In some ways he was the exact opposite of Simon Newcomb, who commanded respect but seemed aloof and dour to his lesser colleagues. Watson was an affable and, perhaps, slightly clownish man; at 240 pounds, his students called him "Tubby."

Less dwelt upon was the fact that Watson was a wunderkind. At age 25, he was Director of the University's Detroit Observatory at a time when an observatory directorship was considered the top of the astronomical pecking order.

Watson also spent his life paranoid about going broke. He seemed to always be looking for profitable opportunities beyond the observatory. He was an insurance agent, and later an actuary. One can see the connection between astronomy and

Fig. 7.2 James Watson. 1855–1871. {Bentley Historical Library, University of Michigan}

statistics. But Watson's venture into the stationery business, selling photos and books, sounds far afield.

Once their presence became known, Mount Pleasant fell over itself to make the eastern scientists comfortable. According to Watson, they were "waited upon by local professors" and "other prominent citizens of the place, by whom hospitalities were extended and a desire expressed to aid in any way possible the successful completion of the arrangements preliminary to the observations to be made."

Watson does not come off as one well-suited to building construction. The Mount Pleasant mayor was pleased to inform him "that the city government would provide any structures that it might be necessary to erect." The astronomers chose to establish themselves 2-1/2 kilometers from the railroad station, on the Henry County Agricultural Society's fair grounds, where their observatory was an extant "hall of fine arts." There is no indication that Watson had to do any heavy lifting.

Moreover, the local (horse-drawn) omnibus line—just becoming popular in Iowa—provided transport for the astronomers. It drove them back and forth between town and observatory, at any time of day or night.

If it was cloudy? "The city council issued permits to half-a-dozen saloons and allowed them to remain open until midnight." None of the other eclipse stations had this kind of luxury.

The Nautical Almanac's Mount Pleasant station was not manned by NAO staff. Professor George Merriman (1834– ; University of Michigan) observed with a Michigan 4-inch-aperture comet seeker by Fitz. Donald McIntyre (1807–1891; treasurer, University of Michigan) was a former University regent and state legislator[1]; he got the 3-½-inch-aperture sighter belonging to Michigan's Detroit Observatory, large pier-mounted refractor.

---

[1] His longtime, cautious lobbying was successful the next year: The University of Michigan became one of the first co-educational universities in the country.

Fig. 7.3 The County Fair Grounds still exist today, near the railroad, just as they did in 1869. {Photograph by the Author}

Fig. 7.4 Contemporary Iowa omnibus. {Iowa Department of Transportation Library}

Ironically, the Detroit Observatory's director did not use a Michigan telescope himself! Watson observed through a 3-½-inch Clark that was normally the finder on Wesleyan University's hardly used 12-inch, also by Clark. It was mounted equatorially for the solar eclipse and must have shone brightly in the Sun.

Fig. 7.5 University of Michigan Fitz Comet Seeker. After 1890. {University of Michigan Bentley Historical Library}

Mount Pleasant, too, had an astrophysical vibe. At Watson's side was Professor John Van Vleck (1833–1912; Wesleyan University). (He was Watson's telescope connection.) For himself, Van Vleck brought to the total-eclipse 'table' an equatorially mounted, 3-½-inch-aperture refractor that sat atop a tripod. Onto it he mounted a single-prism spectroscope. Everything was manufactured by Clark & Sons. Neither telescope nor spectroscope had ever been used before!

Van Vleck had a neutral filter for his telescope. In some hard-to-explain, last-minute bargaining, he traded it to Watson for a green filter. Van Vleck watched the partial eclipse for the beginning of totality with the full aperture of his telescope. This abuse of the eye was certainly of no help when it came time to observe the light of the faint corona spectroscopically. In fact, he saw nothing of the corona through his spectroscope.

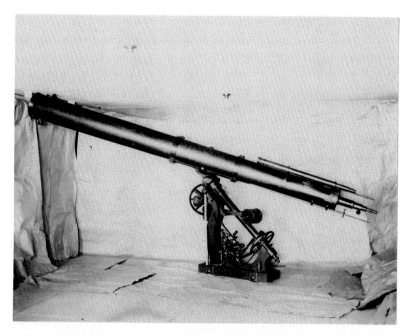

Fig. 7.6 The telescope used by James Watson in Mount Pleasant, Iowa, during the 1869 total eclipse is now stored separately from its parent refractor at Wesleyan University. {Patrick Seitzer}

First Contact: 04:00:10 PM CST
Altitude = 35°Azimuth = 261°

"The sky was beautifully clear." One of the first things Watson noticed was the odd shape of the solar crescent during the total eclipse's precursor partial phase. Others were aware of it, but he discussed it explicitly. When one thinks of crescents in the sky, one thinks of the monthly phases of the Moon. The cusps of these familiar Moon shapes extend (at least, ideally) 180° from each other, all the while to the time at which the crescent becomes so thin as to be nearly unobservable. On the other hand, a partially eclipsed Sun looks more like Pacman. The angle between the cusps changes from nearly 360° at first contact to nearly 0°at second contact.

Second Contact: 05:01:45 PM CST
Altitude = 24° Azimuth = 271°

As totality began, Baily's Beads were replaced by prominences. Included was one the of a "beautiful rose color, that others consistently reported." Watson describe the Moon cover the prominences one by one as third contact approached.

Watson estimated the brightness of the corona to be that of a ten-day-old Moon. Moon brightness is not proportional to the amount of the Earth-facing lunar disk illuminated; the Moon Watson chose for comparison is very much dimmer than the

Full Moon. As the Moon waxes, heterogeneous lunar peaks and valleys produce fewer-and-fewer shadows from which no sunlight is reflected toward the Earth.

(Watson decided that 1869 was a good year for him. He wrote to a German university asking that he be awarded a PhD. *in absentia*. His request was granted, assuming that he paid a certain fee.)

Van Vleck did not see the new, mysterious coronal line in the eclipsed Sun's spectrum. He just noted the expected lines in the prominences. Van Vleck does mention stability problems with his tripod; he may not have had much of a view of the solar spectrum at all.

Fig. 7.7 Visiting astronomers of Mount Pleasant. Front, from left to right: Edward Pickering; James Watson, in chair; John Van Vleck, in chair; and George Merriman. Back, from left to right: Donald McIntyre and William Johnston (Van Vleck's assistant). {Detroit Observatory}

Maximum eclipse: 05:03:12 CST

Altitude = 23° Azimuth = 272°
M = 1.0214

Velocity of the Moon's shadow = 1.51 km/s
Path width = 256 kilometers

Just as Mayer did in Burlington, Iowa, cohabitating in Mount Pleasant was an ancillary troupe from the Philadelphia Photographic Corps. Henry Morton's congregation of observers did so in style compared to those in a Coffin-designed shack. (I will eschew the obvious pun.) Forgoing a quiet rural setting, they situated themselves at the corner of Saunders and Main, near all the conveniences of downtown. Their "observatory" was one of the best hotels in the West, the Italianate Brazelton House (less than ten years old).

Fig. 7.8 Contemporary view of the Brazelton House Hotel, Main and Monroe Streets, Mount Pleasant. {City of Mount Pleasant}

The astrophotographers tried to commandeer the corner of the roof. I write "tried" because they were met by "a small group of religious fanatics [possibly Seventh Day Adventists], 'arrayed in their ascension robes of spotless white,' silently awaiting the end of the world." They were motivated by the Bible, which says that at the Rapture,

I watched as the Lamb broke the sixth seal, and there was a great earthquake.
The Sun became as dark as black cloth, and the Moon became as red as blood.[2]

The assemblies coexisted, both doing the same thing but for very different reasons. It is hard to say which was more pleased with the outcome.

---

[2] Revelations 6: 12–14 NLT

For reasons that are unclear to me, with the Corps was Professor Edward Pickering (1846–1919; Massachusetts Institute of Technology [MIT]), though he took no photographs! In fact, he was a physicist, not an astronomer at all. Soon, though, Pickering would soon become a 'rising star' in the astronomical world. He was an expert in the new physical technique of polarimetry. He spent his entire academic life in Cambridge, Massachusetts.

Pickering started out professionally as a student at Harvard, taught there, moved to MIT, and eventually returned to Harvard as its fourth Observatory Director. Pickering's scientific motto might have been, 'The only thing better than a lot of data is a lot more data.' Perhaps this is why he was willing to leave his beloved Cambridge on the Charles for Mount Pleasant on the Skunk. This, even though he had yet little experience in astronomical observation.

Fig. 7.9 Contemporary Photograph of Edward Pickering. {MIT}

Unlike his colleagues atop, Pickering avoided the apocalyptic mob and chose a comfortable west-facing window inside the hostelry. It was the third-floor corner of his room, through which he observed the eclipse in comfort. Pickering's instruments were high-end, *e.g.*, a very acceptable (for the day) stand-alone polariscope, built after the fashion of polarimetric pioneer François Arago (1786–1853; Director, Observatoire de Paris). (It was a single tube with a quartz plate on one end and a window of Icelandic spar on the other.) But at the same time he went decidedly low-end: a "pocket telescope" strapped to a chair. He jammed a piece of cloth (a handkerchief? Every gentleman had one.) between it and the chair to make the 'off-the-rack' instrument stay put.

Fig. 7.10 Edward Pickering's 1869 solar-eclipse 'observatory' (hotel room) today.
{Photograph by the author}

Pickering observed the chromosphere and then the corona. He described a
"pure white" corona as seen through his small-aperture monocular. To him it was
a four-spiked star, mimicking the points of a compass. A fifth, shorter one, rose
between its north and west counterparts. Yet, somehow, at the same time the
corona resembled to Pickering "a cumulous cloud." His prominence count was
one on the eastern side of the Sun and two on the western.

As for polarization observations, many attempts were contaminated by polarization in the Earth's atmosphere. Unfortunately, Pickering's measurements were
not exceptional. The sky was definitely polarized. The corona? Maybe, but
unmeasurable.

The youngest man among the principal total-eclipse scientists of 1869 was
Pickering. Today, it seems as if he attempted to be the busiest astronomer during
the solar eclipse, too. Beside visual and polarization observations, he made use of
a small, single-prism, stand-alone spectroscope. (It was on-loan to him from Van
Vleck.) Apparently, his room was well-furnished: He mounted the spectroscope
on a chair as well. However, the device was meant for the chemical laboratory and
had no telescopic optics at all.

We can see why Van Vleck had no need for this instrument—it had too wide a
field of view (7°, enough to fit the entire eclipsed Sun and its extended atmosphere) with too little spectral resolution. Rationalized Pickering, "This is probably the best way to obtain the spectrum of the corona, as there is very little loss of

light." He was getting the spectrum from every feature of the eclipse at once. Trying to sort out what he saw, he concluded that the prominences produced bright lines and that the spectrum of the corona was continuous.

Third Contact: 05:04:38 PM CST
Altitude = 23° Azimuth = 272°

Pickering was seemingly a single-man circus act (the one where an ever-increasing number of plates are kept spinning at the same time). He further attempted to measure the changing cumulative brightness of the partially eclipsed Sun, using a photometer. Moreover, he made use of an actinometer. These experiments failed as well. The *three* thermometers Pickering brought with him yielded no consensus on a solar-eclipse-triggered change in temperature. (There is no indication that he ever went outdoors with them during the eclipse.) He even took time to study the background sky. One can construct easily a mental picture in which Pickering's frenzy of solar-eclipse activity all took place without his having to remove himself from his favorite chair.

Pickering came away from his first total eclipse with little of substance. It was not for want of trying. The scientific yield of a bag full of small devices simply did not match that of a single, more capable instrument. On the positive side, Pickering did not have to travel in order to retire after the total eclipse of the Sun.

Fourth Contact: 06:00:55 PM CST
Altitude = 13° Azimuth = 281°

Eclipse Expedition Made Easy

The record for least-traveled, least-inconvenienced astronomer appears to be held back in Burlington. There, amateur astronomer William Pilger (1844–1932; "avid reader") needed only to pull out his 2-inch-aperture portable telescope and set it up in a city park near to his home. He liked the park. Pilger was a member of the Burlington City Park Board.

Pilger observed the total eclipse of the Sun without difficulty. Nevertheless, social circles were hard to cross in those days. There is no evidence that he ever compared solar-eclipse notes with any of the ten professional astronomers visiting Burlington.

Pilger was still in Burlington at the time of his death and does not appear to have strayed far. He is buried there in a (again) nearby public cemetery.

It is true. The NAO expedition was hardly rugged or self-sacrificing (with one exception). Still, everybody enjoys an adventure story. Coffin's official post-eclipse report flew out the door, and 3,500 extra copies were printed. Coffin had to hire his son to help with all the requests for it.

After the total-solar eclipse, the observers retreated east once again by reasonably comfortable railcar. Within a year, the last—and much more adventurous—stagecoach trip ever to traverse the 'Hawk Eye State' embarked. This means of conveyance then became one more element belonging to the lore of the scabrous, wild West only.

8

Meeting of the Grayhairs

The hotels are making a good thing out of the eclipse. Large numbers of gentlemen from abroad have arrived, and are making arrangements for taking a reverse bird's eye view of the eclipse tomorrow.

"In the halls of the hotels we saw meetings between friends long separate, and heard joyous exclamations as grayhaired men met and shook hands and laughed, that neither could recognize in the middle aged other the youth whom he had left, and whom he had since known only through scientific journals."

Astronomers from disparate places in the United States converged to observe the total eclipse of the Sun, but that community was not large in 1869 and was interconnected. Principally, there existed within the total-eclipse expeditions a Washington-Philadelphia Cambridge axis. (It is a bit unfair to say that they were all "middle aged"; in fact, they represented all stages of life.)

The 'founding father' of Philadelphia, Pennsylvania, science was, of course, Benjamin Franklin.

In Washington, D. C., it was his great-great grandson, Alexander Bache[1]. He was raised in Philadelphia and a University of Pennsylvania faculty member until his appointment as superintendent of the United States Coast Survey.

In Cambridge, Massachusetts, it was arguably none other than Benjamin Peirce (Chapter 10). He became most influential there after he, too, went to head the USCS.

Benjamin Peirce went to secondary school with Jonathan Bowditch (Chapter 10). Benjamin Gould's father was Principal when James Clarke attended Boston Latin School.

---

[1] Bache was also head of Philadelphia's Central High School for a time. Thus, the telescope loan (Chapter 10)?

T. Hockey, *America's First Eclipse Chasers*, Springer Praxis Books, https://doi.org/10.1007/978-3-031-24124-6_8

It was Benjamin Peirce who initially hired Joseph Winlock (Chapter 10) for a job at the Nautical Almanac Office. Before taking over the Observatory at Harvard, Winlock joined the United States Naval Observatory briefly and then became head of the NAO. (At one time it was acting-director Truman Safford who seemed poised to take over the HCO.)

Recall that Simon Newcomb's first job in astronomy was for pre-Harvard Winlock's NAO (in Cambridge, before it headed to Washington). Furthermore, Newcomb married into American astronomy when he wed the granddaughter of Swiss-American mathematician Ferdinand Hassler (1770–1843; USCS), first superintendent of the Washington-based Coast Survey.

Newcomb was a friend of the USCS's Julius Hilgard. Involvement of the USCS in the total eclipse of the Sun is the subject of Chapter 10.

I will recount (Chapter 20) Newcomb taking fellow 'inside the beltway' (had there been one!) *apparatchik* Homer Lane under his wing, in an act of 'opposites attract.' Somehow, Lane knew Benjamin Peirce, too.

For reasons that are even less clear, Newcomb was fascinated by Christian Peters (Chapter 12). Newcomb lauded Peters for his science while deploring his morals.

Benjamin Gould knew Peters from the time he spent studying in Germany and got Peters a job, once Peters had emigrated, at the USCS working under him. Other figures who we will get to know better, and who worked for Gould, were George Searle, John Stockwell and Cleveland Abbe (Chapter 12).

Cambridge is, of course, home of Harvard University. Newcomb studied there under Benjamin Peirce, as did Edward Pickering, Gould, Abbe, and Arthur and George Searle. Harvard graduates Charles and James Peirce, of the USCS, were Benjamin's son.

George Merriman once had a position at the USNO. Joseph Rogers would. Abbe rubbed shoulders with those at the USNO. Everybody at USNO likely knew everybody at NAO, once both institutions were located in the same city (1867).

Samuel Langley (Chapter 10) also was employed by Winlock, briefly at the Harvard College Observatory [HCO]. Edward Austin, Asaph Hall, Charles Peirce, Arthur Searle, Searle's brother George, and Nathan Shaler worked for a time there under Winlock, too.

Maria Mitchell (Chapter 12) did piece-work calculations for NAO up until a few years before the total eclipse of the Sun. John Van Vleck picked up extra cash by working parttime for the NAO as well.

Edward Pickering, a life-long resident of the Cambridge area, would have been known at the NAO before it moved. Members of Cambridge's Clark family, who made instruments for three government expeditions, stood side-by-side with the astronomers who used them during the 1869 total eclipse of the Sun.

All the Philadelphia photographers were colleagues (or competitors) of each other. For some like Alfred Mayer, the major connection was through Henry Morton's Franklin Institute.

While still a student at Philadelphia's Central High School, George Davidson (of whom we will learn more when discussing reactions to the total solar eclipse among non-western cultures in Chapter 18) became known to Bache. Hilgard also knew Bache from studying in Philadelphia. It was Bache who hired Davidson and Hilgard. Hilgard, in turn, recommended Winlock for the third Harvard College Observatory directorship.

What is now the Philadelphia suburbs was home to James M'Clune, Samuel Gummere, and Hugh Vail, all of whom, had academic positions. (M'Clune and Gummere appear in this chapter, Vail in Chapter 11). Frederick Bardwell (Chapters 13 and 15) was raised in Philadelphia (and went to Harvard University). So was Edward Goodfellow (Chapter 10), who graduated from the University of Pennsylvania.

Just outside the axis-albeit he was one more who did side work for Peirce and Gould—sat James Watson. (Nonetheless, it was Gould, Winlock, and Benjamin Peirce who recommended him for Directorship of the Detroit Observatory.) There also was Charles Young, invited on the expedition specifically by Stephen Alexander (although their scientific views differed significantly). These two men plus Abbe and Hall were students of Franz Brünnow (1821–1891; University of Michigan), who ran the only other serious American school of astronomy besides Benjamin Peirce's at Harvard. More broadly, Watson and Newcomb shared a background as Canadian Americans.

There is a minor eclipse connection to Albany, New York. Gould and George Hough (Chapter 12) served as Directors of the Dudley Observatory there. Hough championed fellow upstate New Yorker Lewis Swift (Chapters 12 and 15) as the co-discoverer of the famous comet Swift-Tuttle. Before his astronomical career, Harkness had been a newspaper reporter in Albany.

John Coffin, Langley, Winlock, and George Searle all at one time taught at the United States Naval Academy [USNA]—a handy connection when there were not enough eclipse telescopes to go around within the axis. (Coffin effectively ran the Academy during the Civil War, when most other faculty were called up for active-duty service.)

And so forth.

Those Absent from Our Story

Perhaps it is worthwhile to consider who was *not* at the total solar eclipse of 1869.[2]

There were prominent amateur astronomers at the time, of course. Examples include Sherburne Burnham, Seth Chandler, Henry Draper, Henry Parkhurst, William Payne, Carr Pritchett, Lewis Rutherfurd, and Horace

[2] My source for prominent astronomers of the time is the four-volume set, Hockey, T. A. (Editor-in-Chief). *Biographical Encyclopedia of Astronomers*, Second Edition. New York: Springer. (2014).

Tuttle. These men were not Grand Amateurs as defined in Chapter 12. They were not as lucky as Lewis Swift to be invited to join a professional expedition. Each likely had neither the time nor money to take off into the 'wilds' of Indiana, Illinois, or Iowa on their own.

Especially ironic was the absence of those who possessed exactly the skills that this total eclipse of the Sun called for: "… a variety of unfortunate circumstances will prevent Mr. Rutherfurd, Dr. Draper, and other experienced celestial photographers from assisting in the observation of the eclipse coming."

But what of the professionals? Newcomb's assistant, George Hill, seems an obvious choice to go. Nevertheless, he appears to have been something of a hodophobic.

Recall that William Harkness could not be for several weeks without the books of Professor William Chauvenet (1820–1870; Washington University). Chauvenet already was in the 'neighborhood.' Fatefully, during the total eclipse of the Sun, he lay dying in nearby Saint louis.

Edward Pickering's brother, William, would discover a satellite of the planet Saturn as a professional astronomer. Howbeit, in August 1869, he was only eleven years old.

At a stretch, we might count Chester Lyman (1814–1890; Yale College Observatory) as someone who might be expected to come to a total eclipse of the Sun. The rest were there! And the American astronomy axis was exemplified well in the eastern-Iowa portion of the 1869 total-eclipse path.

First Contact: 03:59:32 PM
Altitude = 36° Azimuth = 260°

Second Contact: 05:01:28 PM
Altitude = 24° Azimuth = 270°

Maximum eclipse: 05:02:55 PM
Altitude = 24° Azimuth - 271°
M = 1.0214
Velocity of the Moon's shadow = 1.51 km/s
Path width = 256 kilometers

Third Contact 05:04:22
Altitude = 24° Azimuth = 271°

Fourth Contact: 06:00:55
Altitude = 13° Azimuth = 280°

The oldest scientific association in the country is the American Philosophical Society [APS], based in Philadelphia. Membership is by election. At its Centennial

Meeting, attended by the *illuminati* of American science, Stephen Alexander had read a paper entitled "Physical Phenomena Attending Solar Eclipses." In one strike he established himself as an expert on the subject.

Fig. 8.1 Older Stephen Alexander. {Princeton University Archives, Department of Rare Books and Special Collections}

In person he was small, slight and frail, probably never weighing a hundred and twenty pounds when in his best condition. His countenance was refined, and delicate, and on occasion luminous with feeling; his manner was gentlemanly and courteous, but usually rather reserved until some interesting topic made him forget himself,—then he was fluent and even impetuous in conversation. He was modest almost to shyness, though certainly conscious of his own real merit and ability …

Alexander literally wrote the book[3] on total eclipses of the Sun. (In Chapter 6 we learned that he helped program the whole multi-station NAO eclipse expedition.) Today, we might call him a solar physicist instead of an astronomer. However, Alexander's anecdotal research methodology was anachronistic as soon as it was published. While he did head one of the total-eclipse stations in 1869, it was Ottumwa—far the from the epicenter at Burlington.

One of Alexander's biggest critics was Young. It would not do to have the two observing side-by-side. Young stayed at Burlington.

---

[3] *Statement and Exposition of Certain Harmonies of the Solar System.* Washington, District of Columbia: Smithsonian Institution. (1875)

Alexander was not obviously part of the axis. Despite this—just below the 'surface'—he was, in fact, extremely well connected in the politics of American science. As for Ottumwa, he may have decided—metaphorically—that he rather sit atop a small ant hill than scurry at some lower level of the subterranean maze of large one. Still, the 'optics' is that Alexander was marginalized as of 1869. We might know more, but his papers, archived at Princeton University, have a chronological gap between 1864 and 1875.

"Every inhabitant of Wapello County, save those deprived by misfortune of sight, had ample opportunity to observe the startling phenomena ..." Others who headed to Ottumwa, Iowa, included General N. Norris Halstead. He had contributed $10,000 to build an observatory at Alexander's college; it is likely that he could accompany Alexander anywhere. Four Princeton students and one recent alumnus made the trip, at least two of whom fell in love with young women on the train.

Also at Ottumwa was Professor Charles Himes (1838–1918; Dickenson College). Himes was unusual in that he had missed the Civil War altogether; at the time, he was in Europe earning his doctorate. He represented the Philadelphia photographers, but was atypical among them in that he provided much commentary on the visual appearance of the total solar eclipse.

Fig. 8.2 Charles Himes. *Circa* 1865. {Waidner-Spahr Library, Dickinson College}

Edward Pickering made a point of telling us that Himes wore eyeglasses. Himes is among the only one or two 1869-total-eclipse astronomer for whom we know this to be the case. The comment appears *non sequitur*; maybe it was a shibboleth for some other issue.

"Ottumwa, the city of perseverance or self-will, as the name implies, according to authorities … stands on more hills than Rome can boast, and is beautiful for situation." 111-kilometers west of Burlington, Ottumwa originally was a Native American community on the Des Moines River. A few white settlers began staking their claims in 1843; they came in large numbers with the arrival of the steamboat in 1849. Ottumwa was a mill and (later) coal-mining town. The latter provided it with early streetlighting (coal gas). The city claimed 5,214 residents in the 1870 census.

Unlike other towns occupied by the NAO and the Photographic Corp, there were no local photographers to lend aid at Ottumwa: The visitors were on their own.

Nevertheless, roving astronomers of 1869 were welcome at the county seat, even though a devastating fire had burned much of the downtown less than a year before the total eclipse of the Sun! There is only one recorded local gripe, made by "an Irishman": "Och, Pat, do you know that these men get $5,000 for a few days' work up there, and we are taxed to pay for it?"

As beavers build their lodges upon arrive at a stream, scientists build their observatories upon arrival at a total-eclipse path. About two-kilometers outside Ottumwa, on high, gifted property, the astronomers and crew constituted their temporary observatory on a modest eminence. It was assembled, as guided by Alexander, along specifications drawn up by Coffin before either man left the East Coast.

The geography of southern Iowa is more hummocky than the stereo-typical flat plain of the West. (See the landscapes of artist Grant Wood.) Alexander's company had a commanding view of nearly the entire horizon.

The senior astronomer's principal telescope that he set up for his own use was a 3-inch-aperture achromat, manufactured by the venerable Utschneider & Fraunhöfer firm (later sold to Merz & Mahler). 'Old but good.'

A student correspondent making the trip to Ottumwa had little to say about other eclipse preparations, but did choose to mention that,

> The ladies, as usual, were not behind in their attentions. Invitation to visit the observatory was extended to citizens, and many ladies visited the Hill on Friday … as a return, all who were present received handsome bouquets, with ribbons and cards attached, with the names of the fair donors.

Locals later spoke of "Harvard astronomers" visiting Ottumwa. A gaggle of intelligent scholars from the East—they must be from Harvard, *n'est-ce pas*?

The eclipse. At totality, the astronomers were immediately stupefied by the unexpected appearance of the corona. "Above all, the inimitable, indescribable,

Fig. 8.3 Temporary NAO solar-eclipse observatory at Ottumwa. Stephen Alexander must be the figure standing, looking down, and blocking the telescope eyepiece. Charles Himes would be the figure in discussion with two other men, pointing to the left. The man next to him, and in the center, I am confident is Henry Morton, holding binoculars. {George Eastman Museum}

grand corona, surpassing the fascination, the brightness, the splendor of the won-drous lamp and Genii that came to Aladin in the cave, burst out a celestial crown to the dark Moon."

The prominences " … appeared not unlike pinnacled glaciers or icebergs illu-minated by the setting Sun." Had any of the gathered ever *seen* either a glacier or an iceberg?

Alexander tried to measure the heat given off by the corona during totality. He used a gadget, borrowed from the laboratory physicists' tool kit like all other astrophysical instruments. It was known as a thermo-electric pile, sensitive to warmth. He even cut short some of his other second-to-third-contact observations to use it, assuming that he would be the only one to make do so. (But see Chapter 20.) In fact, it was the first attempt to use an electronic instrument during a total eclipse of the Sun.

The experiment failed. This must have annoyed Alexander immensely because in 1845 he had measured, for the first time, the relative temperature of sunspots and conclusively proved that they were cooler than the surrounding photosphere. Moreover, he had established that the solar limb was cooler than the center of the Sun's disk. Alexander vowed to try again at a future eclipse.

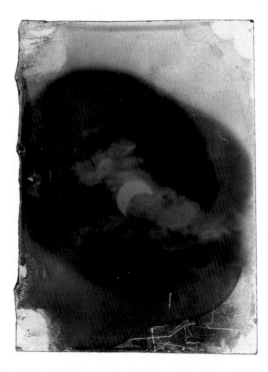

Fig. 8.4 Charles Himes's photographs of the partial eclipse. {Dickinson College}

One of the undergraduates, returning to a theme that permeated his account of the eclipse expedition, wrote, "If the beginning of the darkness was solemn and impressive, equally was the first returning beam surpassingly lovely, a type of beneficent love that guides the universe, that is the center of all moral being." He and the others immediately made plans to attend the 1878 total eclipse of the Sun.

First Contact: 03:58:53 PM
Altitude = 36° Azimuth = 260°

Second Contact: 05:00:55 PM
Altitude = 25°Azimuth = 270°

Maximum eclipse: 05:02:24 PM
Altitude = 24° Azimuth = 271°
M = 1.0217
Velocity of the Moon's shadow = 1.49 km/s
Path width = 255 kilometers

Third Contact: 05:03:52
Altitude = 24° Azimuth = 27°

Fig. 8.5 "Observatory Hill" today. Ottumwa, Iowa. {Photograph by the author}

Fourth Contact: 06:00:31
Altitude = 14° Azimuth = 280°

In what is now the Philadelphia suburbs is Haverford College, home to James M'Clune (1818–1890; Haverford College and Philadelphia Central High School) and Samuel Gummere (1811–1874; Haverford's first President). (The latter's rank likely allowing him to schedule his own leave of absence during a total eclipse of the Sun.) M'Clune originally intended to visit Burlington. Finding that one could not shake a telescope without running into an astronomer there, he opted to move on to Oskaloosa (140 kilometers to the north and not far from Ottumwa). The Haverford College Expedition quickly became part of the NAO/Franklin Institute Expedition.

> Oskaloosa is a quiet place. While it improves gradually, it has not the *drive* and feverish energy of many western towns. Its population is composed mainly of native Americans, the strong temperance sentiment prevailing, making it uninviting to a foreign element, who like their liquor, and a laxity of Sabbath laws.

The city (population 3,204, according to the 1870 census) was originally a small fort (1835) between the Skunk and Des Moines Rivers. The site was selected by Daniel Boone's son. Quakers then established a trading post there (1844). In 1869, Oskaloosa was very near the centerline of the total solar eclipse.[4]

---

[4] My great-grandfather, already an adult, would have seen the 1869 total eclipse had he only emigrated as he did, from England to the vicinity of Oskaloosa, a decade sooner.

M'Clune and Gummere set up at a small brick building at a seminary named Oskaloosa College[5], northwest of town. Here the two observed sunspots while waiting for totality. Once it began, prominences appeared. McClune was struck more by the "orange"—he was the only professional observer to describe their color thusly—streamers of the corona "which gave it an appearance not unlike a Saint Andrew's cross."

Fig. 8.6 Oskaloosa College has become William Penn University. {Photograph by the author}

The party was not 'bothered' much by bystanders. The community preferred to gather atop the new "skyscraper" (three storeys!) on the northwest corner of the town square.

Meanwhile, Coffin spread out still other observers, for additional geographical back up. They went to Monroe City, Missouri, and Kewanee, Illinois. They were not as well placed on the eclipse path as were the Iowa astronomers, but sacrificed eclipse time to outsmart the weather in the unlikely event that all of Iowa was socked in.

The observer in Missouri was NAO "computer" John Wiesner. Weisner traveled with his family. We do not know if the eclipse came out of his vacation time, but if it did, he likely would not himself have chosen Monroe City for a holiday.

---

[5] now Drake University in Des Moines, Iowa

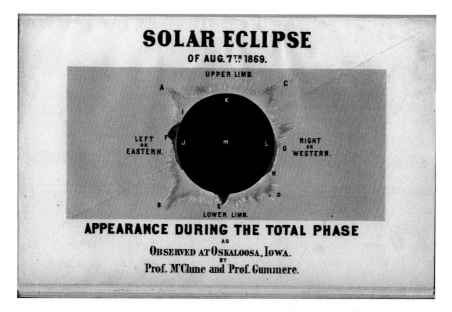

Fig. 8.7 This color rendering of totality in 1869, by Samuel Gummere, became one of the most popular to be reproduced. {State Historical Society of Iowa Special Collections}

Wiesner took with him the seemingly ubiquitous, equatorially mounted, 3-inch-aperture refractor with red-glass filter. This one was provided by an institution where Coffin's name no doubt carried weight: Again, it was the charitable Naval Academy. Wiesner placed the instrument in the front window of the Young Ladies's Seminary, half a kilometer from the railroad station. He probably was not distracted there by the revelry of local eclipse watchers.

During totality, the two Wiesner sons read out a chronometer by candlelight. Weisner tells us that their times are thirty seconds off, because they paid attention to the short end of the second hand and not the longer pointer. We do not know how old the children were.

First Contact: 04:00:53 PM
Altitude = 34° Azimuth = 262°

Second Contact: 05:02:35 PM
Altitude = 22°Azimuth = 273°

Maximum eclipse: 05::03:15 PM
Altitude = 22° Azimuth = 273°
M = 1.0248
Velocity of the Moon's shadow = 1.46 km/s
Path width = 256 kilometers

Third Contact: 05:03:54 PM
Altitude = 22° Azimuth = 273°

Fourth Contact: 06:00:25 PM
Altitude = 12° Azimuth = 282°

Another NAO civil servant, Edward Austin, represented Coffin in Kewanee. One wonders: How many of the extant Alvan Clark 3-inch-aperture refractors made the trip to the eclipse path? In contrast with Weisner, Austin made his stand at a local farm, a quarter kilometer from the railroad depot. (It was owned, somehow appropriately, by a man named August First).

Weisner stopped his telescope down to one-inch. The limb of the Sun began to tremble as second contact approached. Interpreting this observation was above Austen's pay grade; he modestly suggested that it could be an illusion. It was more prone to be a terrestrial-atmospheric effect, as the Sun became nearly a point source of light. The same phenomenon is seen in the twinkling of stars at night.

During totality, "Red flames traveled along the Moon's limb … the whole outline of the Moon was surrounded by a brilliant halo." His "flames" must be prominences. Austin, like many others, was inadvertently giving the phenomenon a cause—combustion—that was, in reality, unknown. Clarifying, Austen saw the two major prominences. Alas, we cannot ask him what a "traveling" prominence is. He probably meant that their appearance shifted from visibility on one side of the Sun to the other between second and third contact.

And then it was back to the conversational halls of the hotels.

9

"Overhanging Monster Wings": The Philadelphia Photographic Corp

"… photography combines the powers of all these observers and furnishes a record admirable for its truthfulness."

What about the photographs, the *raison d'être* of the Philadelphia half of the Expedition? An image of the partially eclipsed Sun was made as early as 1842, the totally eclipsed Sun in 1851. However, this was a daguerreotype.

Useful photography that both did not require extremely long exposures and could be reprinted easily did not come along until the invention of the Wet Collodion Process. The partial solar eclipse of 1852 was registered using this method.

The first such photograph of a total solar eclipse was made in 1860. It was taken by English businessman-turned-astronomer Warren de la Rue (1815–1889; past President, Royal Astronomical Society) and resulted in his receipt of the Royal Astronomical Society's Gold Medal for 1862. It remained the one to beat.

In 1869, the goal was to produce an eclipse image nonpareil. Such a photograph would include the corona.

The 1869 solar-eclipse photographers got the premium telescopes. At the NAO stations, among them were the two borrowed refractors: the equatorially mounted, 6.4-inch-aperture Merz & Söhne and the other 6-inch-aperture instrument.

They needed them. While visual observers still 'wasted' sunlight intentionally, by stopping down the apertures of their telescopes or using filters, the camera of the nineteenth century could not match the efficiency of the eye. It was far less sensitive to light than that in today's average cell phone. The chemically coated, glass plates of 1869 required as much light as possible to capture an image of something as faint as the solar corona.

---

from a Philadelphia newspaper

T. Hockey, *America's First Eclipse Chasers*, Springer Praxis Books, https://doi.org/10.1007/978-3-031-24124-6_9

Still, the light-greedy large achromats meant that the photographers had the option to not place their plates at the very focus of the telescope, for maximum concentration of light. An image produced in this way is exceedingly small. For only the second time in astronomical photography, the 1869 camera operators could afford to introduce an eyepiece before the plate, thereby diluting the light but spreading the Sun's image over a greater area. This enlarged the image on the plate. It still was small, but big enough so that the photographers and astronomers would not have to use a magnifying glass to examine it. The eyepiece was chosen specially so that the eclipsed photosphere, and corona, fit well into its field of view. As for the rest of the instrument, "The stand of the [Oskaloosa] telescope was rested on massive cross-timbers firmly rammed in the ground."

The camera itself essentially was a box with a dark interior that could be attached to the telescope. Inside was a plate holder and spring-loaded shutter for making timed exposures. The key to it was an opaque mask with a hole in it. A triggered spring release, providing smooth motion, pulled the part of the mask with the hole in front of a photographic plate. This began an exposure. Then, another triggered spring release pulled the hole away, replacing it with the opaque part of the mask again. This brought the exposure to an end. Such a method guaranteed uniform exposure across the image. The Corps's cameras and wide-field eyepieces were manufactured by a Philadelphian tradesman from Germany named Joseph Zentmayer (1826–1888; optician).

Fig. 9.1 Joseph Zentmayer. {from a Philadelphia obituary published five years after his death. It was written by the Superintendent of the United States Mint.}

Mr. Zeitmeyer's [*sic*] office in Walnut Street, where he had his lathe close to his counter and near to the cases containing his instruments, was the meeting place of all the scientists of the day. There, at all times, while he was working, professors and physicians and mechanical engineers would meet and discuss problems in optics or in mechanism.

Like their counterparts at the United States Naval Observatory, the Photographic Corp experimented before leaving on expedition by simulating the solar-eclipse experience. They mimicked expedition circumstances right down to constructing a temporary observatory, similar to what they would occupy out west, but instead in west Philadelphia.

Just how long should the aforementioned exposure be? English astrophotography pioneer Warren De la Rue said that prominences are 180 times the brightness of the Full Moon. So, the Philadelphians' experiments involved taking picture of the Moon. But the exposure necessary to render the prize corona? That was to be but an educated guess.

The length of the exposure—for better or worse—would be well known for the benefit of future eclipse photographers, though. One way to log it: During the eclipse at hand, the newly invented chronograph electrically inscribed regularly timed hatch marks onto a paper tape. The photographer interrupted this pattern manually during exposures by use of a circuit-breaking key. The idea was that later measurement of the gap in the tap resulted in the exact exposure time.

Another piece of equipment, vital for all but the shortest of exposures, was the clock drive attached to the photographic telescope. Without it, the camera would not follow the Sun during the exposure. The resulting image would be blurred.

The drive's construction was ingeniously simple: a cone rotating at a constant rate. Its motion was transferred to the telescope by an attached arm in frictional contact with the cone. Moving this arm to different heights on the cone meant that the arm encountered different circumferences, and the telescope could be adjusted to different rates of rotation. All of this was accomplished with a minimum of moving parts.

Photographers a century and a half ago did not travel light. All told, they carried across the Mississippi "five furniture cart loads of material."

The three photography-detailed telescopes were scattered between three observing sites to bedevil clouds. While visual observing could be a lonely affair, each of the photographic stations required no less than *five* people to make everything happen when it was supposed to. Astrophotography was a team sport.

Supporting members of the photographic teams were all enthusiastic volunteers who could abandon their day jobs and disappear into the hinterlands of central North America for several weeks. And, as we will see, some of them actually would never see the total eclipse of the Sun!

Henry Morton divided the men—and they were all men—into three parties, "guided by the desire of securing in each section such a diversity of special ability as might make each self-dependent and complete; also, to leave nothing undone to secure content and harmony of feeling." He chivalrously assigned himself the smaller-aperture University refractor without clock drive and placed himself in the middle: Mount Pleasant. However, he took with him some of the top photographers Philadelphia had to offer: James Cremer (1821–1893), who had a national following for his stereographs; Edward Wilson (1838–1903), editor of the most influential periodical on photography in the country; and John Corbutt, a master at producing lantern slides. (Admittedly, *he* was from Chicago.)

We already know that Charles Himes was in Ottumwa. Carlisle, Pennsylvania, is 176 kilometers from Philadelphia, so, geographically, it might seem a bit much to include him in the axis. Nevertheless, with a passion for photography, he did not consider the distance to be so great as to prevent membership in the Pennsylvania Photographic Society.

It was Mayer in charge of photography at Burlington, using the High School telescope. As mentioned in the previous chapter, Burlington and Mount Pleasant recruited additional assistance from local photographers.

Fig. 9.2 Alfred Mayer. 1861–1865. {American Institute of Physics}

In Burlington, it took two solid days to prepare the camera and test the alignment of the (hopefully) balanced camera/telescope on night-time astronomical objects. From Mayer's 'Diary':

We rose on Friday morning, August 6, to see a driving rain, with an east wind and the same dull murky atmosphere which we had had all the day before. Yet we went on attending to the details of arrangements for the coming day as though we *could* not be disappointed after weeks spent in previous preparation for the work of tomorrow, and after having come over a thousand miles to carry back with us the long thought of eclipse, fixed on our plates in *appearance*, *time*, and *position*. As we retired there appeared a breaking a way in the clouds, and Dr. [Benjamin] Gould told the clerk in the office to "wake as up if the stars came out" We had barely lost ourselves in cloudy slumber when the clerk knocked at the door with "Get up! *plenty* of stars!" We were soon dressed, and Professor [John] Coffin, Dr. Gould, and I were all night, putting our own special instruments in adjustment, and, at the same time, helping each other as we could. When all was finished the sun was rising over the trees on the opposite bank of the Mississippi and the air was as pure and as serene as one could wish.

The final tests had to await Eclipse Day itself. The visual focal length of the newly constituted Burlington telescope was known to be nine feet, but the photographic focus still eluded the photographers. Thus, test plates were taken using the uneclipsed Sun, and then any urge to further adjust the focus had to be suppressed until eclipse time. Stressful, but necessary. The test plates came out fine.

Saturday, at the proverbial moment of truth, it was time to take a photograph of a total solar eclipse, something no one present had ever actually done before. Lanterns were lit so that the men could see and find themselves no matter how dark it got. The photographers took their places like players on a baseball field. Then they waited.

Our example is Burlington, but similar happenings took place in Mount Pleasant and Ottumwa. Mayer was the juggler. He kept the telescope pointed, triggered the exposure, and interrupted the coronagraph circuit.

"H. C. Phillips," of whom little is known to me, stood in the makeshift darkroom, occupying one corner of the building. He coated the plates with collodion (nitrocellulose in solution with ether and alcohol), taking care to avoid bubbles or streaks. He then immersed them in a bath of silver-nitrate solution for several minutes, all just prior to use.

Standing next to Mayer, experienced Philadelphia photographer O. H. Willard installed the still-wet plates in the camera, one by one, with as little time between as possible. He also recommended exposure lengths to Mayer.

"Mr. Montfort," a photographer from Burlington, acted as runner, carrying new plates from the darkroom to Willard and exposed plates to "Mr. Mahoney" (apparently another local) and his assistant "Mr. Leisenring" (also from Burlington). These latter two developed the plates 'on the spot' (meaning back in the darkroom). The process required a bath of dilute ferrous oxide and a dip in potassium cyanide to fix the image. They washed the plates in water and likely dried them over a candle flame. (No doubt they were careful not to expose this source of ignition to the collodion/ether, which is *highly* combustible.) The last step was coating the plates in lacquer.

As a contribution to the pressure everyone must have felt, all the while, there was Miles Rock, of Bethlehem, Pennsylvania, who accompanied Mayer from Philadelphia as a sort of a valet. He loudly called out the seconds off the chronometer so that the photographers knew how much/little time they had remaining. While this duty may not appear to be especially strenuous, nonetheless he was relieved part way through the total eclipse by Burlington resident "Mr. Bonsall." In what may sound like a rather dull opera, after every sixty of Rock's or Bonsall's seconds, O. H. Kendall chimed in with the minutes.

Nine men in all. It is interesting that Mayer entrusted the critical job of developing the plates to lesser-known members of the party. Perhaps he considered this task to be the most routine. Some had commercial expertise.

There is something else to consider: Notice that, except, of course for Mayer, and maybe Willard, it is unclear that *any* of the other party members had time, or, in the case of those trapped in the darkroom, opportunity, to actually *witness* the eclipse themselves. Ironically, these martyrs may have been the only ambulatory residents of Burlington to fail to take in this likely once-in-a-lifetime event. As they were not astronomers, maybe they did not know what they were missing.

Mayer had written the ballet of activity planned for this day, and this day only, well in advance. That plan was to take a machine-gun approach and make:

1. one picture every 4–5 minutes during the partial phases of the eclipse.
2. five pictures starting just before the computed time of first contact.
3. around fourth contact, the same.
4. one picture just before and one just after second contact.
5. one picture just before and one just after third contact.
6. During totality? As many as possible!

At least that *was* the plan.

"… the weather on the eventful day of the eclipse was, at all our stations, perfect …" The partial phase exposures went fairly leisurely. Young called out from nearby that the Moon's limb was about to bisect a large sunspot. The shutter snapped. The resulting image showed the spot's umbra neatly cut in two.

About five minutes before second contact, Willard removed the diaphragm that had heretofore reduced the telescope to two inches of aperture and in doing so increased the light entering camera. The full brilliance of the Sun now was available to the photographers.

A photograph was taken just seven seconds before second contact. Mayer took five more photographs of totality. "… [an exposure of] five seconds was more than sufficient to secure all the details of the protuberances, although it gave no decided indication of the corona."

Things now changed quickly with the Sun, so the photographers did not wish to risk shorter and perhaps insufficient exposures, but they later guessed that the prominences were so bright that an exposure of only one or two seconds might have been all that was necessary to capture the larger examples of these solar features. They further hypothesized that such exposures would yield more detail in the prominences, which they had inadvertently over-exposed. Astronomers and photographers are not immune to rumination.

There were fifty seconds left. Mayer prepared to make a set of exposures, but no one had delivered the plate. "Plate! Plate!" he cried. However, the darkroom was overwhelmed with six plates taken in two minutes. The need to develop the old plates and prepare the new ones had caught up with them. There was no prepped plate for Mayer. In Mayer's impatience to expose, expose, expose, he was left helplessly watching the rest of totality. Photography did not resume until nearly thirty seconds after third contact.

Six more photographs eventually were made, at the scheduled four-minute intervals, before the fourth-contact series. Then the photographers used up the available plates with four more images of the uneclipsed Sun.

As Mayer explained in a rare emotional insight, "... I experienced an indescribable feeling oppression when—I tapped the trigger, and from that instant until the Sun appeared I had nothing but an *instrumental consciousness*, for I was but part of the telescope, and all my being was in the work which I had to perform." Mayer might have been mentally numb, but even he could not master the near simultaneous demand of adjusting the telescope, resetting the shutter springs, registering the time with the chronograph, and taking the photograph. The chronographic record shows that Mayer had interrupted the electric circuit ... whenever he had gotten around to it.

But after all the kilometers and years, a feeling of "oppression"? With throngs along the path cheering and waving during the total eclipse of the Sun, there is something tragic about a man who posterity remembers mainly for his acts during these minutes, feeling—at the time—completely overwhelmed by his burden to perform.

This is all the more so because, by any objective measure, Mayer and his team were successful. They did what they set out to do as a baseline: document the prominences. The corona was 'gravy.' Summarized Mayer himself,

> "Our work was finished ... and of the past, but its history we had faithfully recorded in forty-one perfect photographs. Such good success as attended out efforts speaks for the entire devotion of each one to his allotted duties. Without this devotion on the part of each and all we would have failed, and the harmony and efficiency which pervaded our corps need no other witness."

In his final report, Mayer mentions the tardiness in the darkroom once in passing and never writes of it again.

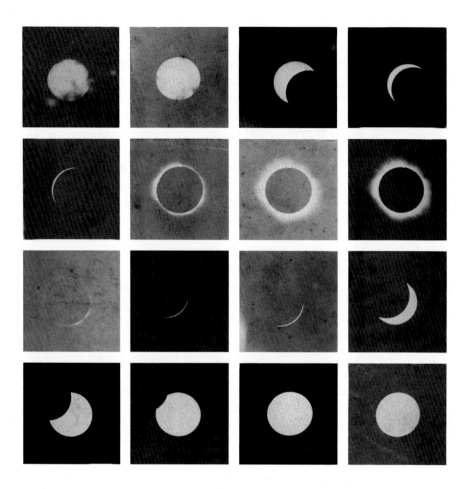

Fig. 9.3 Examples of Alfred Mayer's photographs from the 1869 solar eclipse. {State Historical Society of Iowa Special Collections}

After all the excitement, stowing of equipment, and, perhaps, communal sighs of relief, were over, Mayer had the opportunity to inspect his work through a magnifying glass. The Burlington photographs were so well-focused that granulation appeared in the images of the uneclipsed portion of the Sun. At first, Mayer excitedly interpreted this to be the granulation reported by visual observers of the photosphere, never photographed before. However, on closer inspection, he noticed that the grains were larger than those previously described. And there was space between them. Moreover, the granules on one plate did not match those on another, and different regions on the Sun were granule-less on different plates. Chagrinned, Mayer realized that he was looking at the grains of photographic chemicals crystalized on the plates themselves and not features on the Sun.

Mayer photographs capture sixteen sunspots, some with bright faculae surrounding them. The sunspots moved ever so slightly in placement, from first to last image, as the Sun rotated. Those that did not, Mayer rejected as a much nearer phenomenon: little bits of Iowa, dust spots on the lens.

The sum that is the sequence of Mayer images, all made in like manner, is greater that its individual components. As one looks at the sequence of images, it is possible to notice changes in the individual, restless sunspots—a time-lapse movie, of sorts.

Here is something else from the partial phases. Mayer on the mountains visible at the Moon's limb: "The best method of examining these mountains is to cut a circle of paper of the exact diameter of the image of the Moon, and make its circumference coincide with the general level of the Moon's periphery; the lunar mountains will then appear as black protuberances rising above the white circle of the paper."

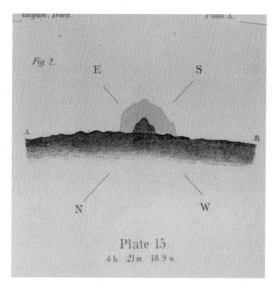

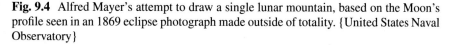

Fig. 9.4 Alfred Mayer's attempt to draw a single lunar mountain, based on the Moon's profile seen in an 1869 eclipse photograph made outside of totality. {United States Naval Observatory}

In the photographs taken just before second contact, the northern cusp of the Sun is seen to be narrower than the southern. That the shape of the solar crescent is asymmetric demonstrates that, from Burlington, the center of the Moon's disk did not pass exactly over the center of the Sun's. But it came close enough.

The time of totality available to the Burlington photographers was just 3 seconds short of the local maximum possible (that on the centerline closest to

Burlington). In photographs of totality in which prominences appear, Mayer noticed them on the eastern limb first, then on the western limb later, as the Moon slowly moved across the Sun's face.

With the advantage of unlimited time to study the eclipse frozen on his plates, Mayer numbered the prominences, gave their position angles around the Sun, and measured their height and breadth using a micrometer. Mayer (as well as others) noted that there were prominences seemingly disconnected from the Sun, "The most elevated and bright of these detached flames float at a height of 20,000 miles above the of the Sun." There is no stated objective to these measurements. They simply are part of the nineteenth-century scientist's perceived duty to quantify nature.

A photographed eclipse is not available only to astronomers in the field. Presumably anybody could look at the plates at any time. Still, Mayer could not help but translate what *he* saw into words. Along the way, we learn that prominence number 8 looks like an "albatross head with bill and head close to the Moon's border." (In a caption for an Ottumwa photograph he says that the bird held some morsel in its open jaws.)

Number 4 on a photograph appears like "… an eagle, with outstretched wings, resting on the trunk of a tree …" Incredibly evocative word pictures! Moreover, we are told that the resemblance is "perfect."

In the outline of his final report to Morton, Mayer separates his description from his results. In truth, all his descriptions of the eclipsed Sun are laden with interpretation, using words like "flame," "surface," "massive," and "cloud."

Mount Pleasant was supposed to be all about the corona, but the prominences took 'center stage.' Alerted by James Watson observing nearby, Morton took the very first picture of the eclipse made by the Philadelphia Photographic Corp, just after first contact. The photographs of the partial eclipse phase, though, were never meant to have significant scientific meaning. They demonstrated that everything in the optical system was aligned and in focus and that the plates were properly prepared. "The serrated character of the Moon's edge is clearly manifest, and the same prominent peak or ridge may be readily identified throughout an entire series of pictures." Morton's favorite photograph of pre-totality shows the lunar limb near the most prominent sunspot with its umbra and penumbra.

As seen in his totality photographs, Morton described the biggest prominence as having an X shape. Upon closer inspection, one appeared "like an ear of corn" (certainly an appropriately Iowan simile). The prominence photographs illustrate the advantage of eclipse photography.

Rather than try to store in one's brain that which appears briefly, photographs allow the astronomer to take his or her time in interpreting what was seen. Compare, for instance, earlier perfunctory descriptions of this prominence as seen through the telescope's eyepiece to Morton's upon studying the photographs:

Fig. 9.5 Henry Morton. {Alamy}

It consists of a solid central mass inclined at an angle of about 45° to the normal at the solar surface, and with three branches from near its end, one sweeping backwards in a direction generally parallel to the solar surface, another forward, concerns, as the direction of the general mass, and a third branching out a little below and running in the same direction as this last. The appearance of the main body, which is of a spindle shape, and with spiral markings, is highly suggestive of a vortical motion which has swept these whiffs of light matter into their peculiar positions.

Assuming he or she stayed awake for all that, the reader will remember that this description is of but one example of one phenomenon (prominences) visible during the three minutes of total eclipse.

Morton goes on to describe two other prominences and then one in which, in description, what is, perhaps, a latent talent at poetry intrudes: "To the left appears a mass of rolling cloud disposed in beautiful streams and curls, like the smoke from a bonfire or burning meadow, swept gently toward side by a light wind." There are other prominences present; a large one seen near third contact "... resembles in shape a great whale with a body made up of dense cumulous cloud matter, with a long tail ... clinging close to the solar edge ..." Morton then recovers and shifts to prosaic quantification: "The length of the entire mass is about 110,000 miles, and the height of its more bulky portion about 28,000 miles, while its length is about 70,000 miles." (All these numbers were based upon apparent angular size and assumed a distance to the Sun.)

Morton concludes with this zoomorphic word picture:

To the right of this, and only showing its entire length in last picture of each series is a caterpillar-like mass of cloud matter, very much like the solid rolls of horizontal vapor which are sometimes seen passing over a sheet of water. At one end rises a projecting bead, but the rest clings closely to the solar edge, and is indented with ring-like divisions, giving it much the aspect of a huge worm … "with two horns."

In the big-game hunt that was photographing the 1869 total eclipse of the Sun, the prize was the subtle corona. However, trying to not over-expose the prominences meant that the much, much fainter corona was hardly seen!

Fortunately, the Mount Pleasant photographers knew this. As third contact approached, they made the photographic equivalent of football's 'hail Mary pass' and left the cameras open wide and long. In the resulting image, the corona is seen. Its variation in intensity with distance from the Sun is proper and asymmetrical structure evident. It was the finest photograph of the eclipse made by any NAO (Franklin Institute) station and approximated what an observer at the time would see with the naked eye.

Yet Morton's photographs were compromised by the fact that he allocated to himself the smallest telescope and the one without a clock drive. Ironically, the outstanding solar-eclipse picture taken at Mount Pleasant, according to Morton's own admission, anyone at Pickering seemed to think large astronomical telescopes overrated. No doubt eagerly sitting at the feet of the great astronomer from beyond, the local enthusiasts sought out his advice. What could they do during the eclipse with the limited, unspecialized equipment at hand? One of these may have been Edwin Hover (1839–1916; Hover & Wyer). Or they included one or more of the Leisenring brothers, also commercial photographers in Mount Pleasant. Pickering convinced them to skip using an astronomical telescope of any kind and to merely point a "common portrait camera" at the Sun, albeit one with as big a lens as one could lay hands on.

Many astronomers would have thought this advice unorthodox at best. Be that as it may, Pickering's new friends did just as they were told and gambled on a single exposure—all or nothing—the length of totality. They could get away with this: They had no one to whom they had to account at the end of the day.

The result clearly shows both an inner and outer corona, with distinctively different visual characteristics. With only a modicum of preparation and investment placed into its realization, compared to that of the out-of-town competitors, this image was as good as anything produced by the Philadelphia Photographic Corp.

In the spirit of inductive science, the scientific mantra for the 1869 NAO eclipse expedition was uniformity. Th last group of Philadelphia photographers set up shop at Ottumwa, in exactly the same way as did those in Burlington and Mount Pleasant. They did so in the same standardized, temporary observatory designed by Coffin. They erected their telescope in the same manner. Their preparations proceeded similarly to those at Burlington and Mount Pleasant. Each member had a duty to perform, each comparable to one performed at other stations.

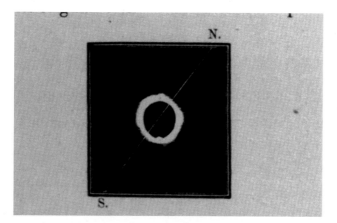

Fig. 9.6 Commercial photographers's photograph of totality, made in 1869 Mount Pleasant. {United States Naval Observatory}

Crisis! The clock drive had been damaged in transit. Fortunately, the ever-handy Zentmayer had agreed to shutter his shop and *accompany* the expedition. He ended up in Ottumwa. So, he was available serendipitously to disassembled, repair, and reassemble the recalcitrant drive 'in the field.' "The trouble anxiety which this cause of delay occasioned, was, however, no small trial of fortitude for the Ottumwa party."

Then it was found that the chronometer had been left behind in Burlington! Ottumwa luck held. Alexander brought a spare. In reality, the Ottumwa party was so busy during totality that they did not have time to pay attention to the chronometer much at all. Things happened when they happened. They extrapolated a time record of their accomplishments later.

Some of the Ottumwa images were ruined. Ultimately, only four plates were exposed during totality—and the result is not very good. Still," One of the Ottumwa pictures, exposed at the very last instant before totality, g[a]ve a photographic record of the curious phenomenon known as Baily's Beads, being simply the last glimpse of the Sun's edge cut by the peaks of lunar mountains into irregular spots." The least experienced team was the first ever to do so. (Some observers were so captivated by the prominences coming into view that this also-interesting display did not register.)

Perhaps realizing that his photographs were no match of those made at the other stations, Himes's report spends as much time on recollecting visual observation as it does the images on his plates:

The corona approached much more nearly in regularity the four-rayed form generally given, and which had always seemed idealized or conventional. The SW. ray was, however, unequally subdivided with the smaller part to the north. The whole seemed of a fibrous, slightly curled or twisted character, somewhat like a cirrus cloud, and of silvery whiteness.

"Idealized." "Conventional." This is the most explicit statement of how what the 1869 astronomers saw—even in their photographs—was influenced by what they expected to see. Expectation was based upon descriptions given at previous total solar eclipses.

Switch scenes to later that Fall. Mayer stands before the American Philosophical Society. He has brought along Zentmayer's camera, the workings of which he demonstrates to his audience. How did he share the photographs themselves? Perhaps lantern slides in a curtained, darkened room?

Before the end of the year, Mayer would send copies of nine eclipse photographs to astronomy epistolarian Reverend Thomas Webb (1807–1885; Church of England), who placed them in what at the time was the ultimate repository of astronomical knowledge: Somerset House in London, home of the Royal Astronomical Society. Mayer's accompanying note read, "The best way to examine the glass photographs is to incline them over a piece of white drawing paper placed before a north window, and use a lens magnifying about eight diameters." (It is, in a sense, ironic that he specifies north in order to avoid direct sunlight, the subject of the photographs in the first place.)

While we have record of Mayer's presentation to the APS, we do not know if questions from the audience followed and, if they did, what those questions were. Nonetheless, the question most on the mind of esteemed assemblage in attendance may very well have been: Did the photographs of 1869's total eclipse of the Sun provide the objective information that science sought?

1860s photographs do not reveal color. For that, the eye was still necessary. But what about coronal shape?

In total, only fourteen pictures of totality were taken by the three Philadelphia Photographic Corp photographic stations. In practice, while the exposures themselves were only seconds long, no team could change plates and prepared for a new photograph more frequently to produce any more. And because the appearance of phenomena such as the corona turned out to be greatly dependent upon exposure length, the images did little to resolve any conflicting testimony of the observers made by eye—much more capable of handling great differences in brightness—as to what the corona actually looked like.

Gould, Morton's adjutant, was not involved with the photography itself; nonetheless, he both observed the corona himself and studied the resulting images. His conclusion? "Almost the first impression given me by an inspection of the photographs, made during the total obscuration at Burlington and Ottumwa, had reference to the limited extent of the photographic record of the corona ..."

In planning—Simon Newcomb for one—envisioned series of progressively longer exposures that demonstrated the radial extent of the corona at different brightnesses. Instead, it was every photographer for himself, using his own knowledge and hunches about photography to be the one to get the superior image of the Sun in total eclipse. Who succeeded is in the eye of the beholder.

Music of the Eclipse

The special connection between Philadelphia and the 1869 total eclipse of the Sun was enshrined in music. An 1862 advertisement explains that that the city's Lee & Walker Company has "… published an extraordinary number of the most popular Patriotic Airs in brilliant styles." The firm eventually printed the work of composers such as John Sousa. In 1869, it released new sheet music from Philadelphian Edward Mack (known mostly for his Civil War marches). The set was made up of the *Eclipse Waltz, Eclipse Galop*[1], *Eclipse Mazurka*, and *Eclipse Polka*.

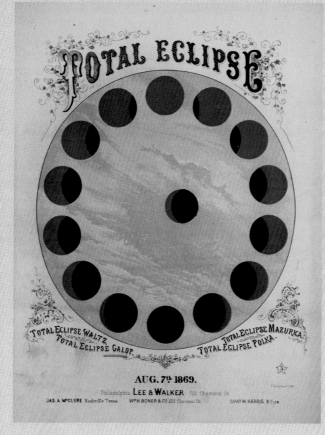

Fig. 9.7 Cover Page on the sheet music for Edward Mack's eclipse-inspired sectionals for piano. 1899. {Sheridan Libraries}

---

[1] popular ballroom dance of the nineteenth century

10

Surveying a Solar Eclipse

> The astronomers, in watching a solar eclipse three thousand years hence, may see and know precisely what the astronomers saw and determined in Kentucky in the total eclipse of 1869.

The very first issue of *Nature,* that publication eventually to become the most prestigious of scientific journals, reported on the 1869 total eclipse of the Sun. It did not discuss the United States Naval Observatory nor Nautical Almanac Office campaigns. Instead, it reported on the enterprise undertaken in part by USNO rival, the United States Coast Survey.

What about the Coast Survey, which has shown up in this narrative seemingly as a guardian angel of other, USNO and NAO astronomers uncertain of the geographic position? The official mapmaker of the United States, the USCS was the oldest scientific agency in the federal government (1807)—the brainchild of Thomas Jefferson. It still exists in a way, subsumed by the National Oceanic and atmospheric Administration.

Simon Newcomb had wanted to secure a job at the USCS, but none were available. He started his astronomical career at the Naval Observatory instead.

Not to be outdone, the USCS co-produced its own solar eclipse expedition, the third organized with the resources of the Navy Department. (The USCS was now a civilian agency but had navy personnel in support roles.) This was *not* to be a physical astronomy enterprise. The goal was to obtain contact times at precisely known locations—the classical astronomy of celestial measurement. Science attempted in the middle of the continent was for the betterment of commerce and naval adventure undertaken in the oceans of the world. "All observations made by the navigator for his position at sea will become more efficient by the improvements in the tables of the Moon, which will arise from the corrections of its elements," wrote the USCS's Superintendent, Benjamin Peirce.

T. Hockey, *America's First Eclipse Chasers*, Springer Praxis Books, https://doi.org/10.1007/978-3-031-24124-6_10

Fig. 10.1 Benjamin Peirce (center). {National Oceanographic and Atmospheric Administration}

The expedition's participants, of course, could not help but note the appearance of the Sun in eclipse. They did so. However, explained one of the station leaders: "Hence the examination of the Sun's disk, its colors and spots, during the different phases of the eclipse, the colors exhibited by the corona, the degree of darkness during totality, and the changes and gradations in the coloring of the landscape before and after totality, the effect of the obscuration upon animals, &c., *were omitted* or only incidentally observed, for the reason that they could not be properly or reliably attended to without interfering with *a more desirable class of observations* [my italics]."

The USCS was equipped with a large assortment of modest-sized telescopes normally used in surveying. Photographs would be taken. Yes. However, they were made in the hope that carefully timed photography might turn out to be a superior means of determining contacts than visual recording.

The USCS obtained the endorsement of the American Academy of Arts and Sciences. In January 1869, it successfully approached Congress for a subsidy to undertake the expedition, using the time-honored 'the other kids are doing it' pitch: "The English government last year sent an Expedition to India to observe

Fig. 10.2 Contemporary United States Coast Survey geodetic transit with telescope by Troughton & Simms of London, England. 1849. {American Museum of Natural History}

the Total Eclipse visible there last August, and France and other countries vied with England … in zeal to observe it." The pitch was successful.

Beginning its preparation earnestly only in June, compared to the other government expeditions the USCS was late in the total-eclipse game. But once committed, Peirce was fervently enthusiastic to the point of existential: "… it is nothing less than a duty owed to civilization that everything in our power and within our means should be done to make observations as complete as possible."

The plan was to establish no less than four stations in four states: Shelbyville, Kentucky; Springfield, Illinois; Des Moines, Iowa; and Bristol, Tennessee. Bristol was considered to be the easternmost location from which both second and third contact could be effectively observed, inasmuch as farther down the umbral path the totally eclipsed Sun would be nearly set.

(Details of the logistical planning are otherwise sketchy: While the National Archives holds a Record Group including text specifically on other USCS solar-eclipse expeditions, the 1869 total eclipse of the Sun is not included.)

Benjamin Peirce asked the Director of the Harvard College Observatory, Joseph Winlock, to mount a joint expedition. Winlock agreed. The HCO and USCS had worked together before, establishing the longitude difference between Boston, Massachusetts, and Washington, D. C. Despite this, the two institutions remained unlikely bedfellows: The USCS was firmly rooted in the Old Astronomy, while HCO was experimenting with the New.

Fig. 10.3 Joseph Winlock. {United States Navy}

It turned out to be clever teaming. HCO provided photographic expertise, USGS substantial manpower. This 1869 total-eclipse partnership effort is reminiscent of NASA's current strategy of relying on the private sector for specific technology.

On a budget of $500, Winlock managed to take ten astronomers to Shelbyville[1] (population 2,000), likely chosen because it was his hometown, and he had taught at Shelby College[2]. He hit up Cornelius Vanderbilt himself for free train-fare and a private car.

First Contact: 05:08:37 PM EST
Altitude = 29° Azimuth = 268°

Second Contact: 06:07:34 PM EST
Altitude = 17° Azimuth = 277°

Maximum Eclipse: 06:08:51 PM EST
Altitude = 17.0° Azimuth = 277°
M = 1.0171
Velocity of the Moon's shadow = 2.06 km/s
Path width = 249 kilometers

Third Contact: 06:10:08 PM EST
Altitude = 17° Azimuth = 278°

---

[1] By apparent coincidence, Shelby County was also the home of the famous thoroughbred race-horse American Eclipse (1814–1847).

[2] After the Civil War, it began to be called Saint James College.

Fourth Contact: 07:04:08 PM EST
Altitude = 7° Azimuth = 286°

Counting up, Winlock's portion of the USCS expedition was swimming in instrumentation: twelve task-dedicated telescopes in all. Winlock operated the normally pier-mounted, Merz & Söhne, 7-1/2-inch achromat belonging to the College Observatory, which he had himself installed. At the time, it was still one of the largest telescopes in the United States.

I write "normally" because the astronomers went to the extraordinary effort to remove it from its dome and remount it outdoors for better view. (When later Shelby College was on the brink of failure, it traded the refractor to the University of Missouri for a smaller instrument and much needed cash.) During the total eclipse of the Sun, Alvan Clark's younger son Alvan G. Clark (1832–1897; optician) stood at the finder of this large refractor.

Shelbyville was the county seat. Eschewing the crowd on campus, George Searle (1839–1918; Paulist Fathers) observed from the roof of the Shelby County Courthouse. Searle was both HCO and USCS at different times early in life but had recently quit professional astronomy to prepare for the Catholic priesthood. Did he intentionally place himself closer to hierophany?

Searle quoted a record length for the Sun's coronal streamers: one-and-a-half times the diameter of the Moon.

It was well known that Winlock's goals for the total solar eclipse diverged from those of the USCS. Supposedly, he would use the great achromat for spectroscopy, the largest instrument dedicated to this task in 1869. Nonetheless, his passion was to take the definitive photograph of the solar corona. As by necessity both tasks had to be undertaken at the same time, it likely comes as no surprise that it was the spectroscopy that suffered.

Winlock's Shelbyville counterpart from the USCS was George Dean. He had with him the USCS's Sub-Assistant F. H. Agnew, who minded the Harvard chronograph. Agnew had injured his eyes surveying Salt Lake City, Utah, earlier in the year. He could not observe clearly the solar eclipse taking place above him.[3]

Jonathan Bowditch, HCO overseer and son of the famous mathematician Nathaniel Bowditch[4], was an honored guest.

Dean used the 4-inch finder no doubt ever-so-gently extracted from the Great Refractor back at the HCO. He covered it with a unique filter made with "London Smoke," which is really an architectural resin. Dean registered timings with a key, breaking the circuit on Agnew's chronograph. Where that mark appeared between

---

[3] Perhaps he had been 'shooting the Sun' with a sextant for too long.

[4] Nathaniel Bowditch observed the total solar eclipse of 1806 as its umbra passed over Massachusetts.

Fig. 10.4 Shelby College, College Street, Shelbyville, Kentucky. Between 1910 and the 1930s. Note the observatory dome. {Kentucky Historical Society Duplication Services}

Fig. 10.5 United States Coast Survey/Harvard College Observatory temporary solar-eclipse observatory at Shelbyville. See figure 17-d, too. {Shelby County Historical Society}

the one-second breaks caused by the clock's pendulum, a contact time was established on paper tape.

The weather at Shelbyville and its various secondary stations (except Bristol) was much the same described elsewhere in the Midwest: Anxiety provoking clouds in the AM, with clear skies in the PM.

Fig. 10.6 Main Street, Shelbyville, ostensibly on 7 August 1869 (Eclipse Day). {Shelby County Historical Society}

At Shelbyville, the business of timing was clearly overshadowed by the work of celebrity Cambridge, Massachusetts, photographer John Whipple (1822–1891; Whipple & Jones), international-awarded photographer of the Moon. Plates were exposed through an equatorially mounted 5-1/2-inch-aperture refractor with clock drive, of Alvan Clark provenance. (It normally was to be found mounted on a pier in the west wing of the HCO.) Together with their darkroom roadies, Winlock and Whipple were able to render *eighty* pictures of the Sun in eclipse, *seven* during totality.

Winlock accomplished this record because, unlike other total-eclipse photographers, he decided to have his camera produce images at the primary focus of the telescope. Without a second eyepiece lens, reasoned Winlock, there was no additional optical element to cause distortion or dim the faint corona. Whipple and he could get results at least akin to those of other astrophotographers—in less time.

Winlock's best-directed picture of totality was an atypical 40-second-long exposure. It left no time for a back-up exposure of similar length. Winlock put all his 'money on one horse.' No other professional solar-eclipse photographer dared go that long.

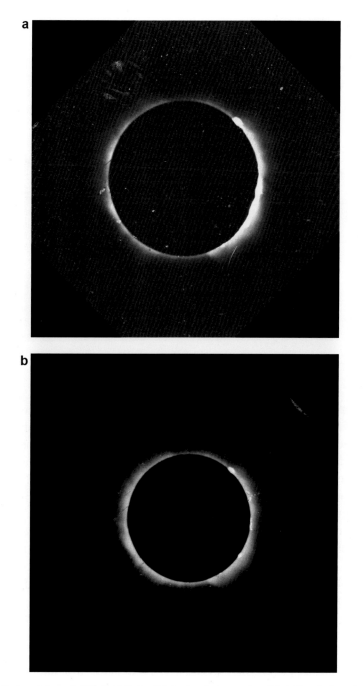

Fig. 10.7 (**a–c**) Joseph Winlock/John Whipple photographs of totality during the 1869 solar eclipse. {National Center for Atmospheric Research, High Altitude Observatory}

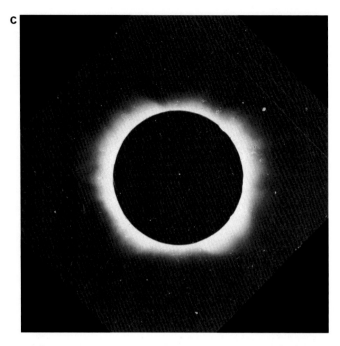

Fig. 10.7 (continued)

Winlock pronounced the result the best rendering of the inner corona ever—better than anything produced during the still-in-the-future total solar eclipse of 1870. Indeed, Winlock's and Whipple's image clearly shows the distinction between the annular inner corona and the streamers of the outer corona, extending a solar radius from the Sun, for the first time. Winlock figuratively or literally grimaced when books and other sources gave Whipple the sole credit.

On the other hand, Winlock also was armed with a double-prism Troughton & Simms spectroscope from HCO and the big refractor of the Observatory was dedicated to its use. It cannot be true that his full attention was given to photography. He observed bright lines in the chromosphere, a continuous spectrum from the corona. He should have seen more, and he might have reconsidered splitting his time during brief totality.

Winlock's and Whipple's work might have received even more attention had the resulting images not been so *tiny*. Nonetheless, they earned a place in the record book for the first photograph of the Diamond Ring Effect. This occurs when one last bit of photosphere peeks over particularly low topography on the lunar limb, just as the annulus of the chromosphere appears. For those lucky enough to see it, the result looks just like its namesake piece of jewelry and is an awe-inspiring sight.

Fig. 10.8 Joseph Winlock/John Whipple photograph of the Diamond Ring Effect during the 1869 solar eclipse. {Harvard University}

Almost as a side note, the Shelbyville deputation thought that they saw coronal polarization. Nonetheless, remember that all physical observations and measurements described here and below were made with whatever little energy was left after the stressful activity of attempting to time the solar-eclipse contacts.

Incidentally, astronomers were not the only ones to travel to Shelbyville for the total eclipse of the Sun. Cashing in on the town's prime position within the umbral path, trains offered $3.00 round-trip tickets for an "Excursion to the Sun."

Charles Peirce (1839–1914; USCS) and Professor Nathan Shaler (1841–1906; newly of Harvard University) established a satellite station at Bardstown, Kentucky. It was even closer to the total-eclipse centerline. They were charged with spectroscopic investigation of the solar prominences with a 1-prism spectroscope from Harvard. (The best expedition spectroscope was probably the one that stayed at Shelbyville with Winlock.) Peirce observed, and Shaler pointed—in the end, to no particular effect.

There was no means for Charles by which to measure the positions of the spectral lines Peirce saw. Without a clock drive, keeping the telescope/spectroscope pointed was difficult, too. Moreover, apparently Peirce and Shaler were startled easily. Admitted the young Peirce, "… the observers were in continual apprehension of some disturbance in the crowd of mostly ignorant spectators …" Shaler further remarked that:

> The crowd was placed, at the request of the observer, beyond a fence, distant about thirty feet from the telescope. At the moment of thirty feet from the

telescope. At the moment of totality a hollow sound, half of fear, half of admiration, called attention to their faces, with dropped jaws and look of horror, which were turned toward the wreck of the Sun.

Charles Peirce ran out of time before he could examine the spectrum of the corona. He complained, "Two seconds more, or a little more privacy, would have enabled me to get it."

Peirce and Schaler ended up using the opportunity to categorize prominences into two types. Categorization is a useful first step in science. Notwithstanding, the USCS men's scheme added little to understanding: 1) "low, long, and yellow" and 2) "high, short, and red."

A Historic Telescope Put to Use

The refractor that Peirce and Schaler—neither an astronomer by inclination—were using is as, or is more, interesting than their total-eclipse observations. It was the Merz 4-inch-aperture Bowditch Comet Seeker (again, seemingly an odd choice for spectroscopy). It was named for the Harvard benefactor who donated it.

Its normal task was this: Whichever over-worked HCO assistant was assigned for the duration was asked to pull this equatorial onto one of the Observatory balconies and "… to complete the sweep of the whole visible heavens once every month." That was the equivalent of 11,500 fields of view! The technique was successful, too, as comets indeed were discovered in the process.)

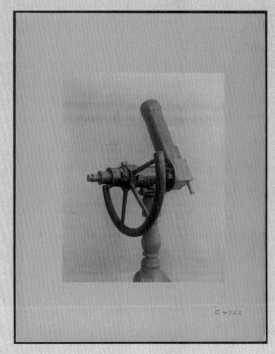

Fig. 10.9 Bowditch Comet Seeker. *Circa* 1890. {Reproduction of a reproduction}

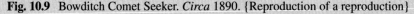

Bristol and its contiguous town of Goodson[5], straddle the Tennessee-Virginia border. Both had once been part of a single, huge plantation[6]. For the solar eclipse, the all-Coast Survey team of Assistant Richard Cutts (1817–1883; USCS), Assistant Alonzo Mossman (1835–1913; USCS), and Sub Assistant F. Walley Perkins (USCS) were assigned there, between the Cumberland and Alleghany Mountains. Though it did not afford the long totality of 'Out West,' the temporary HCO observatory at Bristol arguably was a well-situated place from which to visually observe a total eclipse of the Sun: At 536 meters above sea level, it combined both high elevation and reasonable longitude.

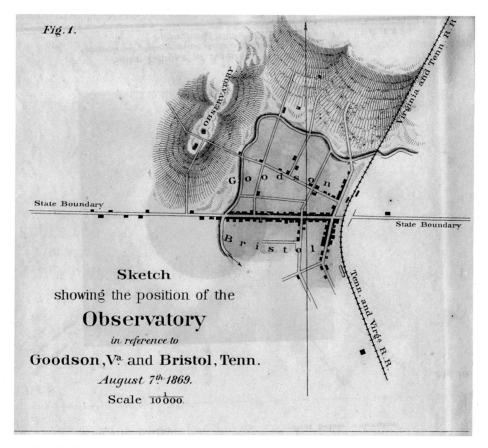

Fig. 10.10 Sketch showing the location of the observing station, Bristol, Tennessee. {United States Naval Observatory}

---

[5] now Bristol, Virginia

[6] This is the traditional term; regardless, I prefer the more descriptive, "forced agricultural labor camp."

On 7 August 1869, though, the Bristol observers were far enough east that they encountered a different weather pattern than that in the Midwest. What was feared and did not happen in places such as Springfield did so in Bristol. The three men dodged clouds throughout their work.

Cutts was equipped modestly compared to others situated by the senior Peirce. He peered at the Sun through a 3-½-inch Troughton & Simms refractor (admittedly, a top-of-the-line British import). He and his colleagues made the effort to cement bricks together, in order to fashion a stable mount for it.

Cutts had no recording chronograph. He called out contact times to Perkins, who looked at a clock and wrote them down. Mossman used a 3-inch "reconnoitering" telescope and checked the chronometer himself.

When the city of Bristol expanded, a commanding view of the railroad station was considered a selling point for real estate. (Today the thought is quite the opposite.) And so it was that the first and, for some time, most fashionable residential neighborhood in Bristol was located along Solar Street.

Fig. 10.11 Marking the location of the temporary Coast Survey, eclipse observatory in Bristol, Tennessee. {Emily Rice}

A HCO/USNO substation was in place at Falmouth, Kentucky. This site was chosen because it was predicted to be just inside the northern limit of the total-eclipse path. The duration of totality there would be short and contrast with those longer, closer to the centerline.

Fig. 10.12 A 1857 hotel used by the United States Coast Survey astronomers in 1869 still exists. Shelby Street, Falmouth, Kentucky. Once named Falmouth House, the building is now known as Pendleton House. {Kentucky Historical Society}

Falmouth observations were managed by George Searle's brother, Arthur Searle (1837–1920; newly HCO), the third astronomer lent to the cause by that Observatory. Before becoming an astronomer, the eclectic Searle had been a broker, professor, and sheep farmer. Another observer failed to show up; a local, Captain W. T. Arnold (USA) replaced him.

The astronomers had along a HCO telescope by Lerebours et Secrétan of Paris and a Dollond from Saint Paul's Church in New York. The idea of a telescope loaned by a church is intriguing, but both the instrument and its biography seems to have been lost.

Along with contact timing, Searle was assigned to specifically study the *partial* phases of the eclipse. It is rather remarkable that he tells us, "During the total phase, no observer used a telescope."

The HCO staff set in an empty lot across the tracks from the railroad station. We have seen how astronomers covet quiet during a total eclipse of the Sun, so that they may hear time signals called out, but also to maintain concentration. That was not to be the state of affairs at Falmouth. Complained Searle, "… just as the total phase was about to begin, the train from Lexington drew up at the station, remained there during the totality, with the passengers shouting, screaming, laughing, and talking all the time, and started again as the sun re-appeared."

Arnold decided to stand on a hillock farther away from the platform than the other astronomers. He alone was able to get trustworthy timings, read off his watch.

Samuel Langley (1834–1906; first Director of the Allegheny Observatory) was at a USCS satellite station in Oakland, Kentucky (near Mammoth Cave). It represented the southern edge of the passing umbra. Again, the idea was an intentionally short totality.

Langley and his traveling companions were disgorged at a train depot surrounded by—nothing. The senior astronomer had two sheds hauled to an adjacent field for use as observatories. Not anticipating much of an exposition, Langley brought for his own use only a 3-inch-aperture refractor from his permanent Observatory; but it and the other instruments had to be protected.

It was not only for want of a roof over their heads and their instruments that complicated matters for Langley: A lineman had to be called in specially to connect the observers, temporarily, with the telegraph system. Nevertheless …

"… our isolated station," sneered Langley with either snobbery, bigotry, or both, "was the resort of all the inhabitants of the adjoining country, white and black, who crowded around the sheds, interrupted the view, and proved a great annoyance." But *these* visitors proved not to be the problem. And there definitely was one.

Could it happen *again*? This time there was certainly no coincidence.

During the eclipse, a special train carrying one-hundred total-eclipse spectators arrived from Bowling Green, Kentucky, complete with a brass band! Somehow the astronomers got their job done, anyway.

It was thought that the Oakland was 3 kilometers within the zone of totality. The USCS men knew that their total eclipse of the Sun would be brief. They did not know how brief. Second contact was called and, before anything further could be uttered, third contact arrived! They had managed to place themselves at the very edge of the eclipse path. Total concealment of the Sun's photosphere at Oakland Station lasted but 1-¾ seconds according to Langley, or 1 second by the reconning of the telegraph man, who had stayed for the show. (His time was determined by "conjecture.") Somehow, Langley was able to 'freeze-frame' the fleeting appearance of prominences in this mind. On the up side, with such a grazing eclipse, those in attendance were treated to 15 seconds of Baily's Beads.

Whether they thought about it in such terms or not, before they left, the astronomers reenacted a rite humans have undertaken for thousands of years. We mark the location of an event that moved us: They placed a heavy stone at their observing site, part-way buried in the ground.

(If the telegraph complicated Langley's life, it also was to simplify it. In 1869, Langley improved the finances of his Observatory by selling standardized, transmitted time to the rail roads. The idea of Time Zones followed.)

Many years later, Langley recalled the 1869 total solar eclipse and in doing so, perhaps unintentionally, chronicled changing technology in simile. At second contact: "… the Sun went out as suddenly as a blown -out gas jet …"; at third contact it reappeared "… like suddenly kindled electric light … "

The coal-mining town of Springfield was now the 'Prairie State's' capital. It was to be a secondary USCS/HCO photographic station. A German émigré, Assistant Charles Schott (1826–1901; USCS), headed it. E. P. Seaver, Robert McLeod, and Charles Fay were among those detailed to Springfield. Eclipse Day was also Schott's birthday.

Fig. 10.13 Charles Schott. {*National Archives and Records Administration*}

All popular methods of dealing with solar brightness before second contact and after third were on display: Schott used a telescope outfitted with a camera. Seaver's 2.7-inch equatorially mounted refractor projected the Sun's image onto a white screen, and McLeod's small-aperture marine binoculars probably engendered little risk for eye damage. (Fay was not even provided with a telescope; he had to scrounge an instrument homemade by a Springfield amateur astronomer.)

Springfield was very easy to get to. Perhaps for this reason, this expedition became an exercise in 'too many cooks.'

Professor Alexander Twining (1801–1884; formerly Middlebury College) and Professor Nathan Dupuis (1836–1917; Queen's University) were at Springfield also; it is unclear why, as they do not appear to have contributed much in the way of observations. Former congressman Alexander Evans [Whig-Maryland] showed up as an honored guest.

As at all USCS stations, the Springfield mission's job foremost was contact timing, of course. The observers here would not rely on telegraph but instead establish their time by the stars and thereby set their clock themselves.

The station had two parts. One was on the grounds of the new State Capital Building, construction of which had begun just the previous year. The astronomers appear to have done their work, simultaneous with the laying of the building's foundation, at a spot 50 meters from the one that eventually would be covered by a resplendent dome. (There was talk of laying a marble observing-site marker for posterity; it never happened.)

So that its complement would be less disturbed, another, more out-of-the-way, site was occupied two kilometers from town—at the Sangamon River Reservoir that supplied Springfield's water. (It did not work *thousands* of spectators showed up!) Here a temporary building was erected to house Schott's telescope.

First Contact: 04:03:26 PM CST
Altitude = 33° Azimuth = 264°

Second Contact: 05:04:11 PM CDT
Altitude = 22° Azimuth = 274°

Maximum eclipse: 05:05:36 PM CDT
Altitude = 21° Azimuth = 274°
M = 1.0233
Velocity of the Moon's shadow = 1.69 km/s
Path width = 253 kilometers

Third Contact: 05:07:01 PM CDT
Altitude = 21° Azimuth = 274°

Fourth Contact: 06:02:32 PM CDT
Altitude = 11° Azimuth = 282°

While waiting for second contact, Schott noticed that the limb of the Moon seemed to connect *via* a cimmerian bridge to the umbra (darkest, inner portion) of a large sunspot. This optical phenomenon is known as the Black Drop Effect. It would plague future observers of the transit of Venus as they tried to catch the moment when the black silhouette of a solar-disk crossing planet completely separates from the blackness outside of the Sun's limb. (See Chapter 20.)

Meanwhile, back in 1869, Boston photographer James Black (1825–1896; a Whipple business partner) started taking 178 (!) images of the solar eclipse. He did so with a 4-inch-aperture, equatorially mounted, refractor by Dollond. As in Shelbyville, these images were made at the telescope's prime focus; they were less than two centimeters in size.

As totality approached, locals were trying to get a little nearer to the sky:

People were hurrying to and fro to secure a good location. Some mounted the cupola of the Statehouse, some the cupola of the Leland Hotel, others sought roofs and the highest accessible points in the city.

And so much for the astronomers' anonymity: Hundreds found the expedition ensconced at the reservoir. It was an example of how proximity to the professionals somehow gave citizens the sense of proximity to the celestial event itself. After totality,

CORONA OF 1869.—SCHOTT.

Fig. 10.14 Charles Schott's drawing of 1869 totality. {Xavier Jubier}

Suddenly the Sun burst out again from behind the beautiful corona, and almost instantly the corona and red protuberances vanished. The people immediately threw away their smoked and shade glasses, which went crashing in all directions with a ringing sound, and the crowd dispersed at once, not waiting for the end, but took up their course for the town more rapidly than they arrived.

A regional reporter was fascinated, not so much by the astronomers at the telescope, but the man tasked with attending to the chronometer. He,

> watched the instrument through the time of the eclipse, never once taking his eyes from the instrument, an act of heroism and devotion to science worthy of more than a passing notice. He gave his time and attention to his important duties while one of the grandest phenomenon the world ever witnessed was taking place.

The decision to take photographs at Springfield was made at the last minute. Captured without a clock drive, the Springfield images quickly drop out of the scientific chronicle. (The promised publication about them never materialized.) They were overshadowed completely by the work done at Shelbyville.

Before leaving the reservoir, the USCS placed a boulder, with a copper bolt drilled into it, in the ground to mark the spot at which total-eclipse observations were made. Today, what is to be found is not that stone, but a similar marker on the Capitol grounds.

Fig. 10.15 An ambiguous engraved stone now marks the location of United States Coast Survey astronomers temporary solar-eclipse observatory outside the Illinois State Capitol building. {Dave Joens}

Each of the principal Springfield observers wrote their own report of the event. The two paragraphs penned by Fay leave much to be desired:

> I saw no … I saw no … I failed to see it. I noticed no … I think this observation at least three seconds slow … My pencil broke while recording this … I stopped to take off glasses; the corona appeared in the interval. I saw no … I saw no …

> I saw one or two flashes across the field before totality… I inferred that they were the effect of nervous fatigue.

Meanwhile, at a Bloomington, Illinois, satellite station, James Peirce (1834–1906; Harvard University) and Joseph Warner (HCO) found a fine observatory: a stone building beside the train tracks. Unfortunately, neither had a telescope to place there. Sitting at the railroad station with the company of "bystanders," the duo accomplished nothing that they could not have done in Springfield.

Not to be forgotten back in Des Moines were two astronomers 'showing the flag' in Naval Observatory 'territory' for the Coast Survey. And they were first tier in the eclipse world. German-american, Assistant Julius Hilgard (1825–1890; USCS) occasionally acted as temporary Superintendent. He was supported by Assistant Edward Goodfellow (1828–1899; USCS) who was a veteran of Stephen Alexander's 1860 expedition. At their heel was former USCS employee J. Homer Lane (1819–1880; more recently, United States Patent Office[7]).

Physician T. C. Hilgard—most likely a relative—made eclipse sketches: "[His] skill in drawing from microscope observations was looked to for enabling him to furnish more precise delineations …" I have never seen the result, though.

These men, too, set up their station in the courthouse square, but at a distance from Newcomb.

Why did Benjamin Peirce send Hilgard so far from the main USCS effort in Illinois and Kentucky? An untreated alcoholic, it may have been Hilgard's embarrassing reputation for public drunkenness—or something more prosaic.

The USCS observers at Des Moines counted as a member of their party the same Sackville Cecil that the USNO counted as a member of theirs. Moreover, it was the USCS that published his 200-word report on the subject. It also gave Cecil a three-inch refractor to use.

Thus, the mystery that appeared in Chapter 4 may be solved. It likely was Thomas Stafford, not Cecil, who manned the USNO transit-circle telescope during the total eclipse of the Sun.

Lane was to be assigned a random 3-inch-aperture refractor belonging to the USCS. Once in Des Moines, though, Newcomb told Lane that he had a glass to spare. Remember the temporary 6-inch-aperture telescope fashioned from the

---

[7] It is unclear why Lane was dismissed.

USNO's refraction-circle objective? It was well-suited to corona observing. Would Lane like to use it? Lane replied, "I gladly accepted this use of so powerful an instrument, though not without misgivings that it might have been more useful in the hands of a more experienced and expert observer." Lane had thought that he might peer through what was now his telescope using a novel smoked-mica filter but was satisfied when handed a piece of "deep-blue glass." What an unusual view—a blue eclipse!

Why was formally unaffiliated Lane and not J. E. Hilgard given possession of this plum refractor? Was it USNO/USCS rivalry? Was it personal? Hilgard probably had more experience with telescopes than did Lane.

In fact, Hilgard hereafter all but drops out of our story. His name is only mentioned in eclipse documents again for having recorded his contact times and those of others. He did this at a modest Fitz 3-½-inch-aperture refractor and without a calibrated chronograph. (He used the traditional method: listening for a member of the Des Moines assembly of astronomers, dedicated to watching the clock, call the seconds out loud.)

Lane was not a 'nobody.' His hobbyist experiments with very low temperatures were significant and were looked upon favorably by other scientists in Washington. Still, he was not a field scientist like Hilgard.

For some reason, *Lane's* written account ended up on John Sands's desk, who included it in his formal report on the 1869 total eclipse of the Sun. So, we read of Lane's impressions in *both* the USNO and USCS post-eclipse publications. In what seems an apt metaphor, Lane 'jumped ship' in the middle of a solar eclipse.

Lane wrote many words about the solar corona, which taken in total convey little information. One rereads Lane. Perhaps that will help. Again, nothing.

It is understood that Lane was excited by two comet-like objects he discovered during the solar eclipse. This excitement lasted just as long as it took Lane to notice after the eclipse that there were two scratches on the filter he had used—the same distance from one another as the "comets."

Everybody knew that the 'bread and butter' of the USCS was cartography. Its scientists were used to portaging equipment to rugged places. So why not go to the now most-westernmost place owned by the United States? The most-distant American soil on the path of the total solar eclipse? Alaska. For the astronomers of the Survey, The 'Last Frontier' was just another day at the office.

So it was that Assistant George Davidson (1825–1911; USCS), newly in charge of Survey operations on the Pacific Coast, headed to the most hard-to-get-to total-eclipse viewing site that any American astronomer would occupy. He was the obvious choice inasmuch as he was already in California. He had anticipated this mission ever since his first visit to then Russian America in 1867.

Davidson was instructed to make solar-eclipse observations including spectroscopy. He was lucky enough to have a single 3-inch-aperture Fraunhofer refractor, equatorial mounted. But as a 'reality check,' there was no spectroscope available on the West Coast. (In a fit of redundancy, he took no fewer than eight chronographs with him.) Davidson's other mission was to undertake the mapping survey associated with the United States' $7,200,000 purchase of Russian Alaska.

GEORGE DAVIDSON

Fig. 10.16 George Davidson. 1875–1885. {National Portrait Gallery}

First Contact: 11:46:09 AM AKST
Altitude = 47° Azimuth = 172°

Second Contact = 12:56:01 PM AKST
Altitude = 46° Azimuth = 196°

Maximum Eclipse: 12:57:49 PM AKST
Altitude = 46° Azimuth = 197°
M = 1.0181
Velocity of the Moon's shadow = 0.80 km/s
Path width = 251 kilometers

Third Contact: 12:59:36 PM AKST
Altitude = 46° Azimuth = 198°

Fourth Contact: 02:07:42 PM AKST
Altitude = 42° Azimuth = 220°

How to get to the centerline? With understated doughtiness, Davidson considered carrying instruments,

> … up the Sutchitna or Kneek Rivers from Cook's Inlet [500 kilometers]; across the mountains that border Prince Williams Sound [[a two-kilometer ascent and descent at the lowest pass] … or up the Alsegh [*sic*] River [Whitewater Class = V+]. Natives do cross the Saint Elias range … but only in winter with snowshoes.

Davidson reconsidered and sailed by Navy ship to Sitka. He then spent eleven days in an open boat, slowly navigating up the unexplored (by easterners) Chilkat River, in order to reach the path of totality.

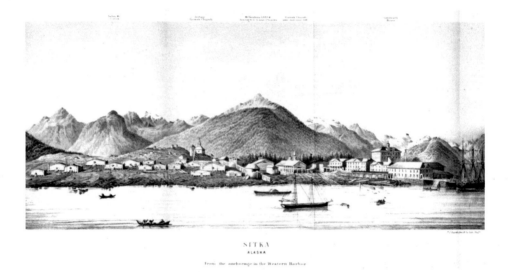

Fig. 10.17 Contemporary Sitka, Alaska. {Chilcott Indian Association/Haines Borough Public Library}

Davidson pursued a somewhat unusual, personal goal for the total eclipse of the Sun: He was a Bailey's Beads denier. He did not believe that such a phenomenon took place and wished to prove his assertion by observation. Proving a negative, of course, is not standard scientific method.

What were Davidson-and-company's solar-eclipse data, after one of the most-exotic trips to reach totality's path? The same as Asaph Hall's: Clouds. A bit of rain. A bit of clear sky. But mostly clouds.

Davidson risked life and limb to see the total eclipse of the Sun. Yet coincidently there was a steamer moored back in Anchor Bay, at the mouth of the Chilkat. Its tourist passengers had made no particular effort to see a solar eclipse. Doubtlessly aware of the irony, Davidson duly reported that, "… the ladies and gentlemen on board had a view of the whole of totality … "

Coincidentally, none other than outgoing Secretary of State William Seward (1801–1872; retired, US Department of State) happened to be touring the recent acquisition that August. Though in bad health, he wanted to spend his last days in travel. He detoured to join Davidson just in time for some fleeting glimpses of the total solar eclipse. This is fitting, because it was Seward who negotiated purchase of Alaska by the USA less than two years earlier. Due to his efforts, we can say that, on 7 August 1869, the Moon's umbra both crossed into and out of North America over territory belonging to the United States.

Fig. 10.18 This drawing was the best that George Davidson was able to construct of the 1869 solar eclipse. {National Oceanographic and Atmospheric Administration}

It is difficult to prove that Benjamin Peirce's promise to improve navigation by observing the total solar eclipse of 1869 was fulfilled. In summary, of the three USA government-sponsored 1869 total-eclipse expeditions, it was the USCS's that spanned more of the eclipse path than any other. It also was the USCS Expedition (save Winlock's/Whipple's photographs) that generated the fewest scientific results.

11

The Canadians: Toques on the Frontier

"For the benefit of those who may undertake an expedition of a similar kind, it may be well to mention a few incidents that occurred during our journey, which, although trifling in themselves, may prove useful to future eclipse parties."

So far, I have related domestic expeditions to see the total solar eclipse of 1869. European astronomers ignored the affair, knowing that the total eclipse of 1870 would soon come their way. Nevertheless, there was one expedition from another country to the fringes of the United States that summer. It was mounted from Canada, a country only confederated two years before.

We have been alerted to the dangers of the Old West, if only from the corpus of movies made by John Wayne[1] and Clint Eastwood. Maybe it was not *that* bad. Still, in the mid-nineteenth century, the America west of the Mississippi was not for the faint of heart. Considering the safety issues accompanying long distance travel in those days gone by, it is amazing that not one, normally often sedentary, eclipse pilgrim was killed or maimed—to my knowledge.

But one ill-starred astronomer did have a much more extraordinary adventure than planned and nearly got himself into great danger. Yet he did it for Queen and country, for he was military man. He turned out to be also a gifted storyteller.

Recently retired from naval service, Commander Edward Ashe (1814–1895; Director of the Quebec [City] Observatory) mounted the single international expedition to the total solar eclipse, representing the Literary and Historical Society of Quebec. (It was Canada's foremost learned society at the time.) Other members of the expedition were chemist James Douglas, Jr. (1837–1918; President of the

---

[1] born, in fact, in Iowa

T. Hockey, *America's First Eclipse Chasers*, Springer Praxis Books, https://doi.org/10.1007/978-3-031-24124-6_11

Literary and Historical Society of Quebec) and Alexander Falconer, a visitor from London, England.

Ashe had been studying sunspots for several years and concocted a theory just published in which they represented, not cool spots in the photosphere as was then commonly assumed, but instead the collisions of intra-mercurial asteroids with the Sun. (See Chapter 15.) He thought he saw sunspot umbras form before penumbras and concluded that the umbra was molten metal, while the penumbra was its dross. There were more sunspots recently because the inner planets happened to be on one side of the Sun, and their combined gravity accelerated the in-fall. These unorthodox ideas made him something of a pariah in solar studies.

This was to be Ashe's second total-eclipse venture, having been in Labrador with Stephen Alexander for the United States Coast Survey's 1860 total-eclipse expedition. As for 1869, at the last minute, the Canadian government approved $400 (a small amount even then) toward Ashe's travel expenses to the total eclipse of the Sun. The naval officer and companions had three days to get ready. This included building a traveling mount for a telescope.

The three men started out by boat down the Saint Lawrence River and sailed through the Great Lakes. They then journeyed *gratis* on the railways from Chicago, Illinois, all the way to Jefferson, Iowa, on the banks of the North Raccoon River (six days and 2,250 kilometers). Jefferson was just a little south of the centerline and had been recommended to Ashe by "the American astronomers."

The mill town had been founded in 1854 and was a county seat. It was now on its way to boasting of 779 inhabitants (1870 census).

Fig. 11.1 Contemporary Jefferson City, Iowa. {Alamy}

First Contact: 03:56:15 PM CST
Altitude = 38° Azimuth = 257°

Second Contact: 04:59:01 PM CST
Altitude = 26° Azimuth = 269°

Maximum eclipse: 05:00:31 PM CST
Altitude = 26° Azimuth = 269°
M = 1.0224
Velocity of the Moon's shadow = 1.37 km/s
Path width = 257 kilometers

Third Contact: 05:02:01 PM CST
Altitude = 26° Azimuth = 269°

Fourth Contact = 05:59:19 PM CST
Altitude = 15° Azimuth = 278°

The Canadian Eclipse Party's 1869 budget allowed Ashe to take along only one major instrument. He made the entire trip with the eight-inch Alvan Clark objective of the new, $1,200 Quebec Observatory telescope in his lap. It was part of the largest telescope in Canada and one of the three highest-resolution objectives used during the 1869 total eclipse of the Sun.[2] (It arrived back home only slightly damaged.)

> I might mention that two of the cases, containing parts of the telescope, were directed "Eclipse Expidition," with three i's in Expedition. This was pointed out to me at Montreal, but the mistake is excusable, for evidently the more eyes we have in an expedition the better. With regard to original spelling, I will relate the following anecdote, which, would have suited "Artemus Ward.[3]"
> The boatswain of a man-of-war has to keep a rough expense book of the different stores that he uses, and this is checked by the master, who on one occasion sent for Mr. Parks, and when he came, he said: "Oh, Mr. Parks, you have expended too much rope for those 'jib guys;' it will surely be found fault with; you had better reduce the quantity;" and on handing him the book, he said: "By the bye, b-l-o-x is not the way to spell blocks." The boatswain took the book very sulkily; and after he had taken two steps towards the door, he turned around, and said "Well, sir, if b-l-o-x don't spell blocks, what do it spell?

This obsession with spelling speaks to snobbery on the part of our hero. In the 1860s, only half of American youth were in school, the fraction for Canada could not have been much different.

> We started on our journey by the evening train. When we arrived at Port Huron [Michigan, USA] our first difficulty occurred; the Custom-House officers would not pass our baggage, although we pointed out the great

---

[2] This mirror is now housed in a telescope at the Collège de Jésuites, Quebec.
[3] the pseudonym of nineteenth-century American humorist Charles Browne

importance of our party, and also, that the Moon would not wait an instant for us. They did not see it; so our baggage was locked up for the night.

One cannot picture Ashe discussing the synodic period of the Moon with an indifferent customs officer and any better outcome. While a man of rank in his own country, Ashe was a mere tourist in the United States.

> We took rooms at a small inn, and then Mr. Douglas and I went by rail to Huron, to see the head of the Customs. After going up two flights of stairs, we were shewn [sic] into a room which two gentlemen occupied. The chief was smoking, with the chair resting on its two hind legs and his resting on the table. We told our story, and shewed him a certificate from the American Consul at Quebec. He looked very hard at me, took the cigar out of his mouth, wrote a pass which he handed to me, and then resumed his cigar and former position. We began to thank him, but as he hid himself in smoke, we retreated down stairs.

This scene is easy to imagine, too. Americans brought cigars back from the Mexican War. Starting in the 1869, they were mass produced in the United States to keep up with demand. Photographs of luminaries such as Ulysses Grant smoking must have helped stoke the fad.

> I never was more struck with the kindness of our American cousins than I was during this trip. On all occasions, they did all in their power to promote our convenience. In the morning we had time to see Mr. Muir, the director of the railway, who kindly gave us a free passage over his line, a kindness that was shewn to us by all the directors of the different lines that we travelled on. I may remark that the cases with the heavier parts of the telescope were broken, and I much feared that the instruments would be seriously damaged. Mr. Muir very kindly had outside cases put on … After we left Chicago, and before going to bed, we left word to be called before crossing the Mississippi. It is not fair to judge of scenery from a view taken through the window of a railway car, but I must say that I was disappointed,—shallow, sluggish, and muddy; but then I ought to remember that I live on the banks of one of the finest and most beautiful rivers in the world.

At this point in the story, it is worthwhile to mention how this sailor-on-land got his job before we continue with his adventure. Due to a shipboard accident, Ashe suffered a fractured thigh which left him permanently *crippled* in his leg. No trained astronomer, he was given the Observatory as a sinecure. All that followed happened to him with but one good leg. Continuing:

> In the morning we were on the prairie, which is not so flat as I had expected to see it, but it is a beautiful undulating country, and if there were trees upon

it nothing more could be desired. It was explained to me by a gentleman who was travelling with us, the reason why trees do not grow on this beautiful land. It appears that on the eastern bank of all rivers and streams only do trees grow; now without entering into the cause of the prairies catching fire, I will only say that in September, when the long grass is quite dry, they do catch fire, and then burn until it is stopped by a river, and as it always burns to windward, and as the wind generally blows in one direction, we have a solution why the trees only grow on one side of a river; and once the primeval forest is removed, it never has a chance of growing again, as the young trees are sure to be burnt, and the beautiful black soil of the prairie is enriched by the deposit of burnt grass.

Ashe fancied himself an anthropologist:

At one station where we stopped to water our engine I saw two children of the soil; they have good reason to complain at their lot. The [American] buffalo and antelope driven away, and if they are hungry they are to go and dig; dig, how can they dig? Let us reverse the picture. Suppose that our cities and towns were by the Indians turned into a prairie, and when we were hungry they told us to go away and catch a buffalo, a pretty hand I should make of catching a buffalo.

I would very much like to leave out the rest of Ashe's paragraph and the one that follows. All the same, at the time his genocidal views were not exceptional enough for them to be edited from publication:

The sooner the poor fellows are shot down or killed by small-pox, the sooner they will go to their happy hunting grounds.

Ashe's racism makes him a less sympathetic character in what is to take place. But first, an ominous portent:

It was pointed out to me that most of the teleqraph-posts were struck by lightning; no wonder, for that king of natural forces, that for so many thousands of years has reigned supreme—splitting the granite rock, and shivering the mighty oak at his will-now to be brought into existence at the will of an apothecary boy, placed in two cups and locked into a cupboard, and then made travel day and night, over hill and dale, and under the vast ocean, to carry messages at the bidding of man,—no wonder I say, that he should try and knock the whole concern into a cocked hat!

Aboard the train the Canadians made the acquaintance of Professor Hugh Vail (1818–1900; Haverford College). One thing led to another, and the American Vail became part of the Canadian Expedition.

"Boonsboro! twenty minutes for dinner!!" Now, then, we shall have something in keeping with the prairie,—I suppose a deer roasted on a stake. Nothing of the sort, I went into a nice dining-room; saw a quantity of pretty girls, or rather young ladies, with short sleeves and low dresses. "Soup, sir! Chicken, sir! peas, sir—" The station at Rugby [North Dakota] is nothing to it. After twenty minutes of capital feeding, we heard, "all-aboard! all aboard!" and as we left, the father of these young ladies was standing at the door, and obliged up by taking half-a-dollar, a great improvement on the English system, where, on asking the waiter for your bill, he asks: "What 'ave you 'ad?" and begins to add accordingly. The next station was Jefferson, 1,398 miles from Quebec. Here the boxes were again thrown out, and the train left for San Francisco. The boxes were left at the station, and we drove up to the hotel, about half-a-mile from the station. As this was Saturday, July 31st, we had exactly a week to select a site and to build an observatory—mount the telescope and take preliminary observations. The American parties were several weeks at their station before the day of the eclipse, and found it not too long to prepare.

Jefferson City is three years old, has about eight thousand [sic; 800] inhabitants, and looks a thriving place. The next day, after church, Mr. Douglas and I rode across the prairie to a station situated about eight miles on the railway from Jefferson. As it was nearer to the central line of eclipse, we wanted to see if it would do for the site of our observatory.

I forgot to mention that the day before I left Quebec, in pulling off my boot I broke the tendon of the plantaris muscle, which made me quite lame. However, the six days' comparative rest made it much better, but still it was far from well.

"Forgot to mention" seems somewhat contrived.

Ashe and Douglas decided to check out a potential alternative observing station. This may have been the nearby town of Grand Junction.

We started for our ride across the prairie about two o'clock, and reached the station in about an hour and a-half. We crossed several streams and some marshy ground, and started several prairie chickens. After examining the place, and finding that it would be very inconvenient to get the material there, we thought that it would be better to remain at Jefferson, and we mounted to return. After we had left some time, and as I was suffering from my leg, and could not ride fast, I persuaded Mr. Douglas to ride on, and get back before sunset to keep an appointment with a carpenter, and not to mind me, as I could ride slowly back. He very reluctantly did so, and when I was left alone, I felt quite at home, steering my horse across the boundless prairie by the setting sun. Now, my horse had crossed many streams, and soft wet places in going out, so I took it for granted that he knew more about the prairie than I did, and would not allow me to get into difficulties, and consequently steered a straight course for that point of the compass in the direction of Jefferson. The sun had just touched the horizon. I was crossing some marshy ground with reeds up to my shoulders, when I saw my horse's nostrils distended, and his ears forward. I immediately put my helm down and brought him round, and just as I had done so, down he sank; I found myself up to my ankles in mud, and up to the calf of the leg in water; the horse was fixed immovable, no struggling, but snorting and dreadfully frightened. I have been in various situations of difficulty; but when I looked up and saw the tall reeds far above my head, and the sun setting, I must confess that I thought my case a serious one. I remembered the fate of a young French officer of the combined fleet that was at anchor at the entrance to the 'Dardanelles,' who went on shore to shoot, and as he did not return that night, we landed in the morning to look for him, and not far from the ship, we found him in a bog up to his waist gun a few feet in front of him, and he quite dead …

At this point the tale sounds quite incredible. So, I consulted Professor Chad Heinzel, an authority on the geomorphology of Iowa. Yes. It is just possible for Iowan mire to act like quicksand. By coincidence, *he* had been trapped fast, in a similar fashion himself! (The trauma took place when he was very young and on foot.) Returning to Ashe's peril:

I knew that if a man once sets up to his waist, it would be impossible to extricate himself; however, when I dismounted I sank up to my knees, and although that was not the place to philosophize, still I did so, and I began to think what is the reason that a man in struggling works himself down, and I immediately discovered that on raising the heel I produced a vacuum, as the mud prevents either water or air getting underneath the foot, and so with 15 lbs. to the square inch, in addition to your weight you soon disappear. That being the case, I did not attempt to raise the foot, but moved it backwards and forwards in a horizontal position until I made the hole so big, that water got under the foot, when I could lift it up with the greatest ease. After extricating myself I tore down some reeds and made a platform round my horse, then I patted his neck, and spoke good-naturedly to him, and then went astern, and by means of his tail worked him backwards and forwards with a rolling kind of motion to let the water well round his feet, and lastly went ahead, passed the bridle over his neck, and sat down with it in my hands right ahead. Now, then, old boy, 'up she rises,' the horse began to struggle, I kept the head-rope taut, and he was freeing himself bravely. If I let go the bridle too soon, he would go back; if I held on too long, he would be upon me, and not only kill me but bury me, so at the critical moment I let go, and rolled over and over amongst the reeds, and the horse floundered past me. When I got on my feet no horse was to be seen, but only the tops of the reeds moving as he was making his way out. I thought I had not improved my situation much, for with my leg I could not walk a mile, and, of course, the horse had shaped his course for the stable. However, when I emerged from the reeds, I saw the dear old fellow standing as still as if he were in his stable. But now came another difficulty with my lame leg, I could not put a foot into the stirrup, perhaps he might have been in a circus and taught to lay down, so I began kicking his forelegs and lifting up one and then the other—but no—he had no idea of it: then I thought I would lash his feet together with the bridle and throw him down, but there might be some difficulty in my remaining on his back when he floundered to get up, well, if the worst comes to the worst, I will lash myself to his tail and make him tow me home; but the idea struck me, I lengthened the near stirrup to about a foot and a-half of the ground, and then lengthened the other and brought it over on the same side, and here I had a nice little ladder to walk up which I did, and then knelt on the saddle and dropped into my seat. I could not help shaking hands with myself and patting my steed on the neck, I then commenced my journey home, which I reached just before dark.

We might think that, after a mangled tendinopathy and serious riding accident, Ashe would consider his worst trials behind him. Nonetheless, in Ashe's mind, the low point was yet to come. And from an unexpected source.

Fig. 11.2 Canadian's temporary solar-eclipse observatory in 1869. {Alamy}

We had agreed to erect the observatory about half a mile from the station, a rising part of the prairie; carpenters were engaged, and an arrangement made with a lumber merchant, who would supply what I wanted and take it back when I had done with it, only charging us for the damage done to the stuff. Early on Monday morning, the instruments were carted out and unpacked; and at sunset the four walls of the observatory were up. Now, as we thought it not advisable to leave all these things open on the prairie, it was agreed that some one [*sic*] should sleep there—and, of course, it was my duty to remain. They sent down a mattrass [*sic*], pillow, and blanket; there was no wood to build a large fire outside, but I collected some chips, and lit a small fire inside, and placed my mattress alongside. A little after sunset a musquito [*sic*] looked over the wall, and then sounded the assembly; on they came, and I with my head in the smoke kept blowing the fire, putting on wet grass to make a smoke; but, after half an hour at this work, I found out the fact that man was not intended for a pair of bellows, and although I assisted the action by compressing my sides with my hands, still at the end of the half hour that I blew I found that I was blown. When once my head was out of the smoke, the musquitoes flew at me; I stood up to fight them, but in so doing I had to fight myself also. Now an army was drawn up in contiguous columns on my cheeks, the skirmishers advancing through my eye-brows [*sic*]; at their first volley I felt as if I was struck with a hackle. I really think that they work their stings like the needle of a sewing machine. Maddened, I struck myself a fearful blow with both hands in the face, and had the satisfaction of making them "leave that," and so I fought myself and the musquitoes for some time: still they attacked me with an impetuosity truly marvelous, and where one fell two took his place. I was getting weak; a storming party had now taken possession of my right ear; I clenched my fist, and with a swinging blow, cleared the ear, but knocked myself down. Exhausted and worn out, I put my hands into my pockets gave them my head. In that half-dreamy state, the long, long hours were passed; and after they had breakfasted, dined and supped; they began discussing me. "Ah," said one, "If you want a good drink, strike between the corner of the eye and the nose." "No, no," said a large party; "if you want a draught of good sparkling astronomer, sink your pump in his temple." "You are wrong," said a dissipated old fellow with frayed wings; "just creep up his cuff, and harpoon his wrist, and there you will drink until you lift yourself off your legs."

The Mosquito Song

Ashe even had the mosquitoes singing. It is the only ballad specifically about the 1869 total eclipse of the Sun. For the benefit of his reader, the multitalented[4] astronomer provided lyrics and (presumably, if called upon) a tune:

♫ The blood of the Indian is dark and flat,
♫ And that of the buffalo hard to come at
♫ But the blood of the astronomer is clear and bright:
♫ We will dance and we'll drink the live- long night.
Chorus: –
♫ How jolly we are with flights so airy;
♫ Happy is the mosquito that dwells on the prairie.

And then they quarreled and fought with each other, and made speeches, –and so the dreary hours dragged along; but when the eastern horizon tinted with beams of light, they staggered off to their respective marshes—some to die of apoplexy, others of *delirium tremens*. Verdict—served them right. From dawn until six, I had a refreshing sleep, and when my relief came, I awoke up, and began to think whether I had heard all this, or only dreamt it. I suppose I dreamt it.

The eclipse? It seems anticlimactic at this point.

The work now made rapid progress: doors with locks, dark room settled, platform support for telescope firmly laid. The next day, began to mount the telescope, but when we came to screw in the object—glass, we found out that the brass seat in tube had been pressed into an oval. What was to be done? No one in Jefferson that knew anything about it; too late to send it anywhere; here was a great break-down. However, a Mr. Kelly said would try; and after some hour's hard work, he got the object-glass screwed home, but could not be unscrewed; so the flats that hold the bolts that secure the object-glass to the telescope could not be put on, but we secured it as well as we could.

Ashe made four "photograms" during just over three minutes of totality. He believed them to be—incorrectly—the most western use of photography during the total solar eclipse. Douglas ran the darkroom. (Alone!) A Mr. Stanton acted as a human camera shutter, by placing a cloth over the objective when not exposing a plate. It was crude, but effective enough.

Falconer described best the eclipse experience of the Canadian Party, in an open letter to Ashe:

---

[4] The nonsensical forced rhyme in the first verse does not bespeak of poetic talent.

To Captain ASHE, R.N., &c., Observatory, Quebec:

Dear Sir,—As requested by you, I now give you the results of such observations as were made by me on the 7th of August last, during the progress of the eclipse.

… Shortly before totality, there appeared on the sun's northern limb several watery-looking globules, which merged into each other as they passed from West to East, and then disappeared. At this instant, also, appeared distinct long, brilliant, yellow, rays of light, running East and West, and far away, and as straight as if ruled; others again ran North and South, and reminded me of the glory ancient painters depict around the heads of Saints. On the Southern limb appeared, just at totality, a large circular opening, or ring of bright silvery light, which assumed the shape of a red-hot crooked bar of iron. This, resting on the dazzling silvery coronal light, gave a strange and wondrous addition to the glorious scene we now beheld. Several constellations shone brightly fourth, and a star or two low down on the Western horizon. I must not omit the strange protuberances seen at this moment: on the Eastern side was one like a tongue bent upwards, with streaks of a reddish hue; the others the shape of knobs, dark and colorless, and rugged in outline.

… when totality took place, all became comparatively dark; every tongue was hushed amongst the groups of persons who had come out on foot, or were seated in their waggons [*sic*], from Jefferson and the country around.

And what did they behold? A wondrous sight! At the moment of totality, burst forth the beautiful coronal light of the brightness of burnished silver! Upon the Southern portion of this ring of light, rested that curved, elongated protuberance, of a fiery redness, rendered more ruddy in contrast with the dazzling silvery light of the corona.

Several constellations shone bright and clear; several stars also were observed above the Western horizon. All these gave the scene a magnificence and grandeur. Wonder and admiration sat upon every face uplifted to the sky. Every voice was hushed. Sublime, indeed, was the scene presented. In reverential awe the groups stood mute. Each one seemed to ponder within himself over the glorious scene in front of him.

Presently, the light of the sun suddenly bursts forth … A murmur is now heard, and voices arise, proclaiming the sublimity of the scene they had just witnessed, one of the most wondrous and imposing sights presented to the human eye, in the firmament of heaven! The words of the Psalmist involuntarily fell from the lips: "The heavens declare the glory of God, and the firmament sheweth his handiwork."

… It remains only for me, in conclusion, to thank you and Mr. Douglas for inviting me to join this highly-interesting expedition, and to congratulate you and Mr. Douglas upon the great success which attended your photographic operations.

I have to thank you for beholding the wondrous and vast prairies west of the Mississippi …

I remain, dear Capt. Ashe, yours very faithfully,

ALEX. PYTTS FALCONER.

GLENALLA, QUEBEC, *August* 28*th*, 1869.

News of Ashe's work made the *London Illustrated News* in October. The 1969 total eclipse of the Sun appears to be the first for which photographs of it were published in popular media.

As you already may have guessed, Ashe was not one to avoid the superlative. He wrote that his fourth photograph (out of four) was "the most remarkable photogram that has ever been taken of an eclipse."

Others were not so sure. Each exposure was the same—ten seconds. This meant that each exposure was far too brief to reveal the solar corona.

There was more. Ashe:

> … it will be necessary to refute some opinions that have gratuitously been given respecting them [Ashe's photographs] … when I found that it would be many months before I could get funds to print my Report, it was agreed upon, after consulting some friends, that the negatives of totality should be sent to England. Unfortunately, I selected Mr. De la Rue as the fittest person to examine them. He never acknowledged the receipt of them, and, after many months,_Mr. Falconer, who had returned to England, sent me a copy of a letter to him, from Mr. De La Rue:

> "THE OBSERVATORY, CRANFORD, MIDDLESEX,
> "Dec. 27th 1869.

Ashe's images seemed to prove that a prominence that had shot up out of the photosphere during the eclipse and then fell back. There were also strange notches beneath the prominences, which Ashe said were reflections from the surface of the Moon.

Across the Atlantic, Warren De la Rue objected and could think of but one explanation. De La Rue:

> "My Dear sir,—I am very sorry to have caused any uneasiness to Commander Ashe; but one circumstance and another have delayed my writing to him. I have received his papers, which I sent to the [Royal] Astronomical [Society], and later on, the original negatives, which arrived safely, although Commander Ashe had neglected the precaution of protecting them with a covering of glass. There is evidence in these negatives of the telescope having moved, or, perhaps, followed irregularly, during the exposure of the plates, and this renders the dealing with the negatives very difficult; more-

over, it contradicts the theory set forth by Commander Ashe in respect to a certain terrace-like formation in the prominences, and also the rapid shooting out of a certain prominence. The American photographs are very much more perfect than those sent by Commander Ashe; in fact, they leave nothing to be desired. To correct the defects of duplication in Commander Ashe's photographs, would entail some expense, and much trouble; and it would be necessary for him to re-write his paper ...

"Wishing you the compliments of the season, I am with best regards,
"Yours sincerely,
"WARREN DE LA RUE."

Bumping into your own telescope is the ultimate rookie move in astronomy. Ashe was outraged by "the crimes I am charged with." He took three pages of text defending his work, concluding that it sharply and truthfully captured the same great prominence that everybody else saw. The verisimilitude of the prominence images was vouched for by those who observed with Ashe. Ashe demanded that Royal Astronomical Society [RAS] return his plates to him.

Presumably it did so. However, there is a committee for everything. The Society also published this:

A Committee appointed by the Council unanimously report that in their opinion, there was a decided movement of the instrument at the time the photograph was taken. This conclusion they arrived at from an examination of the chromosphere close to the Moon's limb, as well as from an examination of the prominences.

The Committee's opinion was unanimous.

The nineteenth-century prestige of the RAS, as final arbiter of all things celestial, was such that its *ad hoc* jury dealt Ashe a professional death sentence. His caustic personality and hetrodox theory of sunspots already had made him *persona non grata* in the astronomical community (and would do so for years). At the RAS's next meeting, the minutes record notice of total-eclipse photographs made by the two USA government expeditions " ... and also by Commander Ashe ... but they are not equal in definition to the others."

There was no controversy involving Falconer. He had considerable artistic talent, which he used to draw the prominences and corona.

None of the other 1869 total-eclipse expeditions saw anything like Ashe's rapidly changing prominences. Ashe insisted they were real (not a blur) but was denied funding to attend the 1870 total eclipse of the Sun visible in Europe. Those who did observe that event saw nothing like what Ashe promised.

It was arguably Commander Ashe, the individual who made the most danger-ridden trip to the 1869 eclipse path, who was rewarded the least for his troubles. As some say in Canada, *c'est la vie'*.

12

Chasing the Umbra through Time and Space

" … *Haec olim meminisse juvabit*[1] [Maybe we will be laughing about even these things in the future]." – Alexander Falconer, after the total solar eclipse

Besides the major, government expeditions to the 1869 total eclipse of the Sun, there were professional or semi-professional freelancers. Like their better-funded counterparts, these, too, were observers—a word implying a particular goal that they wished to achieve under the shadow of the Moon. I contrast this word with watchers, a later subject, people who wished simply to experience the phenomenon. As far as I know, none of the individuals below were invited to join one of the large expeditions discussed so far.

Not all good intentions materialize. Professor Charles Venable (1827–1900; University of Virginia) wanted to make spectroscopic observations—but he did not have a spectroscope.[2] The seeming evaporation of the Illinois College effort is harder to explain insofar as its home in Jacksonville, Illinois, experienced a total eclipse.

Somewhat amazingly, though, most attempts reached the path of totality. Comrades-in-arms with the Navy, Brevet Brigadier General Albert Myer (1828–1880; Chief Signal Officer, USA) and Colonel W. Winthrop (USA) sent the results of their seemingly *impromptu* total-eclipse expedition to Commodore Benjamin Sands. Foregoing the West, they stationed themselves on Whitetop Mountain near Albingdon, Virginia. A peak 1,680 meters above sea level, at the time it was thought to be the tallest in Virginia. (Mount Rogers is a hundred-meters higher.) These two non-astronomers thus presaged the mountaintop observatories of two

---

[1] Quoting from Virgil's *Aeneid* (*circa* 20 BCE)

[2] Why did no one loan him such an instrument? This may have had nothing to do with it. But, in a slightly earlier life, Confederate Major Venable was General Robert Lee's *aide-de-camp*.

© The Author(s), under exclusive license to Springer Nature Switzerland AG 2023

T. Hockey, *America's First Eclipse Chasers*, Springer Praxis Books, https://doi.org/10.1007/978-3-031-24124-6_12

decades later. They reached their destination "after a night's encampment in the woods and somewhat severe march." Told like a real soldier to a sailor.

Fig. 12.1 Whitetop Mountain, Virginia. {Imgur}

The day was overcast. Then, as if by miracle, "At about 3:00 O'clock, however, a breeze sprang up from the northward, which, in the course of an hour, nearly cleared the western sky of clouds, so that a perfectly unobstructed view of the eclipse was had from the beginning to the end of the obscuration …" The bright disk of the Sun disappeared in a flurry of one-or-more sparkling Baily's Beads. To Myer, the Beads made the Sun look like it was "breaking in pieces."

Myer and Winthrop appear to have been prouder of their alpine skills than their totality observations. Their descriptions of the Sun in eclipse are rather bland. But one can infer that the view was spectacular from the summit. Seven prominences were visible. The coronal streamers were so extensive that Myer did not believe that they were real. Incidentally, nobody mentions how safe it was for two Union officers to set up camp, alone, in rural Virginia, inasmuch as the 'Old Dominion' had not yet been readmitted to the Union following the Civil War.

An aside: Again and again, throughout the reports of every expedition detailed in this book, we read foremost about the appearance of the corona. Telescopes stopped down to different apertures, different fields of view, and different degrees of filtering naturally resulted in differing opinions as to the extent of this solar feature. Nonetheless, the descriptions of its morphology are remarkably consistent.

One could argue that Myer and Winthrop were some sort of 'Army Expedition,' though the Army seems to have had nothing to do with it formally. More likely, these were two officers with leave time.

Fig. 12.2 Albert Myer (standing). 1861–1865. {Library of Congress Prints and Photographs Division}

An example of an unambiguously independent expedition was a small group that traveled from Albany, New York, to Mattoon, Illinois. Its most well-known member was George Hough (1836–1909; Director, Dudley Observatory). Lewis Swift (1820–1913; shop keeper) was an avocational astronomer who would eventually discover thirteen comets. Hough had helped him establish priority of discovery for his first, in 1862.

Thomas Hill had just retired as president of Harvard University. As a younger man he had built a mechanical device for calculating the circumstances of eclipses. Hough and Hill were noncontemperaneous Harvard alumni; their connection is unclear.

Still, this total-eclipse expedition sounds like an outing undertaken by a gathering of friends. At least, that may have been how it started. "The whole population of Mattoon was gathered in the vicinity of our station to see the spectacle, vaguely feeling that in some way it could be better seen from that particular spot than from any other."

Then there was Robert Paine[3] (1835–1910; attorney), a Visitor for the Harvard College Observatory. (An observatory's Board of Visitors is similar to a Board of

---

[3] not to be confused with his grandfather, who signed the American colonies' Declaration of Independence from Great Britain; his father, the poet and lyricist; his nephew, President of the American Peace Society; his great nephew, Democratic Party candidate for Governor; his great-great nephew, curator at the Boston Museum of Fine Arts; or his great-great-great nephew, a founder of the science called Community Ecology—all with the same name

Governors.) Paine was known as a philanthropist. It helped that he was very rich. Upon his death, Paine left a small fortune to the Observatory.

Paine was one of the Grand Amateurs in the nineteenth-century tradition: an individual who was free of financial want and who could dedicate himself to his hobby, in this case astronomy. The avocational Grand Amateur had the resources to purchase equipment that put him on par with the research capability of the era's professional.

Paine considered traveling to see the total eclipse of the Sun at Des Moines or Burlington, Iowa. He even noticed that one could take a low-draft steamboat to Fort Union (Dakota Territory), on the Missouri River, which would place him on the total-eclipse path. (A much longer waterway trip to Fort Peck [Montana Territory] would have been closer to the centerline, but few steamboats actually made it that far; see Chapter 18.)

Paine ultimately journeyed all the way from Boston to western Iowa by land, to set up his well-travelled two-¼-inch-aperture refractor at the Boone County Courthouse. The coal-mining town of Boonesboro (which has now combined with the nearby town of Montana and lost its "sboro") was only three-year-old.

Fig. 12.3 Montana, soon to be Boonesboro, Iowa (*circa* 1868). {Library of Congress Geography and Map Division}

Paine reported his observations directly to the Royal Astronomical Society. His prose lacks the ebullience we have encountered elsewhere. It turns out that this was not Paine's first eclipse 'rodeo':

> The corona was good, but, independently of the red or rosy flames, not, I think, as striking or magnificent as the one at Beaufort, S. C., on November 30th, 1834; moreover, darkness was *not* as great. I used a lantern on both occasions to read off the chronometers at the second and third contacts, but it was *not* necessary on August 7th.

Paine did concede that nothing like the naked-eye prominence seen in 1869 Iowa was visible from South Carolina in 1834.

1834? Note how old for his era was this gentleman who alone undertook an even more lengthy and daunting trip than some of those expeditions made up of many participants. He is the only individual I am aware of who observed both of these solar eclipses. Moreover, he spoke to veterans of the *1806* total solar eclipse. From their descriptions, he concluded that it was not as dark as the 1869 event, either.

Winthrop Gilman, Junior (1839–1923; businessman), of Palisades, New York, also was a Grand Amateur who operated a small private observatory. It is said that, "his brain was out of proportion to his delicate body." Nonetheless, he, too, set out for western Iowa, along with a 4-inch-aperture achromat, equatorial mount on a tripod, and regional assistants (all at his own expense). In 1869, it must have been an arduous journey, for a man in his 60s, as well. While not those of a professional astronomer, Sands thought highly enough of Gilman's observations that he included them in his USNO report on the 1869 total eclipse of the Sun. Gilman was the only independent civilian given this honor.

Fig. 12.4 Winthrop Gilman. {published by his distant cousin, Augustus Gilman}

Gilman set some sort of milestone for trekking to the tiniest of venues: Saint Paul's Junction[4], Iowa, was laid out three years before the eclipse in anticipation of the coming of the railroad the next year. At the time of the total solar eclipse, it is imaginable that the place was struggling to maintain a population in double digits. The village sat twenty-five northwest of the bigger Sioux City, Iowa. Saint Paul Junction is the westernmost locale in the US proper from which we have formal 1869 total-eclipse reports.

Gilman could afford fine things. He was equally proud of his Alvan Clark telescope and his brand-new map of Iowa, published just that summer. (With the latter he attempted to establish his exact location.) He was in the position to risk his eyepieces by smoking them.

Gilman expected a little too much from his pride-and-joy instrument. It disappointed him that he could not see the outline of the Moon *before* first contact. Still, one cannot fault his enthusiasm before the solar eclipse actually began.

The New Yorker was among those whose contact times were confused by failure to expect the appearance of the chromosphere immediately after second and before third contact. He, too, anticipated a circular corona. What he got instead was the "lozenge" shape of which we have read much. Gilman wrote little about the total eclipse of the Sun. His medium of description was art.

It was Gilman's local host, N. E. Farrell, who journaled about prominences the color of coals in an "anthracite grate" and a corona "of fine violet, mauve-colored white." It is difficult to picture a corona of three tints at once, still, this passage is to be commended as the only use of the word "mauve," in a scientific report, of which I know.

While it hardly seems possible, Gilman drew for publication fifteen diagrams showing different aspects of the total solar eclipse (starting with Baily's Beads). He rendered a *colored* sketch of the corona. He claims that this work was done just after the fact. Regardless, Gilman's was to become the most popular rendition of the 1869 total eclipse, reproduced even today.

Gilman's drawings show a far greater extended corona than any 1869-eclipse photograph. Indeed, it is more extensive than that as rendered by any visual observer from other observing sites. Either his artwork is exaggerated, or he experienced much better atmospheric-viewing conditions than did other witnesses to the total eclipse. Of course, he benefited from the elevation and longer time between second and third contacts afforded by far western Iowa.

Both Paine and Gilman had no need to accompany someone else's solar-eclipse expedition. Neither did astronomy Professor Christian Peters[5] (1813–1890;

---

[4] now Le Mars

[5] not to be confused with his (also) German contemporary, astronomer, Christian A. Peters (1806–1880; Altona Observatory)

Plate XII.

The Corona of the Total Eclipse of August 7 1869, as seen by the aid of a four inch telescope

Fig. 12.5 Winthrop Gilman's solar-eclipse drawing from 1869. {Princeton University Library Special Collections}

Litchfield Observatory). He unsuccessfully solicited money from college, state, and federal sources. Finally, he received $1,200 from Edwin Lichfield, who had endowed the observatory (hence, its name). Thus, Peters was unique in obtaining NGO monies for his own eclipse expedition. Doing so, he set the precedent for other, privately financed eclipse expeditions in the future.

Of course, there are always *bone fide* loners, too. In describing the efforts made by observers of the total solar eclipse, mention must be made of railway surveyor Jacob Blickensderfer (1816–1899; Union Pacific Railroad). Armed only with John Coffin's eclipse guide, he took it upon himself to make and report observations of the total eclipse from Cherokee, Iowa (founded 1857; the name of the town varied), on the Little Sioux River.

The railroad did not quite reach the settlement. (The next year, the city would uproot itself, buildings and all, to the location of the new depot, a kilometer away!) It may be that Blickensderfer used his influence to ride a train to the end of the tracks. Then he carted, horsebacked, or walked from there to the town site. He set up on a hill just beyond the river.

Born at New Philadelphia, Ohio,
May 9, 1816.
Died at Oakland, Missouri,
February 26, 1899.

Fig. 12.6 Jacob Blickensderfer. {Greg Raven}

Cherokee sat on the total-eclipse centerline. This far west, Blickensderfer experienced more than three minutes of totality. His was also the most isolated expedition in the United States proper. A short note from him appears in the NAO's eclipse report. He promises a more detailed account later. Sadly, Blickensderfer's longer account does not seem to have survived.

Fig. 12.7 Nineteenth-century Cherokee, Iowa. {Photograph of a photograph, by the author}

Perhaps a loner, but not alone. Many years later, a Cherokee pioneer shared from his diary that:

> On August 7 came the eclipse of the Sun … A company of astronomers from the east had come to town several days before the event and were camped on the hill … Everyone in town went to the camp, and many of us got a squint through a good telescope … I was glad to have seen this phenomenon, for millions of people never have a chance to see a total eclipse in a lifetime.

Some total-eclipse expeditions were lengthier than others. Joseph Gardner (1833–1919) of Bedford, Indiana, undertook organized observation of the total solar eclipse. In the sense of "expedition," participants traveled at most across town. Indeed, it sounds more like today what might be a Saturday afternoon, recreational golf foursome. Four gentlemen gathered to photograph the total eclipse using a make-shift camera and telescope assembled "out of odds and ends." None of them were remotely astronomers. Yet, through good luck as much as anything else, the one acceptable image that they managed to make during totality happened to catch the south-limb, naked-eye prominence.

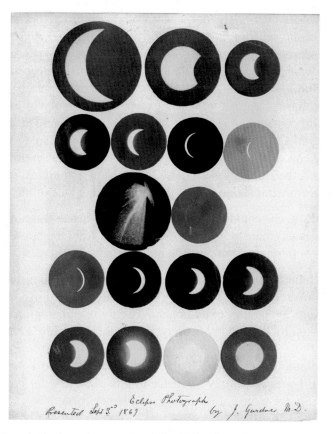

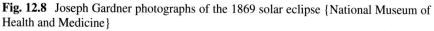

Fig. 12.8 Joseph Gardner photographs of the 1869 solar eclipse {National Museum of Health and Medicine}

At the time of the 1869 total solar eclipse, New Albany, Indiana, was the biggest and richest city in the state. For instance, it had its own Natural History Society.

The total-eclipse 'expedition,' (to just out of town), from the New Albany Natural History Society of New Albany, was organized by the Society's president and another member. On Eclipse Day, the Society first dined, as societies do. It then headed to a knob overlooking Louisville, across the border in the 'Bluegrass State.' Children clutching smoked glass trotted after them, "… with their noses bearing the imprint of dusky carbon."

There's was, arguably, the most crowded eclipse-observing site: thirty-three men (plus crowd) with nothing more sophisticated than a 2-¾-inch equatorially mounted refractor between them, all appearing "with the lividity of death" in the waning Sun. Time was kept track of by a watch apparently hand carried from the Cincinnati Observatory.

Among the members of the expedition were at least three newspaper reporters. Their copy produced the first use of the word "coruscation," to describe the visual arrival of the corona, which I have read. The turnout also included a man and his nephew who had wandered about the solar-eclipse path for two weeks, looking for someone who would share with them a glimpse through a telescope at the total eclipse of the Sun.

The sky of eclipse afternoon was clear except for the smoke of passing steam-boats. After third contact, the larger city to the south still sparkled, with specular glints off pieces of smoked glass in the hands of eclipse watchers.

Louisville, Kentucky, was the most populated city along the entire total-eclipse path by far. Thus, the Hoosiers were poised to make a sociological observation in addition to an astronomical one. In this city, on the other side of the Ohio River,

> … drays and wagons could be seen in the streets as usual. But now and then [after first contact] there would be flashes of light from the house-tops, discovering eager throngs of people with upturned glasses.

At third contact, the New Albany observers heard the roar of thousands.

The umbra hovered over the assembly "like some giant bird in a fairy-tale. The prominences were "mountains of hydrogen." Twenty-one photographs of the solar eclipse were taken by the Albany Society; two were of totality. I have not found them.

Faculty of Miami University of Monmouth College mounted a 'stealth' expedition. I write this because the only reference to it is a single letter posted to a college newspaper. In the words of Professor John Hutchinson,

> [Prof.] McFarland … selected a point in the line of totality lying as far east as could be conveniently reached between Shelbyville, Ky., and Raleigh, N.C.; since that whole line was not occupied by any corps of observers, and being among the mountains they thought that it would give them better opportunities for observation, in a clear atmosphere and commanding position …
>
> … We packed our carpet-sack and … took passage on the Louisville Mail Packet, *Ben Franklin* … We took the train for Mt. Vernon [Kentucky[6]] … and were escorted to our hotel by the whole village nearly who had come out *en masse* to meet us.

---

[6] as opposed to Mount Vernon, Indiana, also on the total-eclipse path

Mt. Vernon is … surrounded by high hills, and one of them … was selected for our observatory … with no greater anxiety did General Bragg, who planted a battery on the same hill during the war, wait for the shadow of Buell's army, than did we wait for the shadow of the Moon.

Most eclipse expeditions considered non-participants to be a nuisance or worse. This one, though, embraced the local population.

Gathered in crowds at our right and left … The whole forming a grand Amphitheater—ourselves on the stage, Mt. Vernon in the arena, the green mountain ranges in the galleries and the Sun hung o'er head as a vast chandelier—soon to have its light turned out for the benefit of the audience.

While the shadow was creeping over the Sun many were permitted to view the Sun through the telescopes … The corona was seen to splendid advantage and the rays of light that pierced the sky from behind the Moon shed a dim, subdued light in the heavens … Prof. Christy had charge of the instrument that was devoted to the good of the public, and many enjoyed a peep through the glass at the Sun while in the shade.

… The Prof. [McFarland] had passed through the same region under Burnside during the war and was recognized by a citizen on the morning we left as a Col. Of an Ohio regiment who had taken breakfast at his house. But the past seems all forgotten …

Note the multiple references to the late war.

Davenport, Iowa, was a pre-Civil War logging town, one of the "Quad Cities" at the confluence of the Rock and Mississippi Rivers. It was established in 1836. The community became more sophisticated: Here, it was that the Davenport Academy of Sciences (founded in 1867) bridged the divide between the professional astronomers and amateurs. The Academy was also the only entity I know of that tried to turn the 1869 total eclipse of the Sun into a profitable corporation.

The 'board of directors' was the Executive Committee for the Eclipse. The business plan was to make use of a specially purposed telescope and camera: two instruments side-by-side on the same mount, allowing photography and direct viewing at the same time. The solar eclipse would be photographed, and the resulting prints would be bound into a book. Copies were to be sold for a subscription.

This all happened; the Committee and others assembled atop the building at Second and Main on 7 August 1869. They succeeded in making 20 acceptable photographs of the *partial* eclipse.

"The entire enterprise was as successful as the most sanguine anticipated, with the exception of not being able to obtain an impression during totality." This self-assessment is roughly equivalent to saying that the surgery went splendidly expect for the death of the patient. Not even a sunspot appears on the images worth printing—all between first and second contact and third and fourth contact.

Fig. 12.9 Davenport Academy of Science members' photograph the 1869 eclipse of the Sun. {Davenport Public Library}

Perhaps knowing that they did not take the best images of the solar eclipse (missing the 'main event,' as they did), the Davenport astronomers could claim that they had taken the largest photographs: Their Sun images appear 10-centimeters in diameter.

The Committee contracted a professional photographer. (with "the most complete photographic gallery … west of Chicago.") Unfortunately, he submitted a bill much higher than expected. Protracted haggling ensued. There was "abuse of a vile tongue," complained the chairman. By the time the delayed book was ready, eclipse fervor had faded. Less than a hundred copies, out of the envision 500, were distributed (some *gratis*). On an investment of $193, the Eclipse Committee grossed $93.00.

(The first president of the Davenport Academy, Professor David Sheldon [Griswold College] did not participate on the Eclipse Committee and chose to observe the total eclipse of the Sun separately. If there is a backstory to this circumstance, I have not been able to retrieve it.)

ECLIPSE OF THE SUN,
August 7, 1869.

Time $\mathcal{5:55'43''}$

ACADEMY OF NATURAL SCIENCES,
DAVENPORT, IOWA.

Fig. 12.10 (**a & b**) Two of the 12 published Davenport Academy of Sciences eclipse photographs. {Putnam Museum and Science Center}

And then there were astronomers and other scientists for whom, on the occasion, for whom zero traveling was required at all. Their observatory already was situated nicely on the total-eclipse path.

Iowa City, located on the Iowa River, served at Iowa's capital from 1841–1857. The Capital Building eventually was occupied by the University of Iowa, which opened for classes in 1855. In 1869, the University sat within the path of the total solar eclipse.

While described as "egotistical, tactless, and mistrustful" by his peers, nevertheless Professor Gustavus Hinrichs[7] (1836–1923; University of Iowa) was a gifted teacher of chemistry who also taught astronomy. Charles Irish (1834–1904; engineer) and he observed totality in 1869 with the University's 4-12-inch aperture Fitz refractor.

---

[7] Perhaps Hinrichs had some right to be a misanthrope. He had emigrated from Germany in order to avoid a civil war. He settled in the United States just in time for its Civil War.

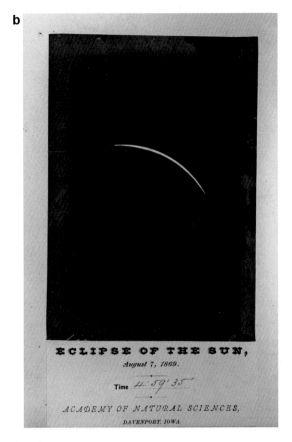

Fig. 12.10 (continued)

Giving up on impassionate testimony, Irish declared: "The phenomenon indicated by this figure was glorious indeed, utterly baffling description." The local newspaper declared that "The eclipse on Saturday was a success," almost as if the circumstances of it were a function of its observation.

Where Truth Meets Fiction

The author of many period novels ('westerns'), C. J. Pettit, wrote *The Eclipse*[8], taking advantage of enthusiasm for the 2017 total eclipse of the Sun across the United States. It is replete with the cowboys, bandits, and inevitable shootouts we might expect. Be that as it may, it also includes a scene in which the hero, named Clay, attends (quite atypically) the University of Iowa. He is present for the 1869 total eclipse. Our protagonist remarks

---

[8] Boston, Massachusetts: Amazon Digital Services. (2017)

that the partial eclipse preceding second contact had an eerie stillness, street-lights turned on, and the corona shone. (Alas, automatic street lighting did not arrive anywhere the world until 1879.)

The time between second and third contacts, as given in the book, is two minutes, thirteen seconds, which is a bit short. Even though embedded in fiction, Pettit's is, apart from duration, a perfectly reasonable description of what Iowa students, as well as Hinrichs and Irish, might have experienced. The only other anachronisms are the well-educated Clay's undergraduate degree in astronomy, as well as his graduate degree from the University of Michigan, which awarded its first PhD. in 1876.

Fig. 12.11 Gustavus Hinrichs. *Circa* 1868. {University of Iowa Libraries, Special Collections & Archives}

Some students 'deserted' Hinrichs and watched the total eclipse of the Sun from the University Cupola, "because of the more extended horizon."

Cupolas were popular architectural features of academic buildings and popular locations from which to observe the total solar eclipse of 1869. On 7 August we would have found in the crowded cupola of Hanover College (Hanover, Indiana) what was apparently a student-led coalition intent to observe the total eclipse. In the carefully distributed hierarchy of tasks, it appears that it was students who operated the available 2-1/2-inch- and 3-inch refractors; the faculty had non-observing roles.

Contrast that effort with the one that took place at a more well-known university in Bloomington, Indiana[9]. Professor Daniel Kirkwood (1814–1895; Indiana University) was famous for his work on the dynamics of asteroid orbits (the Kirkwood Gaps).[10] Yet Kirkwood took a passive role as the Moon's umbra passed over his university. He gathered with a few colleagues on a rooftop to contemplate it. Two colleagues intended to photograph the event, "but when the totality came on, we both broke and ran to see it, and let the important moment pass." Elsewhere in Indiana, Earlham College (Richmond) faculty lectured on the subject of eclipses to 700 people, followed by observation of the total solar eclipse itself. (The Moon's umbra passed over no less than twelve colleges and universities.)

Fig. 12.12 This photograph *may* show the University of Indiana party observing the 1869 total eclipse of the Sun. {Indiana University Archives}

---

[9] The centerline of the 2024 total solar eclipse will pass directly over Bloomington.

[10] Kirkwood was an astronomical autodidact. Even so he was to found a university astronomy program that remains major today. One of Kirkwood's students was Joseph Swain (1857–1927), who stayed on to teach at Indiana University. His colleague, John A. Miller (1859–1946), also remained at Indiana as a faculty member. Under Miller, Vesto Slipher (1875–1869) earned his doctorate degree and went on to become Director of the Lowell Observatory. There he trained Clyde Tombaugh (1906–1997), who became famous for the discovery of dwarf plant Pluto. Tombaugh was a retired professor when I arrived at New Mexico State University. Nonetheless, I learned many things from conversations with him and consider him one of my mentors. Thus, I track my astronomical genealogy back to Kirkwood.

Sometimes it was not about how to observe but where. One amateur astronomer in Des Moines owned a portable Clark telescope—no observatory, temporary or permanent required. On Eclipse Saturday he carried it to the Victorian mansion newly built by Iowa's first millionaire, his friend Benjamin Allen. There, guests beheld the total eclipse of the Sun amidst fountains, arbors, and formal gardens. The house is now the official residence of the state's governor.

Fig. 12.13 The Benjamin Allen House is now called Terrace Hill. Grand Avenue, Des Moines, Iowa. {Terrace Hill, Iowa Governor's Residence and National Historic Landmark}

Let me take us back to Burlington, Iowa, and an unprecedented total-eclipse expedition. It was a professional, institutional expedition in every way except funding. Professor Maria Mitchell (1818–1889; Vassar College), the first female astronomer in the United States, was there. Along with her were eight of her recent Vassar alumni[11]

---

[11] Some sources say "students," but all were from the class of 1868.

Fig. 12.14 Maria Mitchell. {Archives and Special Collections, Nantucket Maria Mitchell Association}

If Mitchell was invited to join the nearby Nautical Almanac Office expedition, she declined. A suffragette, Mitchell was intent on staging a cross between scientific enterprise and calculated publicity event for woman's equality—very much likely the world's first all-female scientific expedition.

Mitchell became a household name while sweeping the skies with her telescope on 1/2 October 1847. On that night she discovered a rare comet. Priority for it was properly—if begrudgingly—given to her by the astronomical community. Some male astronomers spent their entire careers attempting to do what Mitchell did. As historian of nineteenth-century astronomy Trudy Bell explained to me,

> Her discovery was the first discovery of a comet from any observer in the New World. That was a very big deal because it meant that Europe didn't have a hegemony of observational astronomy.

Comet Mitchell brought the New Englander substantial celebrity in Europe as well as the USA. More importantly, she paved the way for American women in science and trained an entire generation of women to become astronomers. She was the first professor hired by the new (1865) Vassar College.

Mitchell was the total-eclipse celebrity of Burlington; she and her crinoline-wearing, hoop-skirted protégés from Poughkeepsie, New York, were the favorite target of local paparazzi. Here were seven unmarried ladies, without male chaperones, facing the vicissitudes of cross-country travel. They carried only tiny parasols for protection—all in the name of science. It had to be seen!

The younger women stayed quietly with two other recent Vassar graduates living in Burlington. Mitchell rented rooms in a boarding house.

On Eclipse Day, Mitchell cleverly stationed her faction away from all the Nautical Almanac Office men, on the grounds of Burlington Collegiate Institute. Neither group would have to share the parochial spotlight with the other. (The president of the Institute provided Mitchell with a "man-servant.")

In her post-eclipse report, Mitchell properly refers to the Moon as "she." All the same, while describing the expedition, Mitchell writes deliberatively in the masculine first person:

> In preparing for an observation of time, the astronomer gives himself every possible facility. He ascertains to a tenth of a second the condition of his chronometer, not only how fast or how slow it is, but how much that fastness or that slowness varies from hour to hour. He notes exactly the second and part of a second when the expected event should arrive; and a short time before that he places himself at the telescope.

Crowds gathered, of course. The president shooed them away to another field.

Before the eclipse, Mitchell placed over the aperture of her 3-inch-aperture, equatorially mounted refractor (the one she used to discover her comet) a circular hole of diameter only 1-½ inches, in order to reduce the amount of light passing through the refractor's optics. (That telescope was her "Little Dollond," which she had named fondly after its manufacturer.) Nobody appreciated more than Mitchell the irony of stopping down an instrument designed to spot faint objects. She also used orange-yellow and red filters with which to protect her observing eye. After second contact, she observed the same two prominent prominences as Coffin did across the city, but her description of one as " … the whorl of a half-blown morning glory" would have read as unusual coming from Coffin's pen. Baily's Beads surprised her.

It seems as if everybody wanted to steal a direct view of the eclipsed Sun, with nothing but clear lens or mirror interfering with its strange light. They sought as direct a connection as possible with the *existential*. Mitchell was no exception and in an unplanned move, removed her filter early to be certain to see the corona in its natural color. Or, at least, she tried to. The filter, which had slid on and off easily every unimportant time in practice, was now stuck. Mitchell wasted ten seconds—an age during totality—trying to get it off.

A democrat, Mitchell allowed each of her alumni to submit their own report on the eclipse to Coffin, along with her own. Mary Whitney and Sarah Glaiser[12] must have come from families of means because each traveled with a Clark refractor the same aperture as Coffin's. Mary Reybold had at her disposal a 2 ½-inch-aperture Dollond refractor.[13]

---

[12] Later that Fall, Glazier moved to Chicago and continued her astronomical tutorage with Truman Stafford.

[13] Coffin's daughter Elizabeth hung out with the Vassar party, too. We do not have a record of what she did during the eclipse, but do know that she collected local fossils and minerals from the region.

With her telescope, especially made for the occasion, Whitney saw the elongated corona—which her teacher fails to mention. Glasier "was surprised by the irregularity of the corona."

Not every woman had a telescope, though. Isabella Carter held a much more modest pair of binoculars, but still managed to pick out the two "ruby-colored leaf-shaped flames" that were the largest prominences. Others were naked-eye observers: One was so perched on the "roof of Mr. Foote's house …" This was a difficult place to visit for a lady, if wearing the corset, hoopskirt, and bonnet common for the day.

Mitchell on the eclipse:

There were some seconds of breathless suspense, and then the inky blackness appeared on the burning limb of the Sun. All honor to my assistant, whose uniform count on and on, with unwavering voice, steadied my nerves! That for which we had travelled fifteen hundred miles had really come. We watched the movement of the Moon's black disk across the less black spots on the Sun's disk, and we looked for the peculiarities which other observers of partial eclipses had known. The colored glasses of our telescope were several, arranged on a circular plate, so that we could slip a green one before the eye, change it for a red one or a yellow one, or, if we wished to look with the eye unprotected, a vacant space could be found in the circumference. In the course of the hour, from the beginning of the eclipse to total phase, this was readily done. I fancied that an orange hue suited my eye best, and kept that in place intending to slip it aside and receive the full light when the darkness came on. As the Moon moved on, the crescent Sun became a narrower and narrower golden curve of light, and as it seemed to break up into brilliant lines and points, we knew that the total phase was only a few seconds off.

Then:

Instantly the corona burst forth, a glory indeed! It encircled the Sun with a soft light, and it sent off streamers for millions of miles into space!

On looking through the glass, two rosy prominences were seen on the right of the Sun's disk, perhaps one-twentieth of the diameter of the Moon, having the shape of the half-blown morning-glory. I found myself continually likening almost all these appearances to flowers, possibly from the exquisite delicacy of the tints. They were not wholly rosy, but of an invariegated [*sic*] pink and white, with a mingling of violet.

Finally, I want to share the expedition of Cleveland Abbe (1838–1916; newly Director, Cincinnati Observatory), better known as an early meteorologist. Perhaps the fact that he was exiting the profession made him an unlikely candidate for one of the government teams. Alternately, his lack of such an appointment was because, after his short stint at the USNO, Abbe worked behind Benjamin Sands's back in an attempt to change the administrative structure of the Naval Observatory.

Fig. 12.15 Cleveland Abbe. {Library of Congress}

Abbe's 'professional travel fund' was effectively penniless. Nevertheless, he was able to cobble together his own, independent Cincinnati Observatory expedition for $540. This was an impressively small sum even then.

After failure to obtain a grant from the American Association for the Advancement of Science, Abbe relied on letters written on his behalf by the Observatory's most influential patrons to solicit free reduced-fare passage, over the long route he planned. He succeeded with each of the railroads that he would travel on. The party nearly got their own "palace sleeping car" from "Mr. Pullman." (For some reason, it fell through.) In a similar manner Abbe obtained free Western Union service. A Cincinnati newspaper chipped in $150 for some additional expedition costs. At the last minute, even the NAO wrote Abbe a small cheque.

The expedition totaled seven. Robert Abbe (1851–1928; soon-to-be physician[14]), Cleveland's brother, would make meteorological measurements. Professor Alfred Compton (1835–1913; College of the City of New York[15]) would attempt spectroscopic observations as well as look for magnetic effects. Robert Warder (1848–1905; Illinois Industrial University[16]) would record contact times and James Haines (Cincinnati Observatory) "by reason of his familiarity with the heavens" would look for meteors. (Such a targeted observational program as the latter, during a total eclipse of the Sun, was unusual; it seems to have reflected

---

[14] He would pioneer radiation oncology and eventually die from exposure to radioactive materials.

[15] now part of the City University of New York

[16] now the University of Illinois

Cleveland Abbe's particular interests.) W. O. Taylor and his brother-in-law would photograph. Taylor was a professional photographer from, of course, Philadelphia.

Cleveland Abbe could have traveled more directly to the total eclipse path, only 210 kilometers south at Lexington. However, he was interested in weather as much as or more so than astronomy. For him, optimum sky conditions over-road all other considerations.

Director Abbe was fixated on points west. For instance, he wrote that totality "swe[pt] southeast near Omaha ..." Sort of. In fact, the path came no closer than 40-kilometers away from that Nebraska city.

So it was that the Abbe party traveled up the eclipse path, nearly as far as was practicable by land. First, it headed northward and then turned west at Chicago (where they conferred with Truman Safford, not yet on his way to Des Moines, Iowa). The Cincinnati observers' goal was farther westward still. They continued to Sioux City, Iowa. (There they ran into James Gilliss (*1840–1898; USA*), whose father happened to be the founder of the United States Naval Observatory.)

> From Chicago to Omaha [Nebraska] the route is too well known to need description. Illinois and Iowa offer to our view a rich rolling prairie, well cultivated wherever it is well watered. The region around the Cedar River being particularly well wooded, hilly, and filled with mineral resources. At Missouri Valley Junction [Iowa], thirty miles northeast of Omaha, we pass from the prairie through the bluffs of the Missouri River, out upon the bottom land of that mighty stream. Those who have no eye for geology will hardly heed indications that surround us, showing that the great muddy river once covered the whole immense area. In those days it was a stream ten, fifteen and twenty miles broad. Its former bed is now dry bottom land, limited by the bluffs on either side, and stretching up and down the river as far as the eye can see—an immense flat plain, covered by a rank growth of coarse grass, in which great herds of cattle are almost lost to sight, so high does the grass overtop their heads. Up this valley of the Missouri, our railroad train takes its way with but a few turns in its whole course. ... our principal diversion consists in watching the herds of cattle, the distant bluffs and the river bed [*sic*], skirted by cottonwood and elm. Frequently the cattle, by running upon our track, necessitates a stoppage, when we jump for wild flowers of which many are new to us, and Prof. Compton, who is our botanist, is kept busy enough analyzing and naming our prizes. A deep cut through Sargent's Bluff [Iowa] brings us suddenly face to face with Sioux City, which his located directly upon the eastern river bank about four miles below the mouth of the Big Sioux [River], and is limited behind by the range of bluffs. From its situation and picturesque environs, we may safely predict for Sioux City a future as bright as that of Council Bluffs [Iowa]; possibly it may some day rival Omaha.

Upon their return to the United States proper, they learned that they had passed longitudes occupied Gilman and Blickensderfer.

From Sioux City the expedition ventured into Dakota Territory—140 kilometers and three days by covered wagon and mule train! (Cleveland Abbe required a special rider on his life insurance policy for such a journey into the hinterland.) Amazingly, only some bottles of photographic chemicals were lost on the trip. Any further west Abbe's team would have trod on 'Indian lands,' according to the Treaty of Fort Laramie signed the year before (Chapter 18) and be decidedly unwelcome.

> ... we soon after ascended to the low bluffs that enclose the Sioux River, and for sixty miles went on ascending and descending the very short rolls of that prairie. Buffalo grass on all sides was the constant and only vegetation. Buffalo hammock dotted the ground everywhere. A house and one hole with a little water was all we saw for twenty miles at a time.

Then there is another digression about mosquitoes.

So it was that in August 1869, Abbe positioned his party at Fort Dakota. Built in 1863 after local settlers had suffered "Indian massacres," this fort had been decommissioned the preceding June. It was now the nascent Sioux Falls City (later just Sioux Falls). The "city" was home to a few-hundred people who had not left when the soldiers did. Many now gathered around Cleveland Abbe *et al.* to watch the total eclipse of the Sun.

Fig. 12.16 Fort Dakota transforming into Sioux Falls, South Dakota (*circa* 1871). {Siouxland Heritage Museum}

The eastern astronomers were rewarded for reaching the greatest western longitude by land to observe the total eclipse of the Sun: They would experience over three minutes of totality at high elevation.

A 3-inch-aperture refractor was deployed on the former fort's parade grounds. The wagon cover was converted into a dark room. Taylor's and Compton's crude actinometer and photometer were pointed upward. The Cincinnati Observatory Expedition would try to do it all. (They forgot to use their polarimeter.)

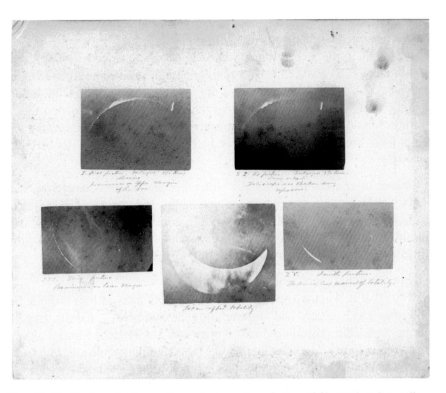

Fig. 12.17 Cincinnati Observatory photographs of the 1869 total solar eclipse. {Cincinnati Observatory}

Cleveland Abbe himself observed with a loaned, prize-winning 6-inch-aperture refractor merely for sketching; Taylor had won it as a prize at the Paris Exposition Universelle of 1867. Everything sounds good, but the Director Abbe's notes are cryptic. Third contact was interrupted by some failure of the telescope mounting and some interruption by an unnamed person. Perhaps he was being discrete.

Conscious that this was supposed to be the spectroscopic eclipse, Cleveland Abbe brought with him a handheld prism and Compton lifted it skyward. His attempt to thusly study the corona failed, but we can appreciate his optimism.

Nothing much came out of Abbe-and-company's punishing trek into the unknown. It seems that the expedition spent all its eclipse-planning time dealing with elaborate logistics. They left their best instruments safely at home. Those assembled specifically for their total-eclipse odyssey were slap-dash. The photographic refractor had no equatorial mount, so consequently there was no clock drive for tracking. Warder used only a 1-inch aperture telescope. For his meteor search, during which a pair of military binoculars might have been of use, Haines had just opera glasses.

Nonetheless, it was Cleveland Abbe who observed the Moon cover the Sun longer than any other astronomer. It was still not enough for him. He wrote, "The totality has passed away like a dream, and no earthly power could recall that shaded Sun to allow us only a moment's longer study of its surface." Alas, the eclipse visitor who had the best view of the event may also have had the worst: The senior Abbe suffered from extreme myopia.

13

"A Darkness That Can Be Felt"

> The day was a most beautiful one, with cloudless sky, bright Sun and clear atmosphere-everything in earth, air and sky seeming to have by common consent determined to give denizens of this lower world every capacity for observing this grand and impressive display of astronomical science.

I so far have described total eclipses of the Sun as if they were mere intersections of dots on the Celestial Sphere. Much more than that, I am convinced that they are the most *spectacular* occurrence in our sky anyone is likely to see.

I posit that few people have ever witnessed a total eclipse and been unmoved. There are stories of clueless motorists merely turning their lights on and proceeding down the highway, but they are a rare exception.

What is it that tantalizes you and me? While Simon Newcomb would remind us that there are important scientific reasons for studying a total eclipse of the Sun, one cannot discount the sensorial experience outside the boundaries of physical science. And this has little to do with the *position* of the Sun and Moon on the Celestial Sphere. As far as conveying this, the pens of the astronomers, reflecting after the great event, also proved equal to the task.

We are told correctly that, "There is no premonition of a solar eclipse …." Eventually, once well past first contact, the sky takes on an unusual appearance. The Sun is eventually reduced to nearly a point source of light. Shadows cast by terrestrial objects have seemingly unnaturally sharp edges.

Near total solar eclipse, the fuzzy penumbra that normally surrounds the darkest umbra of any object's shadow is gone. One George Carter, of Chicago, wrote an entire report to Benjamin Peirce on this subject alone. Raconteur professional

T. Hockey, *America's First Eclipse Chasers*, Springer Praxis Books, https://doi.org/10.1007/978-3-031-24124-6_13

Fig. 13.1 1869 solar-eclipse photographs taken by Joseph Winlock/John Whipple at Shelbyville, Kentucky. Upper right, after first contact; upper-left, near second contact; lower-right, totality; and lower-left, after third contact. {University of Navarra}

Benjamin Gould described it this way: "The light grew cold and weird, the heavens assumed a grayish and, finally, an almost leaden hue, and the outlines shadows became preternaturally sharp and distinct."

Depending upon the angle of illumination, pools of water looked black. Distant rivers resembled "line[s] of silver."

A bit visually disturbing is James Watson's comment, " … persons standing near had a peculiar ghost-like appearance." NAO fellow-traveler Darwin Eaton, who seems to have been unnerved by everything to do with the total solar eclipse, was agitated by the "Dark, weird and ghastly appearance of landscape."

Luckily, on top is this, there was not so much as a tremor. Tremor? The year before, in South America, there had been both an eclipse of the Sun visible and a major earthquake. Tens of thousands were injured or worse by the latter. Some amateur natural scientists wondered if the two events correlated. Were they even, perhaps, predictive, one of the other? No.

Nearer second contact: As the dark silhouette of the Moon encroaches on the brilliant Sun, whether or not the approaching shadow is visible has much to do with the appearance of the local, surrounding terrain and how obstructed the view is. General Myer instructed his "guides" specifically to watch for the shadow;

even though they knew where to look (west), they "saw nothing of which to give notice."[1] On the other hand,

> Mrs. Farrell [Winthrop Gilman's hostess] distinctly saw the Moon's shadow rapidly approaching in the air. It appeared to go upwards from the western horizon like the lifting of a dark curtain.

Richard Cutts referred to the "cold dark shadow" and recorded, "… it appeared to strike me bluntly—forcibly, rather than to pass softly, quietly by." Eaton—writing as Eaton was wont to write—tells us that

> Just before totality, as the shadow approached from the west, an undefined horror seemed gathering in the sky—a cloud, and yet not a cloud … All nature seemed to hesitate, and everything was contradiction. I can well understand how such an event, not understood nor expected, would make the stoutest hearts quail and fill the sternest mind with fear and dread.

Fig. 13.2 Richard Cutts. {Anonymous}

---

[1] Having watched the total-eclipse shadow, I can say that it moves faster than any first-time observer might expect.

Nevertheless, it is hard to top this, which made its way into the local history books: "fearful as a procession of spirits in the lower circle of the 'Inferno.'"

The locals simply said that the shadow, "… embrace[d] thousands of acres of as fertile land as the Sun ever spread his life giving beams upon."

Of the final seconds of the partial phase, *uber*-enthusiast Gilman expressed that, "The feeling was as if one was about to enter a cavern." Then, "just as the last visible ray disappeared, … it [was] like a flicker of a candle going out."

During totality, blue sky all but disappears. But a *tour d'horizon* reveals 360° of variegated colors. In the words of abecedarian professional Edward Pickering, "The sky presented the appearance it has during the morning twilight, the western horizon, however, being bright yellow." Why morning? Recall that his view was through the window! Perhaps more useful is actually the description by Eaton:

> The appearance of the heavens was peculiar. While the stars were shining overhead, a beautiful twilight, tinted with hues of sunset extended entirely around the horizon.

For some 1869 solar eclipse observers, this effect began a "few minutes previous to totality," and the "horizon changed color, as it is frequently seen to do on the approach a storm." Not all agreed on what color, though. Maria Mitchell's protégé Sarah Blatchley (1844–1873; Vassar College Expedition) called it that of the "Mississippi of a leaden color …" for the young Miles Rock, it was "a bright orange-purple light [that] encircled the horizon, shading into the dull, ashen hue of the zenith." Another observer saw the " … pale orange color of vegetation." The sky was even called "brimstone." Maybe no one or two colors did the eclipse sky justice? One person on the scene found those clouds that were present as "arrayed in every color of the rainbow" and the rest of the sky "as if bands of broad ribbon of every conceivable hue had been stretched in parallel lines half round the universe."

During the few minutes of total solar eclipse, the dark of night comes to the day, or, at least, that of deep dusk or dawn. "The sky was also sensibly darker in the vicinity of the Sun than it was near the horizon," explained Gould mentee John Stockwell. Not everyone could agree on *how* dark, though. That " … the light diminished enough that the steam from the ships on the river appeared to darken from white to brown" is not very helpful.

For Gould, "Throughout the period of totality the daylight was amply sufficient for making and reading pencil-memoranda without copious effort …" The ability to read was frequently used as a photometric measure. In the case of Mitchell,

> The darkness was at no time so great that print could not be read. My assistant used a candle for the chronometer, but I am not sure that it was necessary.

Eaton "could read print on common newspaper" (which, I learned, through perusing many of the day, was very tiny).

Though not part of his USGS assignment, Robert McLeod undertook a poll: "Of six gentlemen, all of perfect vision, who experimented during the totality, with cards prepared for the purpose, three could read distinctly *Diamond* type [*i.e.*, 4-½ point = 1.5 millimeter]; one nothing smaller than *Pearl* [5 point = 1.8 millimeter]; one nothing smaller than *Nonpareil* [6 point = 2.2 millimeter]; and one nothing smaller than *Bourgeois* [9 point = 3.2 millimeters]."

Others used more direct astronomical analogies for sky brightness, either qualitatively like Professor Frederick Bardwell's (1832–1878; Antioch College) "… nearly the same as when the Sun is hid behind a high hill just before setting" or quasi-quantitatively like John Eastman's (1836–1913; USNO), "… about equal to that on a clear moonless evening, at the time when third-magnitude stars can easily be seen." The Moon was a common reference, too, as in Blatchley's "… Sun visible to [the] naked eye without pain … cloudy moonlight sky." I find fascinating astronomer Eaton's continued description, here negating itself: "Appearance of an approaching thunder storm [*sic.*], but without clouds, lightning, or thunder."

Ultimately, the subjective darkness of a total solar eclipse is based upon how literally one takes the idea of the disappearance of the Sun and day becoming night. It is an exaggerated expression. Still, as total eclipses go, 1869 was dark.

The astronomers at Plover Bay, 64° North, experienced sky brightness during their entire stay. Their latitude and the time of year conspired such that, while the Sun did set during the brief night, twilight never disappeared completely. So, the darkest sky they saw for a week was likely during totality!

There was very much a sense that one was under a shadow. Observers high in the cupula of the Cedar Falls Orphan Home could look out the window and note its extent: "… trees could be seen basking in the dim sun-light in the vicinity of [the town of] Janesville [Iowa], eight or ten miles distant."

It was aphotic enough that stars and planets 'came out.' Eclipse doyen Stephen Alexander made a point about " … the peculiar brilliance with which the stars seemed to spring out without the long struggle through twilight."

Perhaps only astronomers would take the time to count the nighttime stars visible to the naked eye *during* a total eclipse of the Sun, but this they did. The record was seven. Many noted red Arcturus, with similar first-magnitude Regulus in second place. Some mentioned Vega. There was at least one claim for each of Antares, Altair, and Alkaid. Several observers spotted planets as far away from the total eclipse of the Sun in the sky as Saturn.

Eventually the shadow departed. "And when the light returned, it seemed like morning coming from the wrong direction."

There is no objective reason why a total eclipse of the Sun should affect sound. And yet, astronomers pushed their expository writing to the limit in order to describe what they sensed during the solar eclipse, "a profound silence." At totality's beginning: " … a death-like plainness gradually diffused itself over the countenance of all, and when the last faint crescent of the solar light was obscured, gloom and silence seem to have taken possession of the Earth and the living beings around." And it its end: "The world seemed to breathe again, and at once life and animation took the place of what was an oppressive stillness." Even at the ocean, cartographer Joseph Rogers found that "there was no sound except that of the sea on the beach ..."

The temperature fell. Newcomb's "the suffering chill was appalling" was probably an overstatement but refers to a real phenomenon. Some observers reported that their hands became cold. "... Coats were buttoned closer, and the lady of our party almost shivered ..." Still, when we read that, "... the temperature was forty-two degrees cooler than one hour before," we must conclude that we have reached the point of embroidery.

Eastman was the only full-time astronomer whose assigned job during the total solar eclipse was to make meteorological measurements, aided by his wife. He documented temperature very precisely with professional-grade "alcoholic thermometers." His result was simply a steady decrease in temperature as the afternoon waned ($73°$ F to $69°$ F).

Others used household thermometers; N. E. Farrell recorded $70°$ F. at first contact, $68°$ F. eleven minutes after third contact, and $71°$ F. an hour later. He may be excused for missing the temperature immediately after second contact. An even 10-degree eclipse drop was claimed at Bardstown, Kentucky. Far to the north, Asaph Hall just wrote, "$46°$ F. partial, $42°$ F. total." Pickering purported to notice a dip in temperature inside his hotel! James M'Clune told of a total solar eclipse, during which "dew was deposited on the grass in the public square of the village, and a chilly sensation was felt, which seemed to indicate a greater change of temperature than the three degrees shown by the thermometer."

Fig. 13.3 John Eastman. {National Oceanographic and Atmospheric Administration}

The following report from an amateur meteorologist is suspect: "The eclipse yesterday caused the thermometer suddenly to fall with great rapidity, and the weather turned so cold that frost was visible this morning ..."

Does a total eclipse of the Sun affect the wind? Many included a change in the breeze within their solar-eclipse reports. However, what that change was appeared to be very location specific. E. P. Austin wrote,

> The day had been very hot, with considerable wind, which died out almost completely about the time the Sun was half eclipsed. At this time ... the heat of the Sun was no longer oppressive, while near the totality there was a slightly chilly feeling.

At Alfred Mayer's location, "A low moaning wind now sprang ..." Eaton acknowledged his own 'over-the-top' response to the total eclipse of the Sun when he wrote of the wind, "a tornado, and yet only a slight breeze!"

Christian Peters registered, using a barometer, a short-term peak in atmospheric pressure at the time of totality. The barometers of Burlington, Iowa's Harold Thielson (1814–1896; railroad engineer), Professor J. M. Mansfield (1843/44–1894; Iowa Wesleyan College); R. C. Tevis[2] at Shelbyville, Kentucky; and others showed either a rise or dip. John Coffin's barometer was "deranged in transportation, and rendered useless."

It was a long shot. Nevertheless, scientists checked to see if the Earth's magnetic field varied during a total eclipse of the Sun. It did not.

In a more villatic America (e. g., Illinois), the effect of a total eclipse of the Sun on animals was well documented by the astronomers. Nathaniel Shaler informs us that:

> The cocks all began to crow, at several points, with the sleepy crow of early morning and not the exultation of full day ... Cattle were evidently much alarmed, and ran, with tails up and heads erect, across the fields almost in stampede. At the close of the eclipse a hen was found, with her chickens under her wings.

> Elsewhere it was disclosed that "... doves flew to their cotes ... [domestic] geese marched in haste to their night quarters ...," and "... turkeys [were] surprised that night had found them so far from their accustom roost ..."

"... pigs squealed in resentment over being sent to bed hungry ..." "Farmers reported that sheep sought repose." Albert Myer's "horses continued to feed quietly during the increasing darkness, as at an approaching sunset," but in the city they exhibited uneasiness, "... growing so restive and alarmed as to compel their owners to unhitch them from the vehicles to which they were attached."

---

[2] All that I know about him was that he was a Mason.

Fig. 13.4 Enhancement of a twentieth-century cartoon, illustrating the 1869 total solar eclipse. {B&W original by W. L. Purcell; this modern watercolor version by Kristyn Blaylock.}

Despite being downtown, Pickering heard a commotion: "… cats made the afternoon hideous with their screams …" Someone "noticed a bat flying around." From Alaska we learn that "Even the everlastingly howling curs seemed to be quieted down."

As for wild fowl, Hall quoted a naval commander saying, "before the Sun was entirely obscured the sea birds around us had gone to roost." Back in the United States,

> The birds ceased their music, and began to flit and flutter about with an uneasiness painful to look upon. The pigeons no longer flew in lonely pairs, but in compact groups tried to escape from they knew not what, circling in low flights, slimming close to roof and ground. Before the extinction became total, a large flock that had been flying round and round the building on which we stood, dropped themselves upon the roof of the church immediately in front of us, and clung to the shingles in huddles till the darkness was

gone. Smaller birds separated from their mates, and flew singly and in wild disorder, frequently beating themselves against walls and other obstructions. A group of martins flew in among the people standing on Exchange Block [Des Moines, Iowa], evidently seeking protection from a power it is in their nature to fear.

Of the birds, "… one of them [struck] one of our party in the face in its distracted flight." Students at the University of Iowa reported that, " … the swallows that nest in the cornice of the University came twittering homeward and circled in clouds around the dome, preparatory to retiring, and a quiet flock of geese that fed in the campus talked the matter over as the dark became denser …" "Jaybirds became very boisterous, either from fear, or singing a royal welcome to the celestial show."

And then there were the insects. "Flies nested on ceilings." "Crickets chirped," said J. B. Reybolds of Vassar. "Fireflies twinkled in the foliage," added Mitchell. Mayer was bothered by moths. M'Clune noticed that "Bees came swarming to the hives …" "… gnats, and other heralds of the night, fluttered in the air." "… butterflies fell to the ground in a state of stupor …" "… katydids chirped their nocturnal notes, but hushed into silence as soon as they found that they were 'victims of misplaced confidence.'" Despite complaints about mosquitoes before the total solar eclipse, there were none made during the solar eclipse.

Controversially, many saw streaks of light through their telescopes during the 1869 total eclipse of the Sun. Colonel Winthrop watched a "shower of bright specks." Charles Himes and Joseph Zentmayer were more forthcoming: "… bodies like meteors crossing the dark image of the moon from cusp to cusp."

A bit of groupthink set in: Austin saw 'meteors,' too, and thought them to be a local phenomenon. Upon hearing of other, similar reports for disparate places, he reevaluated and concluded that they were real.

As the objects observed were seen crossing the Moon at several different, widely spaced locations—from Iowa to Illinois to Indiana to Kentucky to Virginia--parallax would demand that they be very away. This argument applies if every observer were watching the *same thing*.

Ignoring the distance problem for a moment, the astronomers did not see the known August Perseid meteor shower. While it does peak mid-month; at eclipse time, the radiant constellation Perseus is ill-placed in the daytime sky. The eclipse phenomenon, then, would be an all-too-coincidental newly discovered shower.

A few observers realized that there was a simple explanation for what they were seeing: insects or airborne seeds.

Astrophysicist Professor Bradley Shaefer (Louisiana State University) is also an expert on optical phenomena in the sky. He tells me that,

What is going on is that when seeds or insects pass nearly across the Sun, the remaining light is very strongly forward scattered, and these extremely nearby bodies appear bright. I have often seen such a phenomenon while looking close to the Sun with binoculars, as part of a program seeking sun-grazing comets.

After totality, we are told that "… flowers opened again their petals, to greet the new born [*sic*] day …" All returned to normal. "…. the eclipse was gone, and there was not a single mark on Earth or in sky to indicate that the greatest event in the history of astronomical science had just occurred."

In an attempt to take an unique spin on a topic that was becoming well-worn after the total eclipse of the Sun had passed, newspaper editor John Mahin undertook this bit of speculative science fiction:

> Few people trouble themselves to think what the effect would have been if the eclipse of Saturday had lasted any length of time, and the sun been blotted from the heavens. [Natural] Philosophy declares that not only would a horror of darkness cover the Earth, but the moisture of the air would be precipitated in vast showers to the Earth, and the temperature fall to a fearful point of cold, nothing less than two hundred and thirty degrees below zero, Fahrenheit. The Earth would be a seat of darkness, and more than Arctic desolation. Nothing could survive such freezing cold a moment more than one could live in scalding water. In three days after the cooling process began nothing created would be alive but the monsters that wallow in deep ocean, and the eyeless reptiles that make their haunts in the caves which penetrate far under ground [*sic*]. The thought of this possibility cause[s] a shudder even yet, though the danger is passed. By the by, was there not clearly a perceptible and unusual dampness and stillness in the atmosphere during the eclipse?

No Creature Stares at the Sun Except for *Homo Sapiens*

It is amazing that no one we know of suffered eye damage while observing the 1869 solar eclipse through a telescope, considering how little regard was given to safety. One close call appears to be Edward Goodfellow. "… Mr. Goodfellow's sight had latterly been affected by over-use, and he was obliged to abstain from using his eyes for some time after the eclipse." Amazingly, Goodfellow wrote, "… but, limited as was the extent of my view, the impression of beauty and sublimity produced by the total eclipse was one to endure for a life-time." In other words, he felt that it was worth it.

Professor Ralph Chou (University of Waterloo) is the internationally recognized expert on eclipse eye safety. He tells me that:

> Dark green or grey filters would have attenuated the visible light to a level that ensured that a direct view through the telescope would not result in an eye injury. Goodfellow, on the other hand, was unlucky with his choice of a yellow filter because the attenuation of visible light

with such a filter would not have prevented him from sustaining some degree of photochemical injury when looking through his telescope.

Newcomb had a yellow filter, too, but his objective called for *avoiding* looking at the partial (and potentially harmful) phases of the solar eclipse. Goodfellow was undertaking contact timing. Those assigned to this task were vulnerable especially, because the job *required* observations before second and after third contact.

Ultimately, the total eclipse environment, as it affected all the senses, is in the eye of the beholder. There was Charles Young's experience when he abandoned his spectroscope:

I turned my head, and for perhaps 10 or 15 seconds beheld the spectacle, the most beautiful and wonderful upon which the human eye can ever rest. The impression was overwhelming, and so nearly blotted from my memory all that I had observed before I can now recall it with certainty only from the written notes kept by my assistant …

But there was Eaton's: also

I was completely surprised, and for a moment almost bewildered. The whole scene—the heavens above and the Earth beneath—had an appearance of sad awful gloom, as if some strange and fearful calamity were impending.

Either way, a total eclipse of the Sun is a visceral experience. In 1869, its variety of effects were recorded minutely, in keeping with the scientific philosophy of word-renown naturalist Hermann Ludwig Ferdinand von Helmholtz, that "Each individual fact, taken by itself, can indeed arouse our curiosity or our astonishment, or be useful to us in its practical applications."

14

Standing on the Edge Looking Up

"I request that you will place the circulars in the hands of competent parties who may be willing to make the observations."

Simon Newcomb could have organized a very different expedition to the total eclipse of the Sun. While his team and others diligently kept track of contact times at sites near the centerline, these intervals are of less value than those extremely near the limits of the total-eclipse path. If he had spread out those under his direction along the edge of totality, he could have mapped with great resolution the umbra's width as a function of time. More precise information about where the Moon and Earth were in their orbits about the Earth and Sun, respectively, would have resulted.

One might imagine stationing all available astronomers along the predicted northern and southern limits of the path, as far west as they were willing to go. The required measurement could be made without a great deal of instrumentation, thus obviating the need to stay close to the railroad arteries. Given a good source of time, all that was needed was the naked eye and the experience astronomers of the day had in making instantaneous judgements. For instance, a nineteenth-century astronomer knew his or her personal equation, the delay time between recognition of a phenomenon (*e.g.*, the complete disappearance of the Sun) and his or her reaction to it (*e.g.*, pushing a button, glancing at a timepiece). It would have taken all of Newcomb's power of persuasion to convince a great number of observers, including some not under his direct supervision, to forgo longer observation time near the centerline. A *coordinated* operation between the United States Naval Observatory and Nautical Almanac Office (plus, perhaps, the Coast Survey) would have been ideal.

T. Hockey, *America's First Eclipse Chasers*, Springer Praxis Books,
https://doi.org/10.1007/978-3-031-24124-6_14

However, experienced personnel had different goals for the 1869 total eclipse of the Sun. Newcomb chose to assign eclipse-limit monitoring to avocational astronomers. He realized that people already living in the 'sweet spots' for path-limit observation (within 10 kilometers of that predicted) need only answer, at a minimum, a simple 'digital' question sufficient for the amateur observer: 'Did you see totality? Yes or no?'

Relevant postmasters were mailed an open letter, addressed to communities far from the centerline but still likely to experience a total eclipse. Instructional content was written by Newcomb, above Benjamin Sands's signature.

The inquiry asked for volunteers to time totality. These pleas often ended up in the local newspaper. The call for "intelligent observers" was answered, largely by professional people, whose time was their own to take off in the middle of the workday. Additionally, among these were individuals who saw science as another rung on the ladder of social ascension.

Haphazard attempts to refine the Moon's obit by using an array of observers during a total solar eclipse had been tried as early as 1715 (in England). Nothing had been undertaken on this scale, though.

The northern limit of the 1869 total eclipse of the Sun was explored in Linn County, Iowa, by no less than three groups of people. The county surveyor, observing the eclipse through his transit, recorded totality as lasting 1 minute, 3 seconds according to his watch. He sat atop the county courthouse in the cupula.

In Mechanicsville, Iowa, nascent total-eclipse observers stood "eighty-five rods northwest from [the] Chicago and Northwestern railroad depot." They experienced a mere 25 seconds of totality, a minimum of those northern-limit durations recorded.

Samuel Yule (1815–1888) had a "spyglass." He positioned himself for the solar eclipse, south of Stanwood, Iowa, also "on" the Chicago and Northwestern Railroad. Presumably, standing on the tracks was less dangerous than it is today. One minute, 4 seconds.

Ordnance officers of the United States Army, posted to the Rock Island Arsenal, timed the span of the total eclipse from the bank of the Mississippi. An unusual setting, the tiny Rock Island was a cemetery for Confederate troops. The "Arsenal" still exists today, a rather lonely place attached by bridge to what is now the city of Rock Island, Illinois. One minute, 17 seconds.

James Bell lived near the thirteenth mile post east of Rock Island, in Colona, Illinois, along the Chicago, Rock Island and Pacific Railroad easement. Bell, the Reverend S. H. Wood, and their unnamed wives followed the total solar eclipse from this house. Afterward, when he wrote the Naval Observatory, he made sure to point out that he had an "Elgin watch." Fifty-nine seconds. Newcomb writes, "Original record not sent, but the descriptions of the observations and phenomena,

in establishing the umbra's width. These included one from English emigrant Professor and Reverend Charles Phillips (1822–1889; University of North Carolina[1]), whose higher-education journey could have caused him to come into contact with both Benjamin Peirce and Stephen Alexander.

Clearly, Newcomb was left with a large stack of letters with numbers included, the kind of mail with which he was very comfortable. Nothing much *came* of it.

Newcomb's citizen science was ahead of its time. Nevertheless, it ultimately failed. Many participated; one might have anticipated at least some sort of *statistical* application of the data. However, there were so many opportunities for systematic error.

The irregular lunar limb spends proportionately more time near the solar limb during a brief, acentral total solar eclipse. This can confound particularly measurement of the event's duration. As the Moon's limb travels nearly tangential to that of the Sun, the points of second and third contact are vulnerable to the Black Drop Effect.

Moreover, it was not clear always where enthusiastic observer-recruits were located, or that they fully understood the importance of 'calling' an edge contact quickly. And, of course, there was the simple unfamiliarity on the part of the participants with observational astronomy and what exactly it was they were supposed to accomplish. We can assume that virtually none had ever seen a total eclipse of the Sun before.

How many had never spent much time looking up into the sky at all? Nor worrying about the second hand on their watch? By the same token, after one sees the sheer alienness of the solar corona for the first time, how easy is it to return to the prosaic act of examining a clock?

The spirit was willing. Nonetheless, the inadequate, conflicting, and sometimes simply wrong results were essentially worthless for determining the exact location-as-a-function-of-time for the Moon. Even if this were not the case, it turned out that hardly any of Newcomb's newfound (self-selected) corps lived on, or happened to station themselves at, the *exact* path limit. In reality, otherwise was too much to expect.

Given the effort Newcomb put into the project, the result must have been, if not heartbreaking, at least frustrating for him. He did not try it again.

The United States Coast Survey undertook a similar experiment, but one with important differences in methodology. It was orchestrated by Julius Hilgard.

Unlike, Newcomb's get-what-I-can-get approach, Hilgard specifically targeted two population centers predicted to be sitting on the eclipse-path edge. He then had USGS personnel professionally survey a reference point central to these locations, ahead of, or immediately after, the total solar eclipse.

---

[1] though it was temporarily closed in 1869

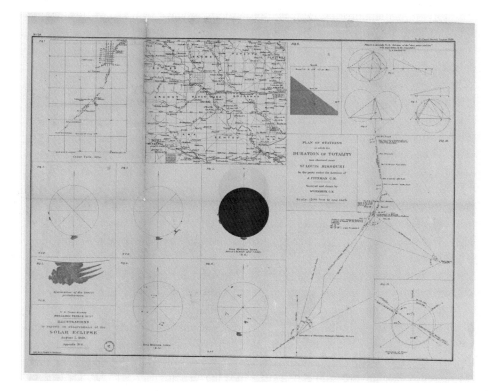

Fig. 14.1 Julius Hilgard's maps for regions on the predicted path's edge of totality, during the 1869 solar eclipse. {National Oceanographic and Atmospheric Administration Photo Library}

There was one actual town that met the criterion, Cedar Falls, Iowa, calculated to be on the northern limit. What we would now call the Saint louis, Missouri, suburbs marked the southern limit. (The large city itself was too far south; a lot of people missed out on a total eclipse because of this unfortunate geography.) Hilgard solicited—and trained—volunteers from just these two venues.

If you will permit me a digression: It was in Cedar Falls that Iowa's homeless waifs, left without parents by virtue of recent warfare, grew up. Once empty, the Civil War Orphan's Home was repurposed as Iowa Normal School for the preparation of teachers. Later it became Iowa State Teachers' College, which, coincidentally my father attended. I never lived in Iowa myself until well into adulthood. But in another, literally cosmic coincidence, it was this institution under yet another name (the University of Northern Iowa) that asked an astronomer and self-proclaimed eclipse junkie to join its faculty. Me.

On 28 June 2004, the Nautical Almanac Office arrived at my landlocked doorstep in Cedar Falls. It did so in the person of Simon Newcomb. Sort of. In fact, my visitor was Alan Fiala (1942–2010; Chief, Nautical Almanac Office [retired]). Fiala, the

NAO's eclipse expert, was an intellectual descendant of Newcomb. Fiala had himself organized ten total-eclipse expeditions and edited circulars with predictions.

Fig. 14.2   Alan Fiala. {United States Naval Observatory}

Did I know that my very neighborhood had played a role in the 1869 total eclipse of the Sun? I did not. I never had heard of the 7 August 1869 total eclipse of the Sun.

Fiala's goal was different than Newcomb's. Now that a robust theory for lunar motion was available, he wanted to use the total-eclipse shadow size from long ago to determine if the Sun's physical diameter was changing. Only in the latter twentieth century was the lunar limb profile, which had confounded Newcomb and others, known well enough such that it could be subtracted or added to observations in order to obtain corrected times for a mean spheroid—greatly improving the accuracy of shadow-size calculations.

Though confined to a wheelchair by childhood polio, since retirement Fiala had crossed solo the states of Kentucky, Illinois, and Iowa in his quest. In Iowa, he had already visited Marion, Mechanicsville, Stanwood, and Red Oak.

Recall that, as Hilgard's USCS enterprise, long-ago residents of my town had been encouraged to attempt determination of the 1869 total-eclipse path's northern limit through timing of the disappearance and appearance of the Sun's photosphere. They agreed. (He had had similar success in Missouri.)

Now, Fiala was in Cedar Falls. The result was an entertaining day driving about town, in search of the *exact* spots from which path edge timings had been made, replotting their locations using the new technique of Geographic Position System (GPS) measurement. It was not an effortless undertaking. The geographical descriptions from 1869 no long fit the layout of what was now a larger, modern, midwestern city.

Once Fiala departed, I took up the search. No fewer than four different observing stations were manned (though even more were planned) by the enthusiastic citizens of 1869 Cedar Falls, thus setting something of a record. Because of Hilgard's targeting, we know more about these 'civil' total-eclipse observers than we do about those in other municipalities and across the rural USA.

Fig. 14.3 The only extant building and one of the few remaining railroad depots (important to our story) from the era can be seen in Cedar Falls, Iowa. It is now a bank. {Photograph by the author}

Asa Horr (1817–1896; physician) was astronomically inclined with his own observatory; in 1869 he was one of the founders of the Iowa Institute of Science and Arts. William Anderson (1814– ?); formerly, county surveyor and Justice of the Peace), W. Wormwood, and Horr sat in the cupula of the only-just-built Civil War Orphans' Home on Eclipse Day, a fine perch from which to watch for the umbral edge.

Fig. 14.4 Civil War Orphan's Home. {University of Northern Iowa Donald O. Rod Library, Special Collections and Archives}

Photographs show the building to be a lovely piece of architecture; alas, it burned down in 1965. The University's student union now sits above its remains.

The address of J. H. Stanley's store, mentioned in the Prolog, was a nicely situated locality, though his reported observation "across the street" from the shop leaves a lot of available street and, hence, uncertainty.

One still can locate the small hill west of town on top of which we can imagine eclipse observers were pacing, awaiting the big event. At the time, it would have been accessible by horseback. Now, it is surrounded by the streets of a housing subdivision.

Dempsey Overman (1820–1874) and his brother John (1817–1906) were among the founders of early Cedar Falls. In fact, it was they who named what originally was but a hamlet. Dempsey was one of those movers and shakers who donated to every civic cause, the sort who cities name parks after. In Cedar Falls? They named a park for him (after the younger Overman donated the land).

Just two and a half months before the total eclipse of the Sun, the junior Overman bought the lot for a new house, three blocks from the one to be built fifty-five years later and which eventually became my home. From there he, his sibling, and another gentleman "recently returned from Germany," observed and reported upon the total eclipse of the Sun.

But where was the Overman house? It only could have just been finished on 7 August 1869. The county auditing plats indicated to Fiala that the deed to the property on Twelfth Street had gone through many hands since. With my (rather limited ability to) help, he narrowed its location to that on a large lot, the kind one would expect Overman to have procured. There, he was confronted by an ill-tempered dog and a homeowner who said, "I advise you to not come any closer." Fiala took his GPS measurement from the safely of his car.

Seventeen years later, I determined that we were off by one address. Fiala encountered his canine nemesis at the wrong house.

I doubt that Fiala foresaw such adventures when he became a PhD. astronomer. I may be wrong. His attempt to suss out time variation in the solar diameter established a null result within the precision of his measurements. The quest for measuring a variation in the Sun's diameter continues.

With no suitable town on the total-eclipse path's *southern* limit, the second part of Hilgard's USCS experiment was to situate five volunteers from the vicinity in the countryside, along a line radial to Saint Louis. The observers' line segment crossed the predicted path. They stood at one-mile intervals. The gentleman who ended up closest to the umbral edge experienced a 10-second-long total eclipse.

(This exercise anticipated the total solar eclipse of 1925, visible across New York City, in which dozens of potential observers stepped forward to station themselves on Manhattan rooftops. Depending upon who saw a total eclipse and who saw a partial eclipse, the southern limit of totality was established to have been between 230 Riverside Drive and 240 Riverside Drive, on the Upper West Side.)

No such luck for Hilgard. All his observers placed near the eclipse's projected path limits saw the Sun in total eclipse! While this was likely edifying for those present, the fact that there was no one who missed out on totality meant that the path limit still remained undetermined experimentally.

Of course, there is always serendipity. Dr. Henry Ristine of Marion (1818–1893; chief surgeon for the Burlington, Cedar Rapids and Northern Railway) was visiting the countryside; his reputation for conscientiousness suggests that he was making house calls. He inquired. Yes, the residents of one house did see a solar eclipse on 7 August, though totality lasted but an instant. So, he asked the neighbors. Why, yes. The residents were on the look-out for a total eclipse, but the bright Sun (photosphere) never completely disappeared. The two properties were just 4/5ths of a kilometer apart. This story was repeated to Hilgard, who must have immediately realized that to his chagrin he did not have the means to survey the locations of remote farmhouses.

The Curious Case of W. H. Pulsifer

William Pulsifer (1831–1905; businessman) was a classic Grand Amateur astronomer in the American Tradition. He grew up in Boston, but made his fortune in the West, the transportation and economic powerhouse that was the city of Saint Louis. (In a conflicted state, Pulsifer chose the winning side in the Civil War.) He was President of a raw materials company—Pulsifur literally wrote the book on the history of lead—and later an officer of an insurance company and a bank.

In middle age, Pulsifer turned to astronomy as a pastime. He built his own observatory which housed a high-end, 4-inch-aperture Clark refractor. What is more, he took the unusual step of surpassing mere observation and made more complicated spectroscopic observations. Grant Amateur territory.

A total eclipse of the Sun was the perfect opportunity to indulge his hobby. This he tried with very limited success. His home observatory was twenty-five kilometers *outside* of totality's path.

Twenty-five kilometers! While others crossed half a continent to see third and fourth contact, Pulsifer abjured a morning's horse ride to leave his swanky neighborhood of Lafayette Square. I suppose that we will never know why. He did not make the same mistake twice; he and the same telescope traveled with Harvard astronomers to Texas for centerline observation of the 1878 total solar eclipse.

Totality was not always necessary to make an emotional impact in 1869. Sycamore, Illinois, was not *quite* on the umbral path. It experienced a magnitude 0.98 partial eclipse of the Sun. This did not matter to Charles Marsh (1835–1918; businessman), who co-invented a forerunner of the modern farm implement, the combine. He was very ill in August but crawled out of his sick bed to watch the event through the window. In his autobiography, Marsh writes more about the solar eclipse than he does about the death of his wife, three months earlier than the celestial event.

Eleven-year-old James Keeler (1857–1900) saw a most unusual solar eclipse from La Salle, Illinois. It was 99% total. The sight made a great impression on him: Keeler would become a professional astronomer. He was among the first to apply the physical principle known as the Doppler Effect to spectroscopy. Doing so made possible the measurement of radial motion in astronomical objects. Keeler was to succeed Samuel Langley as director of the Allegheny Observatory.

One voice from the edge was silent. Lexington, Virginia, home of the Virginia Military Institute [VMI], experienced a magnitude 0.98 eclipse. Moreover, VMI's new Chair of Physics was renown as an astronomer. Matthew Maury (1806–1873; former USN) had been the first Superintendent of the USNO. He left that position and resigned his commission in order to fight with the Confederate Navy. In 1869, he only just had been pardoned. There is nothing whatsoever about the solar eclipse of 1869 in the Maury Papers stored digitally at VMI. The correspondence nearest in time concerns an order for carpet. To my knowledge, no professionally credentialled astronomer from the former Confederate States of America observed the United States's most famous celestial event of the decade.

15

Vulcan

Lede:
Paris, KY., Aug 7 – No Scientific observations made here.

The Sun was *not* Simon Newcomb's principal target during the 1869 total eclipse of the Sun. If this book were fiction, I might write that he was on a 'secret mission,' the Moon's position as a function of time only a cover story. In reality, he made little effort to disguise his true quest. He simply understated its importance to him.

For instance, he attached opaque screens to the courthouse of known angular diameter based on his pre-eclipse measurements with a sextant. Looking at the eclipsed Sun past these screens was supposed to assure that the brighter, inner layers of the Sun would not interfere with his view of the corona. In fact, while observing the corona, he mistook the Moon itself for one of his screens. They *were* useful, though, for eliminating light from the Sun during the second part of his observing plan, the one for which he intentionally tried to minimize expectations.

Notwithstanding that he was their organizational leader, Newcomb left the photographers and spectroscopists alone. Officially, Newcomb's personal goal was to study the corona. Known for his work on the orbits of Uranus and the newly discovered planet Neptune, he was in the process of commencing his more difficult project to improve the mathematical model for the orbit of the Moon. Such an exercise in applied mathematics required exact measurements of where the Moon was and where it was going, at a very specific time. A total eclipse of the Sun theoretically helped with this requirement. Nevertheless, Newcomb likely knew ahead

T. Hockey, *America's First Eclipse Chasers*, Springer Praxis Books, https://doi.org/10.1007/978-3-031-24124-6_15

of the event that the complications of making contact timings in practice would not provide the precision he needed.

So why would Newcomb himself interrupt his beloved calculations and travel through six states to obtain this datum? In fact, he did not bother to put himself in charge of the timings at all.

History loves the discoverer. The bigger the thing discovered, the better. Many of us still celebrate Columbus Day, even though: a) Christopher Columbus did not intend to discover a New World – he wanted to take the back route to Asia; b) He was a bit dimwitted inasmuch as even the ancient Greeks knew that the globe was so large that sailing from Europe to Asia was suicide by thirst, hunger, or mutiny (unless one was outstandingly lucky enough to unexpectedly run into an uncharted continent before that fate); and c) his appalling treatment of the Native Americans set the stage for centuries of abuse.

Now, who discovered/made the tiny SARS-CoV-2 vaccines that have saved untold lives. When is their holiday?

Astronomy is no different. All the same, we have seen that the professional astronomer throughout most of history was not charged with making discoveries. He or she refined positions, and other physical properties, from the existing compendium of astronomical objects.

It makes sense. Unless you already have reached a peak in the profession, why risk your time (or career or paycheck) on unknown celestial bodies that may or may not exist? Especially when there is so much in the heavens that we do know enough about, and which needs attention? However, this no longer applied to Simon Newcomb.

Per contra, from the United States and France respectively, Newcomb and Urbain Le Verrier (1811–1877; Paris Observatory) spoke each other's astronomy. Le Verrier, too, did his most important work with pen and paper, not telescope and star chart. He achieved lasting fame by correctly predicting the existence, and calculating the location, of a new planet—Neptune—based upon irregularities it caused in the orbit of Uranus.

Le Verrier's current goal was to refine the orbit of all eight planets. Frustratingly, Mercury was not playing well with the others. It turned out that it did not follow the pristine elliptical path dictated by Isaac Newton.

Could this small world's orbit be perturbed by the gravity of a missed planet, orbiting between it and the Sun? Le Verrier believed that there was such an object. The planet-in-waiting was to be named Vulcan. Once Le Verrier announced his conclusion, there were individuals who proceeded to 'find' the extra planet (but with no verification).

The problem was that there is another way to make a dark spot on the face of the Sun—the ubiquitous sunspot. Sunspots come and go, change their shape, as well a change their position on the Sun as the Sun rotates. Who could say whether

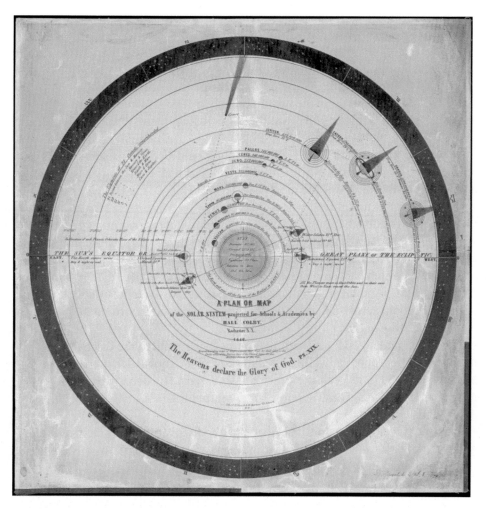

Fig. 15.1 Nineteenth-century map of the Solar System. Note the inclusion of planet Vulcan. {Library of Congress Geography and Map Division}

any given dark outline seen on the Sun was caused from within or something backlit beyond?

The trick would be to catch Vulcan when it was not in front of the Sun, when it could be seen illuminated in at least partial sunlit against the black of sky. Yet Vulcan always would appear near the Sun, trapped in a small orbit that kept its angular distance from our star small.

It is hard enough to observe Mercury. Known planets with orbits inferior to that of the Earth can only be caught just after sunset, when blue sky turns to higher-contrast dusk, but before planet sets itself. Or briefly in the morning, when the planet rises above the horizon and the Sun has not yet done so (brightening the sky so as to make the planet fade out of visibility). It would be worse for Vulcan.

All the same, there is an exception to this rule of visibility. During a total solar eclipse, the Sun may be far from the horizon, and along with it its erstwhile nearby attendant, Vulcan. Plus, the blinding glare of the Sun is missing. Vulcan would appear just as the stars do in the now inky welkin. The stars are well mapped. An extra one would be immediate cause for suspicion: an 'new' planet has wandered into the field of view.

Newcomb knew all of this. He eventually would become jealous of his colleague at the United States Naval Observatory, Asaph Hall[1]. This was true even at a time when Newcomb was the most famed astronomer in the United States and a scientific celebrity around the world. Why?

In my (admittedly unscientific) survey of ten introductory astronomy textbooks, with publication dates sprinkled throughout the last thirty years, Newcomb's name is nowhere to be found. Instead, we read in each who in 1877 used the very USNO telescope for which Newcomb lobbied successfully, to discover the two moons of the planet Mars. And those were not major planets; they were just satellites. What if Hall had come across a major planet, a first-time discovery from the USA? He would be a national 'hero.'

Newcomb understood very well that immortality went to the discoverers. As a mathematical astronomer, while he had improved our knowledge of the empyrean, he had not added to its census of interesting citizens. His deeds were more intangible than a rock in space. Therefore, they offered little *excitement* in the eyes of the American public.

In 1869, Newcomb saw his chance to check 'discovery of a planet'—the then-greatest professional accomplishment in observational astronomy—off his career 'bucket list.' It was Newcomb's goal to pursue Vulcan.

There was another point of view, though. Observing near Newcomb during the total eclipse of the Sun was Christian Peters, also a newly minted citizen. Yet Peters's background hardly could be more different from Newcomb's. Peters received a formal, classical education in astronomy while studying in Europe. He worked in Sicily, but when he chose the losing side in a revolution, he was forced to emigrate to the United States and ended up at the Lichfield Observatory.

Peters was a successful asteroid hunter, with 42 discoveries eventually 'in his belt.' He would spot (109) Felicitas in October 1869.

Withal Peters was a Vulcan skeptic. There had been an 1862 report, by an amateur astronomer, who saw what he thought could be Vulcan transiting the solar disk. Peters was observing the Sun at the same time and believed that this 'discoverer' had confused two sunspots, thinking them to be the same object moving at planetary speed. He was merciless in his anti-Vulcan rhetoric.

---

[1] I selected books that mention Hall.

Fig. 15.2 Christian Peters. {William Sheehan}

On another occasion, a solar observer claimed to see fast-moving silhouettes move across the face of the Sun. Peters dismissed the sighting as likely that of a flock of migrating birds. So, when Newcomb casually asked Peters if he wished to look for Vulcan during 1869 totality, it pushed a button. The 'gruff' Peters replied that he was there "to observe the eclipse and that, he would "not go on a wild goose chase after Le Verrier's mythical birds." Doing so, he echoed a number of other astronomers.

The 1869 total eclipse of the Sun seemed to be an excellent opportunity to locate one or more vulcans. Nearly every observer commented on the naked-eye appearance of Mars to the left of the eclipsed Sun and Venus to the right, "… as if herald and handmaids …" These bright planets conveniently marked the plane of the disk-shaped Solar System, the ecliptic. This was presumably where an inter-mercurian planet also would orbit the Sun. One did not have to seek a much fainter interloper everywhere, 360° near the eclipsed solar disk, during less than three minutes of totality. Moreover, while Mercury itself was placed closely in the sky with respect to the Sun, Newcomb knew the greatest angular distance from the Sun that Mercury could achieve. The same was true of Venus. He need only look between these limits, a segment of the distance between the two brighter planets.

Indeed, Newcomb could narrow the hunt still further. His calculations showed that, to effect Mercury, Vulcan's orbital radius about the Sun must be only half or less that of Mercury itself. Based on assumptions about its size and mass, it would be four times less bright than Mercury. (He assumed a single body in this scenario; he appears to have expected an albedo similar to that of Mercury.) He was looking

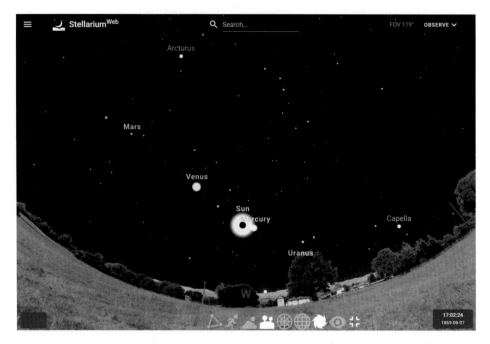

Fig. 15.3 Appearance of the sky during the total eclipse of 7 August 1869. {By the author, using Stellarium® software}

for a second magnitude object[2]—maybe slightly dimmer if it was to be seen in front of or behind the corona. Newcomb's screens and box time were odds-on foremost to allow his eyes to quickly see objects of this brightness through one of the two telescopes he used.

Newcomb acknowledged an alternate version of the Vulcan hypothesis—what Peters was referring to—that the precession of Mercury's perihelion was due to the gravitational influence of more than one unknown body. In other words, the intra-mercurial planet could instead be a fleet of asteroids. Each of these would be smaller and dimmer than a single planet. If there were a great many of them, they would show up *en masse* as an unresolved belt wrapped around the Sun. Newcomb was confident that he could see this phenomenon, too. "… the main object I kept in view was to determine whether there was anything at all visible outside the usually assigned limits of the corona, and yet so near the Sun as to be invisible at other times."

With these various parameters in mind, after second contact, Newcomb quickly convinced himself that the supposed planet or planets were not to be seen in his available telescopes. He abandoned his plan partway through totality to gaze at the corona.

---

[2] the brightness of the star Polaris, the North Star

If one is looking for the intimate or, at least, revelations of emotional state, the reader will be dissatisfied with Newcomb's diary. It is largely a list of the day's appointments. Still, he went so far as to write privately that he "was disappointed in not finding intra-mercurial planets."

Benjamin Gould was also on the hunt for Vulcan. His formal report makes it sound like an afterthought—something he undertook for twenty-five seconds after he had completed his survey of the corona. Nonetheless, if not on the lookout for a faint planet, why did he bring along such a large-aperture, wide-field-of-view telescope? And why did he equip it with an occulting disk, designed to purposely *block out* the corona and its obtrusive light?

Gould previously memorized the star field that would be the total solar eclipse's backdrop. He even rigged the motion of his telescope so that it would stop when he hit the angle that he thought marked the farthest distance from the Sun that Vulcan could be expected to reach. Clearly Vulcan was on his mind from the beginning.

Said Gould of his search,

> … I devoted 25 seconds to a careful scrutiny of the vicinity of the Sun, in order to ascertain whether any other luminous body might be visible there … a slight motion of the unclamped telescope sufficed to throw [the star designated] π Cancri into the field. This I succeeded in detecting, although with difficulty, thus fixing an outer limit for the magnitude of any planetary body with my sweep …

Any scientific measurement has error bars. Gould was not saying that Vulcan did not exist; he was saying that if it did, he was confident that it was dimmer than the magnitude 5.8 star, π Cancri. It is a fair statement. Because the star was close to where the inter-mercurial planet ought to exist, there would be, as a practical matter, no difference in atmospheric extinction or the sky glow caused by the nearby totally eclipsed Sun effecting both. "Had I not in advance made myself familiar with its position, and with the motion necessary to bring it into the center of the field, I do not think I should have detected it; but it seems highly improbable that any star surpassing it in brightness by one magnitude could have failed to arrest my attention, even in that close proximity to the Sun's limb …"

Less than half a minute may seem like little time to afford a task with such a huge potential payoff. Gould swept along the ecliptic outward from the Sun on the Mercury side, then repeated the process on the Venus side. It took him *longer* than expected. He finished a few second before third contact.

Gould's report stops abruptly there. While a null result is important in science, ofttimes just as important as a positive result, it lacks 'sex appeal.'

(Even today, null results often are the subject of footnotes. Recently, the American Astronomical Society launched a new journal.[3] In a crowded market, why yet another professional periodical? Part of the explicit mission of this journal is the publication of null results, which might otherwise fail to find a home in print.)

Fredrick Bardwell, "Aid" at the United States Naval Observatory, was Benjamin Sands's 'sleeper cell.' Too dramatic a metaphor? Sands's report mentions him in part of a single sentence. He was sent to Bristol, Tennessee, accompanied by one other serviceman, not affiliated with the USNO, and with a minimum of equipment. It is unclear if the USNO/Harvard personnel already there knew of him. Nobody references Bardwell in print (except Bardwell).

One can imagine Sands's thinking: Asaph Hall was liable to experience bad weather. What about Des Moines, Iowa? Had he put too many of his espial 'eggs' (men, materiel) in a single 'basket'? What if Des Moines was clouded out, too? (If the solar eclipse had taken place one day earlier, this easily could have happened.) Bristol was far away from Des Moines. In the worst-case scenario, perhaps Barwell could say that he saw the total eclipse, thereby avoiding Sands having to admit that USNO was completely 'skunked.'

Wherever did Sands get Bardwell? Bardwell served in the Army (not Navy) during the Civil War. Back then, his wartime posting was not considered a 'plum' appointment: He was an officer in charge of one of the few African-american units, as portrayed in the movie *Glory* (1989).

Bardwell appeared in Bristol the very day of the eclipse. He traveled light: no plans for a temporary observatory and only a small telescope to be lashed to a post. He and his traveling companion simply climbed a hill with an unobstructed horizon that was (described with a dose of self-importance) "… far enough from the public highway to be free from the intrusion of idle visitors." After the total eclipse of the Sun, townspeople—probably unaware of the slight to the major party—rechristened Bardwell's as "Solar Hill."

Key to Bardwell's hoped for success was a chart he drew, showing the positions of stars and planets near the Sun for 7 August 1869. It turns out that he was an (apparently self-appointed) Vulcan chaser.

Bristol ended up being a hub of interest for the illusive Vulcan. Even so, Bardwell has little to say about the particulars of his quest. His attention seems to have drifted from a patrol of the ecliptic to the eclipsed Sun itself sometime during totality. He was the only astronomer to describe the prominences as "… a rosette of bright purple…"

With regard to the search for an intra-mercurial planet, the evidences so far seem merely negative, and considering all the difficulties necessarily atten-

---

[3] *Research Notes of the American Astronomical Society*

Fig. 15.4 Frederick Bardwell; others unidentified. 1863. {Library Company of Philadelphia}

dant on the search, is by no means conclusive against its existence. It is quite possible that this interesting question will yet be definitely settled by new methods which were suggested by prosecuting the recent search.

No Vulcan, though.

Bristol assignee A. T. Mossman (USCS) looked perfunctorily for Vulcan, too. He wrote sounding apologetic: "I had no time to examine the space between Mercury and the Sun for the supposed planet, and I saw no stars except Venus, as my attention was almost wholly directed to my telescope, watching, as I was, for the instance of the Sun's appearance." *Almost*. Mossman wanted to make certain that his boss, Benjamin Peirce, and others who would read his report, knew that he did his job. Notwithstanding, he obviously had Vulcan on his mind.

Meanwhile, George Searle searched for an intra-mercurial planet from Shelbyville. James Haines did so from the Dakota Territory, 1,200 kilometers away. Again, nothing. Vulcan would be forever a null result.

Most astronomers classically trained in positional astronomy (measuring more precisely what you know instead of searching for that which you do not) rebelled.

The was something even a little embarrassing about planet sleuthing. They would not waste precious totality time on Vulcan.

Setting Vulcan aside for a few moments, what are we to make of a small group of observers who *did* see a new 'star' during the total eclipse of the Sun? Their description was taken seriously enough at the time to be published in the premiere scientific journal, *Nature*. Remember the *ad hoc* Winthrop Gilman expedition?

Gilman seems to be the only member of his party with any serious experience in astronomy. For example, Leon Vincent (–1916; Iowa Falls and Sioux City Railroad) was a civil engineer who had access to a 1-inch-aperture telescope, small even for the day. It had no use in observing the nighttime skies. The owner of the property at which Gilman was staying, N. E. Farrell, had only field glasses with him, which, admittedly, he was well accustomed to using inasmuch as he had been a member of John Powell's expedition to explore the Rocky Mountains the year before (Chapter 20). Others had no optics at all.

Said Gilman: "At no time during the eclipse did it occur to me to look for intra-mercurial planets." Be that as it may …

The unusual phenomenon witnessed at Saint Paul's Junction was seen with no more help than the *naked* eye. "A few moments after the corona [became visible], a small but exceedingly bright point, like a star, was noted independently by four of the party …" (but not Gilman). It came to be called the "Little Brilliant." An unexpected bright star is easy to dismiss. Still, four people saw it! "… each of the observers felt quite positive that what he had seen was truly a star [or planet]." Unfortunately, Gilman, his eye glued to his eye piece to watch the total eclipse he had come to see, missed it. Baily's Beads? Apparently not. "It [the "Little Brilliant"] formed near the limits of the corona" about [1-]1/3 of the Moon's diameter from the center of the eclipsed Sun. Three people even were able to specify its position as being in the same radial direction from the center of the Sun as the now well-known "anvil-shaped" prominence.

Gilman took depositions from those who had seen the "Little Brilliant." He did so just after the total eclipse of the Sun had concluded, at which time he realized the phenomenon's potential significance. The witnesses' stories matched, even though Gilman deposed each out of earshot of the others. All attempted a judgement of dimension—for instance, some fraction of the apparent diameter of Mercury—though it is well known that estimates of extent among astronomical bodies is skewed by brightness. The true angular size of such a solar-system object as Vulcan should have been beyond the resolution of those without a telescope.

Gould dismissed the Saint Paul's Junction observations as that of a known star, 82 Cancri. However, this star is at the limit of naked-eye magnitude; there is nothing "brilliant" about it at all.

This was an important point to Gould because, if the Vulcan suspect was brighter than 5th magnitude, he was fairly certain that it would have been picked out previously. Or, at least, observed by other 1869 total eclipse parties. He thought that that it might even have showed up in the photographs. Looking at the congregation's photos of the eclipsed Sun, he saw nothing. Therefore, Vulcan does not exist. (To be fair, Gould never demonstrated that the solar-eclipse photographs would record a star.)

It is useful to know where the Saint Paul's Junction contingent was coming from when they insisted the mystery star was, in fact, "brilliant." For this reason, it is instructive that they happened to describe the famously bright star Arcturus, visible in the sky during totality, as merely a "glimpse" of a star!

Nonetheless, John Hind's (1823–1895; British Nautical Almanac Office) *Nature* paper concluded that the folk in Saint Paul's Junction had seen something important.

They were intent only on an outing to watch a total eclipse of the Sun, but, in fact, ended up discovering a new planet.

It was the other way around from the way Gould saw things. Hind felt that it was Gould, the professionally equipped astronomer on a crusade to ferret out such a thing, who had mistaken *Vulcan* for the *star*. He had allowed—in the words of the appropriately named Freddy Mercury—"fame and everything that goes with it" … to slip from his hands.[4]

Could things get any stranger in a town where the number of a total-eclipse expedition's dilatant members threatened to exceed the number of permanent residents? Yes it could. Just *before* totality, Vincent came (running?) up to Gilman,

> "… exclaiming that he saw a miniature crescent-shaped star under the Moon, and that I must come and verify his observation. So interested was I in my own work that I paid little attention to Mr. Vincent's announcement, but on his returning a second time, more urgent than ever to have me look at the object, I did so in a hurried manner, using his glass. I detected nothing in the few seconds I gave to the search … During totality the object was forgotten, but shortly after third contact he [Vincent] readily picked it up in the same locality.

Vincent was so excited that—like another contemporary Vincent in the Netherlands, who also reproduced the starry sky—he drew his new foundling. Seeing the artwork later, Gilman realized that he probably had not looked at the proper spot. Vincent's crescent was far from the Moon.

Clearly, though, Gilman was not having it with Vincent's crescent-shaped star. Ironically, stories from Saint Paul's Junction, of celestial objects that did not belong, had greater traction than Gilman's total eclipse observations. In the words

---

[4] Mercury, F. "We are the Champions." *News of the World* (LP). Westminster (UK): EMI. (1977)

of a popular turn-of-the-nineteenth-century author, "It may be that out from the floating islands of space, two processions marched across the Moon."

Not likely, though. Just as Gilman did, today we feel safe in dismissing the Vincent crescent. We know too well, as Gilman no doubt did, too, that an out-of-alignment telescope in the hands of someone not well versed in its optics can render easily a non-existent crescent. A crescent implies an object of great angular extent. Even the crescent Mercury is hard to pick out through a modest instrument. That there should be a stray object of such apparent size, but which—for the first time in history—appeared only to Vincent, stretches incredulity. As the late-great Carl Sagan was fond of saying, "extraordinary claims require extraordinary evidence."[5] And it just is not there.

The crescent-shaped-whatever tarnishes the observations of the "Little Brilliant." The fact that it also was observed *with the naked eye* by multiple witnesses means that relegating it to a mere optical effect does not 'wash.' What exactly were they seeing outside of Saint Paul's Junction?

It was not an inter-mercurial planet. We now know that there is no such thing. An exploding star, a nova, especially one that drew no comment anywhere else, sounds awfully coincidental.

An orbiting Iridium satellite, with its shiny hull, can turn just the right way with respect to the Sun that the glint is both brief and spectacular. Knowing when such an event would occur from our latitude and longitude, I once told a class of students that I would conjure up a new star before them. Timing my mumbo jumbo to the event, I pointed my trusty laser pointer into the vault of heaven and … behold! . . . a new star.

Anyhow, we are discussing the nineteenth century, not the days of *Harry Potter* novels. No artificial device in the sky, not even an anachronistic airplane, could have produced the "Little Brilliant."

It was a clear sky. We quickly run out of potential explanations. Must we fall back on mass hallucination? The excitement of the eclipse? One sheep following the lead of another? Would it have even been necessary to have a total solar eclipse in order to see this strange sight? Of course, without an eclipse to behold, why would a group of people be starring simultaneously at the heavens on an otherwise random Saturday afternoon? What good is a "Little Brilliant" with no one to document it?

Not a word about the "Little Brilliant" appeared in print again. The philosophy of science always has had trouble with anomalous results. Induction, a favorite scientific method, involves repeated experiment or observation. But what if the observation is specific to a given place and point in time? The 1869 total eclipse of the Sun will not come again.

---

[5] after French polymath Pierre-Simon Laplace (1749–1827; École Militaire)

Meanwhile, back at the observing sites populated by professional astronomers, one suggested a typical post-eclipse conversation went as follows:

"How was the photography?"

"...successful..."

"And the observations with the spectroscope?"

"Very successful."

"And did you see an inter-Mercurial planet?"

"No."

An Eclipse Is What You Make of It

Using the brief time of totality during a solar eclipse for other than gazing at the eclipse itself may seem odd. But it is not unique to the search for Vulcan.

Charles Coale (1806–1879) was an acquaintance of Albert Myer, and the General allowed him to append a total-eclipse report to his own. Observing the eclipse late in the day, at Abingdon, Virginia, Coale did so between clouds. In describing it, he set some sort of record for use of colorful language:

> In giving to the clouds "*all* the colors of the rainbow," I was probably bordering on the extravagant, though not more so than is allowable in country journalism. I distinctly remember, however, that there were distinct bands of pink, purple, yellow, orange, and fiery red, and each slightly tinged with different shades of its own color. One of the bands, I remember, had, to my vision, a slight lilac tinge. I do not remember to have observed any green or blue, but I remember that the lower edge of the purple had a very faint blue tinge. All these resting against a dark background gave them an indescribably gorgeous appearance—the lines of color seeming to be divided by strips of black. They all lay in horizontal lines, one above the other.

132 words just on eclipse clouds! Elsewhere he describes them as "indescribable." Certainly, his unique psychedelic prose gives lie to that statement.

What of Coales's impression of the total solar eclipse itself? "Having no astronomical knowledge or aspirations, and no instrument, my whole attention was directed to the clouds, and my whole soul absorbed in their ... grandeur."

Guiding many groups of first-time solar-eclipse viewers, I never have seen anyone forgo the blackened Moon, glowing prominences, and ephemeral corona to watch the nearby clouds.

1869 marked the death nell for Vulcan. It should have been the end in reality. While perhaps not intentionally looking for a misbegotten planet, a dozen trained observers had looked the vicinity of a total eclipse of the Sun and seen nothing that did not belong.

I say "should," because 1878 brought with it the similar, transcontinental eclipse described by Ruskin in the first chapter. Newcomb made one more try for Vulcan during a total eclipse of the Sun; he was among very few to do so. Maybe the corona was too bright in 1869? Maybe this time it would be dark enough to see vulcanoids?

Newcomb was accompanied in Colorado by James Watson, who had decided that discovering Vulcan would be a nice cap to his asteroid-collecting career. Newcomb again saw nothing and surrendered. Watson thought he *did* see something –*two* new planets—and did not give up.[6]

Lewis Swift was the only other astronomer to claim that he eyed Vulcans, and Swift's Vulcans were not Watson's! Watson went to his grave—though under continued skeptical attack from Peters—believing that he had found the intra-mercurial planet. And his prestige kept the chimera alive decades longer.

A failed search was undertaken during the total solar eclipse of 1883 … and 1887 … and 1889 … and 1900 … and 1901 … and 1905 … and 1908 … Regardless, after 1869, Vulcan hunting was a 'niche sport.'

Finally, it was none other than German physicist Albert Einstein 1879–1955; Berlin University) who killed Vulcan outright. Yes, Einstein. World-renown scientist, of course, but not one known for his prowess at the telescope. The most famous physicist of the twentieth century whacked Vulcan with a pad and pencil. His watershed General Theory of Relativity explained Mercury's anomalous motion without resorting to an extra source of gravity in the neighborhood.

Vulcan. What a waste of a great name! It finally was given to one of the asteroids (as Vulcano), but only discovery number 4,464.

By the 1960s, the missing planet Vulcan was forgotten totally in the popular imagination. However, if the Solar System cannot have a full-sized Vulcan, maybe some other place else can. Arguably, the imaginary planet was immortalized best, as an alien world orbiting a star elsewhere within the Galaxy, in American television's 1966–1969 series, *Star Trek*.[7]

---

[6] Charles Young *about* Watson: "… one of the most energetic and able men I ever knew … extremely self-confident (but perhaps no more so than his abilities justified), selfish and unscrupulous in advancing his own interests."

[7] Viacom/CBS

16

Americans in Totality

"Well, it proved one thing … and that is that the papers don't always tell lies."
(an "old lady")

Benjamin Peirce called the United States of 1869, "… an intelligent population greatly interested in the scientific aspects of the event [(an eclipse)], and ready to render aid in procuring the desired results." Indeed, we have seen locals eagerly standing ready to assist visiting astronomers on their quest.

As we have discovered, municipalities jockeyed for the attention of eastern total-eclipse expeditions. Some advertisements were well thought out. Others were not:

Those astronomers … ought to come to Indianola as it is as high a point as can be found in Iowa. This will be a big eclipse—will be nearly [as] dark as midnight.

A "big eclipse." To what were the citizens of Indianola making a comparison? "Dark as midnight." Was it meant that the town had some special dispensation from the Sun?

The article is correct in that elevation might mean something to the visiting astronomers. But Indianola is near Des Moines, Iowa, far from the northwest corner of the state, which is its highest point. Indianola is perched 200-meters short—not even close. No professional observers visited Indianola in 1869.

In nineteenth-century rural, agrarian America, through which most of the path of totality traversed, life was hard. And routine. A total eclipse of the Sun came to be an unlikely source of entertainment, too.

There was great public interest in the total eclipse, often inspired by the local newspaper. From Sioux City, Iowa, to the Wilmington, North Carolina, editors published "eclipse editions." Even organs with circulation as limited as the *Kentucky Freemason* and *American Phrenological Journal* wrote extensively about the upcoming event.

T. Hockey, *America's First Eclipse Chasers*, Springer Praxis Books, https://doi.org/10.1007/978-3-031-24124-6_16

Fig. 16.1 Typical newspaper article about the upcoming 1869 total eclipse of the Sun. {From the collection of the author}

Solar-eclipse articles provided a diversion in a way that political coverage, written in anticipation of the upcoming election, did not. So, we see, nestled in second-page columns between headlines describing a racially motivated "murderous riot on the steamship *Dubuque*;" local drownings; county land sales; church services; county fairs; the curse of public drunkenness; "Personal Habits of the Siamese Twins"; ads *e.g.*, velocipedes, a reward for finding a lost revolver, and cosmetics ("If you would be beautiful, use Hagen's Magnolio Balm"); and unabashed gossip, announcements about the forthcoming total eclipse of the Sun.

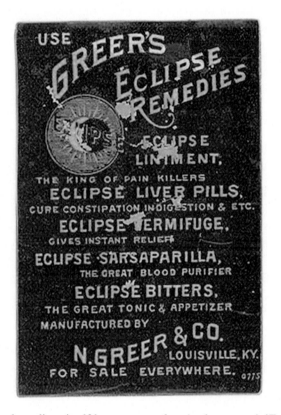

Fig. 16.2 The solar eclipse itself became a product 'endorsement.' {Ferdinand Meyer (Peachridge Glass)}

Well ahead of time, these publications composed stories about what the total solar eclipse was and when, where, and how to see it. The Council Bluffs, Iowa, *Nonpareil*—not even in the path of totality—devoted two columns to the subject. Sometimes the press filled space with enough eclipse-related *bon mots* that the overinformed "old lady" was initially "… afraid it would never 'go off.'"

The dailies and weeklies said that "thousands" would witness the total eclipse of the Sun. This is surely a great underestimate.

Print media began to prepare its readership for the total eclipse of the Sun as early as June. Most paper editors started out understanding little of solar eclipses themselves. One hedged his bet: "The *proposed* eclipse of tomorrow…" [my italics]. They reprinted descriptions of what was to occur that appeared in other papers or had been prepared by experts.

"The best we have seen" was that written by "Prof. White, of New York." This summary appeared frequently. Notwithstanding, I can find nobody by that cognomen who actually stood in the path of totality. Apparently, he was an armchair total-eclipse observer.

The Hampton, Iowa, *Reporter* was reassured and concluded,

This exhibition will come off promptly at the advertised time. No postponement on account of weather.

Inevitably, misinformation was spread, as well, in the process of preparing a readership for the " … one great event of the century … " (We do not know where the Civil War fits into this hierarchy.) Examples include,

- Christopher Columbus predicted a solar eclipse and used the information to re-establish his authority over a mutinous crew. (This story is 'whitewashed:' Refer to Chapter 2.)
- The only two total eclipses of the Sun previously visible over large parts of the United States during the nineteenth century were those of 1807 and 1831.
- "The Sun will rise eclipsed in the interior of Liberia …" (This is over seven-thousand kilometers from the point in the Atlantic Ocean at which the eclipsed Sun *set*.)
- "The shadows of Earth will commence crossing the Sun's disk about four o'clock …" (Shadows plural and "Earth" are problematic.)
- Totality should be observed through a filter (instead of the partial phases).
- Large stars will be observable in the daytime. (Stars have no apparent size; the author means "bright.")
- It will be the last total solar eclipse of the century. (The next would take place in December 1870.)

Still, these local bulldogs of erudition mostly got it right. Despite their efforts, "We learned that on that day of the eclipse several families who don't live far from this place, not having *any* previous knowledge of the affair, got badly frightened when the eclipse occurred, and in their terror yelled and screamed so as to be heard by those living over a mile distant." [my italics]

When mentally transporting oneself back to the latter nineteenth century, it is easy to be unsurprised by folk who not only did not know the date of the total solar eclipse but were even unaware that a solar eclipse was to take place. It is only slightly more difficult to imagine people of the time who did not fathom what such a phenomenon *was*.

Contemporary newspapermen were not sympathetic. It does not seem fair to me to go so far as to turn the tale of poor terrorized families into a cynical advertisement: "People who are too penurious to subscribe for their home papers ought to be scared a little now and then."

The problem with generic pieces composed elsewhere and reprinted was that they did not always cover basic information such as, for a given locality, when the eclipse would take place! "We have seen no general statement of the exact time at which the eclipse a week from Saturday will commence here ... it is impossible to tell how long the eclipse will be total here ... have your smoked glass ready ... you will not have chance to see another this century."

Eclipse circumstances of sufficient exactitude for anybody other than the professional astronomer were fully predictable and available well ahead of time beginning early in the nineteenth century. The requisite information was to be found in an almanac by a person who knew how to read such a source, but almanacs were considered the truck of sailors.

(Even if an almanac was available: In Nashville, Tennessee, use was made of the "Cumberland Almanac," which certainly sounds like an appropriate reference. Lamentably, and unbeknownst to Nashville, all the times listed were for Washington, DC.)

It is credible that no citizen of a small midwestern town had access to an almanac nor the expertise to make use of its tables and charts. The citizens would have to rely on the authority of an astronomer. If no such individual passed by prior to the total eclipse of the Sun, this avenue for instruction was closed, too.

Still, it is unexpected that the above-quoted passage comes from the *Iowa City* newspaper. Gustav Hinrichs, among few residents of Iowa who could be called an astronomer lived there. Why did he not offer up the time? It may have been that he did not feel that informing the public was within his purview.

Parenthetically, that last phrase in the article above, "this century," was in common usage. Indeed, for most locations on the 1869 total-eclipse path, the wait time between such eclipses was to be much, much longer.

Residents of Keokuk, Iowa, knew when to observe; it was not clear from where.

The scientific inclined of Keokuk of course took much interest in preparing for elaborate observations. Taking our instruments, we started out to the bluff southwest of the city. On reaching this, refreshed by the fresh clover fragrances, we saw a higher one, beyond, which we determined to reach. And still meaning 'excelsior,' we went on from hill to hill till the 'totality' was preparing to occur ...

Of course, sometimes there *were* outspoken, local adepts. A summary of social occasions in the Marshalltown, Iowa, *Times*, includes:

> Sunday evening Rev. Mr. Percival delivered an eloquent and beautiful discourse, rich with gems of thought, and clothed in almost poetical language, on the subject of the Great Eclipse religiously considered. The audience listened in breathless attention, as if fearful of losing a single idea out of this constellation of gems.

There is no wonderment in the public learning about a forthcoming solar eclipse from the pulpit. The eclipse affirmed the natural theology popular in that age:

> A total eclipse, of the Sun, is so rare and surprising a phenomenon, and its occurrence can be correctly predicted many years in advance, to the very day, hour, and minute, and almost second, that all who come within its range will be compelled to acknowledge the unvarying nature of the laws for astronomy. The infidel may, indeed, point to those laws as evidences that when one shuts his eyes to spiritual truth he can see no God in nature; but one of the greatest intellectual triumphs of true Christianity is to show that the laws of nature are intelligible to man only because the Lord has made us after His image and likeness, and that He has made them in their own limits unvarying for an analogous reason that He has made the swiftly-moving earth firm under our feet; to give us something on which we can establish ourselves on a firm, unyielding basis of thought and action, and on which our mental edifices can be raised, trusting in His laws.

Then there were ordinary Americans took it upon themselves to become eclipse 'experts': "So it happened that a general enthusiasm on the subject was produced which manifested itself in much street conversation, interspersed with rather obscure jargon about the Penumbra, the Umbra, Mean Time, Local Time, Total Obscuration, &c., &c., which had it fallen on the ears of a casual visitor would have created the impression that this quiet country town was the seat of some great University and that all its doctors, lawyers, preachers, merchants, tradesmen and mechanics were learned professors or ambitious undergraduates."

Finally, there were those who were well-informed about the total eclipse of the Sun yet were skeptics. In Effingham, Illinois, "One old gentleman, over 70 years of age, and one of the earliest pioneers of Illinois, insisted that there would be no eclipse, because he said it was impossible for any person to foretell such things." A peer, after the eclipse was well progressed, insisted that it would not be total because "there was not stuff enough to cover the sun."

Assuming it was real, where would a person situate themselves to view a total eclipse of the Sun? A suggestion:

We know of no better place than the top pf the steeple of the Catholic cathedral on Fifth street [in Louisville, Kentucky]—one of the highest in the United States—but as comfortable seats are scarce up there, we would recommend any point in the city from which an unobstructed view of the Sun can be had. The middle of any street running east and west would do, if it were not for the danger of being run over by the cars.

Or:

The editor of the Indianapolis, Indiana, *Journal* has become confused on the eclipse question and don't [*sic*] know where to go to see it. One day he talks of one place, and the next another. Come to Greencastle, [Indiana,] Berry. The eclipse will be total here. By the Way: you now discover the inconvenience of living in such an out-of-the-way place as you do.

".... at exactly the time calculated a party of gentlemen, including many of our citizens and several members of the press, and gentlemen from other cities, together with a party of observers from New York, who came prepared to make accurate observations, through the courtesy of Dr. Grissom of the Lunatic Asylum, were stationed upon the roof of that noble institution near our city, provided with smoked and painted glass[1], and other necessary instruments." I do not know that these "other necessary instruments" were.

Certainly, there were those willing to travel to see the total eclipse of the Sun. "A party of Cincinnati [, Ohio,] gentlemen have engaged the sidewheeler General Lytle, to leave that city on Saturday morning, at 7 o'clock, so as to reach a point near Louisville in time to see the total eclipse on that day." They were undeterred by the fact that the boiler of the General Lytle had exploded several years earlier at the cost of several dozen lives.

The bourgeois of the cities and towns tended to remain there. This was also where the journalists who documented the total-eclipse were. We have a better understanding of townspeople's experience than that of the greater rural population. "... and as the hour for the great obscuration approached, men, women and children sought elevated positions, even to climbing upon the roofs of the highest buildings ..."

"Dr. Hazen," of Davenport, Iowa, charged 50¢ to view the total eclipse of the Sun from his office atop the local pharmacy. (Profits were part of the plan to support the Davenport Academy of Sciences.) The sight was captured safely by a telescope and projected onto a screen.

In Wheeling, West Virginia, "The entire population was in the streets ... ," which is perhaps an exaggeration. Perhaps not. "The solar eclipse ... appeared

---

[1] Sometimes asphaltum varnish was used.

promptly on time, and unlike most exhibitions of man's arranging, more than 'filled the bill'."

Of course, eclipse watchers in different places saw different total solar eclipses. In the West, the phenomenon occurred high in the sky and interrupted the day. In the East, it was a horizon event and appeared as a premature onset of evening: "The Sun sets on the 7th at 58 minutes past 6, so the eclipse ends but a few minutes before sun set."

Most citizens regarded the 1869 total eclipse of the Sun with the naked eye. The Davenport *Daily Gazette* made some attempt at urging eye safety:

> Great preparations have been made by the various scientific organizations of the country for accurate observation and exhaustive investigation. All that most people will command for observation will be a piece of smoked or tinted glass. [These materials are now known to be ineffective solar filters.] A common opera glass [now known to be dangerous to the eye], screened, will render the phenomena more distinctly visible—those wishing to see all the features must not let their attention wander, as the appearances are visible but a few moments.

Notice the implication again that filtering is necessary during totality, a misconception widespread even today.

The Burlington, Iowa, *Hawk-Eye* distributed the following promotion the day before the total eclipse of the Sun:

> All persons wishing to view the Eclipse will do well to call at the Post Office Exchange this day from 8 am to 1 pm, and procure Smoked Eye Glass. No postponement on account of the weather. With these glasses the eclipse can be seen *through the heaviest clouds or rain storms.* [my italics]

They cost from 5¢ to 10¢.

[The Louisville *Evening* Express] "observed one enthusiastic eclipser mounted on a chimney of a six-story building, looking at the phenomenon through a whole window, sash and all, smoked entire for the occasion."

Still,

> We regret to announce that Alderman Riggs received a serious injury to his eye-sight [*sic*] during the eclipse. He was on a visit to the country, and traveling on the road when the obstruction began. He ha[d] no smoked glass at hand, or other protection for the eyes, and in his efforts to observe its different phases seriously injured his eye-sight.

(The deleterious nature of observing the total eclipse of the Sun with unprotected eyes was sometimes delayed: " … weak eyes began to feel the evil effects of starring at objects so bright" the next day.)

Fig. 16.3 For slightly more money, a consumer could use this product while waiting for the total eclipse. *Circa* 1871. {Library of Congress Prints and Photographs Division}

On the other hand, the sheet explicated another solar-eclipse-viewing method:

Quite a number of enthusiastic gazers, eager to find an original and complete way of beholding the grand spectacle, tried looking through the bottoms of tumblers and bottles. No matter if they did find them full of a hardly translucent fluid—such was their zeal that with the coincident of the movement which brought the vessel before the upturned eyes, its contents would be emptied into the open mouth. The result of this plan seemed to be everything, if not more, than desired, and not only did its followers see more in the same time but the eclipse lasted them at least four hours longer than it did anybody else.

By Total-eclipse day, "… In every parlor, at every dinner table, on every street corner, beside every counter, within every counting room, between drayman and steamboat-Captain, merchant and customer, lawyer and client, physician and patient, master and servant, mistress and cook, hostler and chamber maid the staple of conversation was the eclipse … "

Afternoon, Eclipse Saturday. "The Sun looked cheerfully down without a single thought of the humiliation he was about to suffer." "… the whole population became astronomers for an hour or so."

Fig. 16.4 The eclipse generated its own brand of drink: Greer's Eclipse Bitters. {Ferdinand Meyer (Peschridge Glass)}

Not everybody was content merely with watching the total eclipse of the Sun. There were those who wished to participate, somehow, in it. It was as if action during the solar eclipse made it more real. But what to do if no proper optical instrument was available? Simple measurements still could be made. Benjamin Sands had put out a public request for local weather measurements during the total eclipse, so we read of viewing parties at which "They were furnished with thermometers, barometers, and improvised telescopes …" (I would like to learn more about an "improvised" telescope.)

Others found ways to monetize a total eclipse of the Sun: "Clothier Benjamin advised everybody to buy a linen suit in which to celebrate the occasion …" "Before viewing this sight, go to the Spurrier House Saloon, Sixth 10:30 o'clock, and partake of their Eclipse lunch." 'Now that the solar phenomenon is passed, remember that there is one confectionery which totally eclipses all others …'"

When the moment came,

> … all business for the time was suspended … Every person of any considerable age in this county who was not unfortunate enough to be blind, viewed this wonderful phenomenon in the heavens.

And,

> Men walked more softly and spoke in lower tones of voice. And for some minutes before the appearance of the Moon, the hum of conversation had died away, the members of the party were all at their respective spots, and expectation was on tiptoe …

During totality, some fell to their knees in prayer. Said a "very pious … lady,"

> I would just like to know why the "good man" sends these things; it must be because the people have done so many wicked things. I do wish they would behave themselves.

Ora Williams

Ora Williams was the only historian I know of to have watched the 1869 total eclipse of the Sun. However, at the time, he was a barefoot seven-year-old. At an even younger age, he had witnessed the first train to arrive in Des Moines.

Williams came from a very literate family farming near Adel, Iowa, 30-kilometers west of Des Moines. On weekend trips to town, scarce money was spent on magazines and newspapers as well as supplies. Williams considered himself to be a "product of the *McGuffey Reader* and *Ray's Arithmetic*." His prize possession was the *Atlas of Astronomy*.

On the afternoon of the total eclipse, Williams filled a tub with water in the barnyard. He and his seven brothers and sisters were able to follow the eclipse's progress reflected in the water, with little risk to their eyes.

As an adult, Williams was curator of the Iowa State Department of History and Archives. In a 1948 recollection, he posited that, "… I have thought a thousand times about that celestial spectacle and am profoundly convinced the solar eclipse August 7, 1869, widened and colored the thinking of my lifetime."

Williams lived to age 93. He is the most recent living observer of which I know, he and his first-person story dying in 1955.

Very soon it was over. "All quiet this week on—the eclipse will not appear again … "

How long had totality lasted? Subjective impressions varied from, "a moment of inexpressible terror" to "a minute passed as if an hour."

Then,

> … we were all back on Earth again. The show was over.

After the fact, reporters had another problem: how to describe a phenomenon that nearly all of his or her readers had seen for themselves? As little 100 words might be spent on the post-event recapitulation. Other wordsmiths, though, struggled to find a unique 'hook.'

A typical example of 'totality journalism' comes from the Fairfield, Iowa, *Ledger*:

At 5 minutes before 4 o'clock … the contact of the Moon's shadow was first detected; a shout went up from the street below, and hundreds of eyes were straining through pieces of smoked glass, in the direction of the Sun. Slowly and steadily the shadow of the Moon covered the face of the Sun … each moment now appeared more weird-like, more appalling. Suddenly from the far northwest, there fell upon the Earth a shadow so deep and dark that it seemed like thick black cloth hung from above and covering all beneath and behind it … looking up, we saw that the eclipse had reached its total phase. The grandeur and sublimity, the wonder and all of the moment can only be imagined.

"Weird" was a commonplace adjective used to describe the total eclipse of the Sun, as in, "A dark shadow hid the Sun, darkness like night, and yet unlike it, fell upon the Earth and pervaded the heavens, and the strange, weird appearance was greatly intensified."

There were publications that took a more religious approach to the total eclipse of the Sun. "It seemed to speak directly to our spirits, with assurance of protection, of gracious mercy, and that Divine love which has produced all the glorious combinations of matter for our enjoyment."

But for others on the opposite end of the emotional spectrum, "This great occurrence on Saturday last, was shrouded [sic] with gloom and a desolate foreboding …" The total eclipse "… portended evil …"

What about after a total eclipse of the Sun? Sometimes a rather parochial view might be circulated. "The observations made in Brighton [, Iowa,] will be made the foundation of a new theory regarding the corona." [my italics]

On the other hand, newspapers did not fail to mention total solar eclipse expeditions to other locations. The editor of the Sioux City Journal wrote about the total eclipse as seen from Siberia.

Occasionally, professional astronomers provided a testimonial for the local press. Stephen Alexander opined in the Ottumwa, Iowa, paper, "The succession of phenomen[a] during the total eclipse of Saturday last, and which all could observe was very accurately that indicated in the Daily Courier of Friday."

*   *   *   *   *

There is no event so magnificent that it is free of its huckster and jokers. Certainly not the opportunity of a total solar eclipse:

There are in Sioux City, as there are everywhere, some very good fellows, who enjoy a very good joke. We are glad this is so. This would be a dull world indeed if there were no spice mixed up with its every day [sic] affairs. The occurrence of the great solar eclipse gave the que to some chaps about the Northwest Transportation Company's office for the realization of no small amount of fun at the expense of enthusiastic and credulous seekers after astronomical truths. The ball was started by enlisting the services of the press—the press up the street. On Friday the wags were gratified in seeing, in that intelligent and enterprising sheet, a lengthy account, of the hifalutin order, descriptive of a 'historical instrument'—'a telescope *six hundred years* old, possessing unparalleled magnifying power'—which was being forwarded to Secretary Seward, 'commissioned by the Government to establish an astronomical bureau at Sitka.' Then followed some 'facts of its beautiful history.' 'It originally belonged to Galileo,' who became its possessor 'at the time Hydia Alda made his decent from the Himalia Mountains,' 'in whose family it remained about four hundred years.' By the aid of this instrument, it further appeared, the sunken 'Russian fleet' was successfully raised 'in the harbor of Sebastapol [*sic*];' and later, it was 'used to recover the three million dollars sunk on the steamer Carter, at Pampa Bead, on the Mississippi.' In conclusion, it was gravely announced, that, for ten cents each, and for the benefit of a certain Sunday school, our citizens might observe the eclipse through the wonderful instrument. There were about twenty-five victims, who paid their ten cents each, and found themselves sold and feeling quite cheaply in consequence. Prominent among the number was the writer of the 'history' of the 'beautiful telescope.' The instrument consisted of a carpenter's saw-horse, set at an angle and covered with rich damask cloth and tarpaulin. The observer was instructed to take a certain position and put his eye to a certain little hole. That little hole was covered by a 'lense' [*sic*] of common glass, which in turn was covered by the letters *s-o-l-d*! The editor in question, it is said, occupied about fifteen minutes time in endeavoring to get a good view, and even to the last was unsuspecting, imagining that the characters he observed were peculiarities of the Moon's surface, which the wonderful magnifying power of the historical instrument rendered visible.

Cincinnati, Ohio, just missed being on the path of totality. Still,

On Vine Street, below Third, some enterprising merchants managed to get possession of a huge pipe, and into the extremity of the monster they fitted a piece of tin, mounted it on some tripodal arrangement, and then covered the great fellow with a piece of cloth. All who passed were invited to 'take a free look" through the telescope, and, of course, with the laudable desire which all had, of seeing everything that was to be seen, the invitation was generally accepted. But every one [*sic*] who looked through the deceitful instrument saw only a tallow candle shining, and flooding with its dim wick light the word "sold."

Like every other human experience, the total eclipse of the Sun was the subject of literary humor. From the puerile: "A couple … who insisted upon viewing the eclipse through the same glasses, were so delighted with the affair, that they

continued to gaze long after the Moon had taken its departure, and at midnight …
they were still investigating." To the more subtle:

> Mr. J. S. Moore took a photographic view of the Sun during the total obscu-
> ration on Saturday. He could not get his camera sufficiently near the Sun to
> take a life-size view, and will be obliged to enlarge the view taken if he has
> a picture worth inspecting.

Or with some pathos:

> Last week two or three parties bought tickets for the eclipse, of some irrever-
> ent wag, and came several miles to town to see it. They stood around until it
> began to grow dark, waiting for the 'show' to come in, and then started
> home, denouncing the whole thing as a '… Republican swindler,' originated
> by the Radicals to get a crowd to their convention."

Two themes come together in the following satirical passage:

> The hole in the Moon observed during the eclipse, is supposed to be caused
> by the tunneling of one of the mountains for the purpose of building a
> Railroad. The Board of Supervisors up there are under arrest for refusing to
> ley a tax to pay off the Railroad bonds.

We guess that some local politics is being lampooned.
More obvious political satire appeared, too:

> Before the next [total eclipse of the Sun] comes off all of the present popula-
> tion of Kentucky—except perhaps a few babies that have already had the
> measles—will pass away. The generation which comes after us may see
> another [total eclipse of the Sun], unless Radicalism should abolish all such
> things by a seventeenth amendment to the Constitution …

Occasionally, type setting led to odd juxtaposition:

> As the Moon began to move away the flood of light rushed out again, as if
> glad to escape from a forced imprisonment, and revealed the Sun as gradu-
> ally as it had covered it up.

> At Shakopee, Minn. Two girls aged 15 and 16 years, daughters of a Swede
> named Anderson, each gave birth to an illegitimate child within a day or two
> of each other…

> And, immediately following without break:
> Hon. O. A. Allen died to-day in the Insane Hospital, at Sommerville … The
> eclipse here was obscured by clouds.

Meanwhile, the rituals of life did not shut down entirely during the solar eclipse. In Quincy, Illinois, mid-eclipse, a

> … funeral procession found itself on Main Street, nearing Sixth, at the moment of totality… As the light faded, the solemn procession halted in the street, and while yet the city was enveloped in partial darkness, a prayer was offered up … As the light burst upon us from the sun the procession moved slowly on, while those composing it began singing a funeral chant. The scene will not soon be forgotten.

And there is this warning:

> Our Republican friends in the county, who meet at 3 o'clock Saturday August 7, for the purpose of appointing delegates to the county Convention, will need to imitate "The Wise Virgins," and go with their lamps well filled, because the eclipse will cause total darkness to prevail over this county. Do not be found like the "Foolish Virgins" without oil in your lamps, but have them filled and well trimmed.

Between celebration and sorrow, there were wonders:

> Last evening we were shown an egg at D. W. Miller's grocery store, which is decidedly a curiosity. The egg was laid by one of a quantity of chickens belonging to Mr. J. T. Bradford, on the last Sunday evening, and has a very correct delineation of the eclipse of the Sun of Saturday last upon its surface. The side upon which it appeared is partially flattened. The portion representing the eclipse seems partially imbedded in the shell, while around the edges diverge representations of light as seen when the Sun was in its total phase. It is not the work of art, nature, as it is often the case, seemed to have done the work. That hen was badly frightened in those dark moments.

The overall view that the Fourth Estate provides of a total solar eclipse is that it is a source of *amusement*. The Mount Pleasant, Iowa, *Journal* put it in perspective:

> Two notable things visited our little town on Saturday last—the eclipse of the Sun and Forepaugh's Menagerie and Circus.[2] This singular conjunction was not anticipated long before, and we never expect to witness the like again. The people's attention was diverted. Some preferred seeing the eclipse, others, the larger share, the show.

It may be true that for every *yang* human beings expect a *yin*. There was a darker side to the total eclipse of the Sun 'in the air,' too. Even though the eclipse was during the summer,

---

[2]Originator of the three-ring circus, Adam Forepaugh (1831–1890) was Phineas Barnum's (1810–1891) major business competitor.

> We have been enjoying splendid October weather during the past week. The general belief is that the approach of the total eclipse of the sun is the cause of all this unusual frigidity. If so, we have no desire that this kind of an eclipse comes oftener than once in a century at most; and less frequently if it can be got along with.

A farmer reported that " … his wheat was all 'spiled' by the rains, caused by the eclipse." An individual, curiously described as a "sleepist" [=insomniac], predicted "that the Sun would never shine clearly after the eclipse …" The Vinton paper in which his forecast appeared admitted that it had been hazy or cloudy for ten days since, though this kind of sky during a prairie August is hardly an outlier condition.

7 August 1869: Elizabethtown, Kentucky (on the total-eclipse path), caught fire and burned!

> The great eclipse of the Sun occurred just about the time the fire was subdued. The eclipse gave a somber hue to all nature, well befitting the calamity which had destroyed one-half of the business houses of our town.

People died during the total solar eclipse. People expire during any three-minutes span. But only in regard to the eclipse did theologians, and those who considered themselves to be, wonder: If you die in such a darkness, *are you doomed to Hell*?

Less conjectural, two women reportedly *were* killed *by* the 1869 total eclipse of the Sun.

> A Mrs. Gifford of Marion County [Iowa] died Saturday from effects of fright at the Eclipse. She had no knowledge of its approach, and being alone at the time, as the eclipse commenced she fled to her nearest neighbors. When she arrived there, her reason was gone, and she fell down in a fit. A doctor was called, but pronounced the case hopeless.

Both sadly ironic, as well as more plausible, is this account of fatality:

> A young woman … lost her life in consequence of the eclipse. She lighted a match to smoke a piece of glass, and unfortunately set fire to her dress in throwing the match away.

The word "unfortunately" seems understated at a time when the eclipse itself was described as "The King of Terrors."

The story of sixteen-year-old Thomas Bowman is more mysterious:

> When trying to make his way to his brother's homestead in 1869, and upon the day of the great total eclipse over the Sun, in August, he was lost. At about 4 P. M., when the Sun was fully overcast he became bewildered and lost his bearings. It seemed like night and the prairie wolves howled savagely over the trackless waste.

My attempt at analysis of the populace's reaction to it assumes that everyone paid attention to the total eclipse of the Sun at all! Surly the "gaze of all humanity" in Wilmington, North Carolina, must be an exaggeration. Maria Mitchell recalled that, during totality, a woman in a veil wandered ghost-like through Mitchell's observing site, seemingly, from Mitchell's words, oblivious to what was going on above.

And what are we to make of this gentleman? "Prof. Collins left town last week for a tour in the western part of this state and Nebraska, where he will rusticate until after the eclipse …" Is he traveling to observe the total eclipse of the Sun, or to avoid it?

In 1869, publishers realized that astronomy sells. The Cedar Rapids, Iowa, *Times* wrote, "State papers that have *recovered* from the eclipse are now talking about the comet that maybe seen these pleasant evenings." [my italics.] On the same page as a fairly accurate report on the total eclipse of the Sun, we find in the *Brighton Pioneer and Home Visitor* an article headlined, "The Comet is Here."

> For more than ten years past the most scientific astronomers of the world have told us, through publications in the magazines and otherwise, that during the months of July, August and September, this year, 1869, the most wonderful comet the world has ever known would re-appear.—They have also assured us that it would approach nearer the Earth than any comet ever did before—and that either the Earth or the comet would have to change its course, or a collision would be inevitable. As this comet is said to be many thousand times larger than the Earth, and as it is a solid mass of fire, with a tail of fire that would reach round the earth more than a hundred times, it is not at all unlikely that a collision with it would proves as disastrous to the Earth as the late accident on the Erie railroad did some of the more unfortunate passengers. According to the astronomers, it was this comet that preceded the terrible civil wars in Greece. It was immediately followed by a terrible contagion in Persia and the Eastern countries—a most dreadful plague that in a few weeks swept from the face of the earth more than one-half of the people of the countries visited by it. Some years after, this comet appeared again, and was preceded by a most terrific civil war in Rome—and was followed by a plague, or scourge that piled the dead up in heaps in the streets of that proud but corrupt city—until there were scarcely enough persons left alive to bury the dead. This comet is now visible—having made its appearance on time—thus verifying the prediction of the astronomers …

The comet is totally fictional. As luck would have it, Comet 11/P Tempel-Swift *was* discovered in 1869 (not a decade previously). But later that Autumn. It is a periodic comet of only telescopic brightness and lies in an orbit that will prevent it from ever coming close to the Earth.

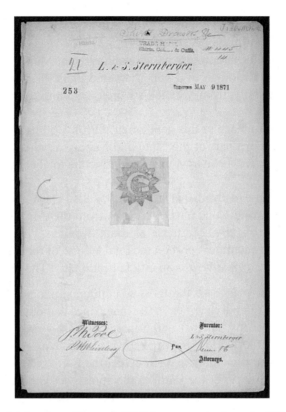

Fig. 16.5 The total solar eclipse of 1869 became a trademark. 1871. {Library of Congress Prints and Photographs Division}

Did the total eclipse of the Sun result in a greater public appreciation of astronomy? Maybe. There is no straightforward means with which to measure such a hypothesis. Immediately below the comet article, this is typeset—no doubt tongue-in-cheek—as a supposed excerpt from a tabloid:

> The journal says that the eclipse has developed such an enthusiastic love for astronomy among the young folk of Muscatine [, Iowa,] that a great many have taken to sitting up nearly all night, watching the movements of the heavenly bodies. The young ladies very generally become so much exhausted in their scientific pursuits as to fall over in the arms of their companions.

Anyhow, the immediate hoopla soon ended: "Children resumed their play—men and women returned to their accustomed pursuits—but the grand eclipse is still, and long will be, the theme of thought and conversation."

Newspaper articles about the 1869 total eclipse of the Sun did appear as late as 1948. And the event grew in drama with the retelling.

Ever heard about the big eclipse us old timers had back in the early seventies? [*sic*] Well, that was a whale of a big show, and they ain't been nothin like it pulled around here since. It came along about three o' clock in the afternoon, and everything went *dark as pitch for half an hour*. Schools closed so the kids could run home before dark, and folks had ta light the gas[3] ... [my italics]

*   *   *   *   *

In 2017, approximately 175 residents and visitors crowded onto that tiny piece of Iowa that lay in the total eclipse path of that year. A watcher explained her presence: "It's something different, something new for me to do." The same words likely were heard 148 years before.

---

[3] We are told explicitly that, in Terre Haute, Indiana, at least, "The gas company will not light the street lamps [*sic*] this afternoon during the total eclipse of the Sun."

17

In the Shadow of Benjamin Banneker and "Even Thoughtless Women and Children Hush …"

Note: Historical quotations may contain harmful language that reflects attitudes and bias of their time.[1]

> "Why, Samba, how black you are!" said a gentleman the other day to a negro waiter at a hotel, "How in the name of wonder did you get so black?"

> "Why look here, massa, de reason am dis: de dey dis chile was born dere was an eclipse."

I have described Maria Mitchell's total-eclipse expedition and reaction to it. According to stereotype, women, if involved in the sciences at all, were supposed to be unimaginative and not at all adventurous. So, ignoring the ingenuity and field work of a team of scientists (and totally missing the point), what did the local Burlington, Iowa, newspaper editor write about the newly elected member of the American Philosophical Society and her Vassar Expedition?

> The beaux at Des Moines who plumed themselves on their supposed brilliant prospects of captivating these young lady astronomers, must, as soon as possible, pack their carpet bags and make their appearance in Burlington or miss their opportunity. Of course, a lot of sprightly and handsome young ladies, coming "out west" wanted to go to a handsome town, with handsome scenery adjoining, and handsome young men by the score, and of course came to Burlington. Des Moines may have an old observer or two, but the wide awake and handsome ones of both sexes are all in this city, we judge.

---

[1] slightly modified statement made by the United States National Archives

T. Hockey, *America's First Eclipse Chasers*, Springer Praxis Books, https://doi.org/10.1007/978-3-031-24124-6_17

Fig. 17.1 Vassar College senior astronomy class. 1868. Mary Whitney is seated, looking at a book. To her left (standing) is Mary Reybold. Sarah Glazier is next to her (seated). Sitting across the table (with hands folded) is Sarah Blatchley. Both Blatchley and Reybold would die tragically young.

The repeated interjection of the word "handsome" now seems *non sequitur*. However, any description of young women during the era would likely focus on their marriageability.

> ... we should meet [Mitchell] and her class with open arms. Good-looking girls, twelve to fifteen hundred miles from home, must be looked after and treated with distinction ...

Mitchell was always "Miss Mitchell" in the contemporary literature. Sexism was common in publications about the 1869 total eclipse of the Sun, in the writing of both laymen and astronomers alike.

Winthrop Gilman felt obliged in his after-eclipse report to mention that one of his observers was "... by the way, a lady ..." Benjamin Peirce found it necessary to qualify a feminine solar-eclipse observer as someone who "conversed intelligently on the matter." So did Asa Horr, when referring to a Susan Simons who even owned her own telescope. With Darwin Eaton's comment, we see a pattern: "Some five or six *intelligent* observers (among whom were … Mrs. Judge Roach and Miss McLean) …" [my italics]

The same word keeps popping up. It does not in descriptions of male counterparts.

Were any female observers designated 'unintelligent'? They would not appear in the reports.[2]

With just a few exceptions, we likewise do not know the names of the married women who observed the eclipse in totality: They usually were documented as, *e. g.*, "Mrs. Stockwell." That these women had the opportunity to observe assumes that they were not merely traveling companions: "Our party now consisted of Prof. McFarland and Lady and two daughters, Prof. Stoddard and Lady, and ourself, minus the Lady."

Outside of the Vassar Expedition "co-eds", we possess the names of even fewer *unmarried* women. (Joseph Winlock's 12-year-old daughter, Anna, accompanied him to the total-eclipse path and grew up to become an astronomer ["calculator"] at Harvard College Observatory.)

In 1869, a lone "female college" was installed within the long path of totality: Saint-Mary-of-the-Woods (near Terre Haute, Indiana). Of the twenty-two United States' colleges and universities mentioned in this book, only one besides Vassar (the public, University of Iowa[3]) had granted any degree to a female student by August 1869. This statistic does not just include science, for which womankind was not customarily thought to have an aptitude.

Mitchell does not choose to comment on the issue of intelligence. By the next year, she had read about what actions did and did not take place during totality. So, instead, she took on the general issue of aptitude.

In an article she wrote about the 1869 total eclipse of the Sun for a widely read magazine, Mitchell twisted the female stereotype. Her prose is gender neutral until the last word of the last paragraph:

My assistants, a party of young students, would not have turned from the narrow line of observation assigned to them if the Earth had quaked beneath them. Was it because they were *women*?

---

[2] Reports: At Earlham College, "… the ladies were requested to be on the look out [*sic*] for the planet *Venus* [my italics] …"
[3] Female students at the University of Wisconsin were relegated to the Normal School.

Change in the Air?

Eclipse expeditions always collect a camarilla of local celebrities. ("… a large number of prominent citizens with many ladies assembled;" notice the two, seemingly mutually exclusive, classifications of attendees.) The year 1869 was no exception. The Mount Pleasant astronomers, because of their luxurious (by comparison) digs, invited such spectators.

These included Arabella Mansfield (1846–1911; attorney), who ended up taking notes at the spectroscope for John Van Vleck. She was the first female lawyer in the United States. After winning a lawsuit that would allow her to be, she was admitted to the bar the very year of the total solar eclipse.

When the National Woman Suffrage Association formed in the Fall of 1869, she served in its leadership with Susan Anthony. The organization of the Association for the Advancement of Woman also began in 1869. Its first president was suffrage-journalist Mary Livermore. However, for the next two terms, that office was held by … Maria Mitchell.

Meanwhile, a young Mary Whiting (1846–1927; Ingham University alumnus) was inspired by a lecture on the topic of the 1869 total solar eclipse. A recent college graduate, she went on to attend additional lectures by Edward Pickering and eventually founded the departments physics and astronomy at the new Wellesley College. Wellesley quickly became the United States's leading institution for the education of female students in the sciences.

And then there was Mary Newson. Newson was *born* on 7 August 1869. She would become the first female American allowed to earn a PhD in mathematics from a European university.

The only other published account of the total solar eclipse apparently written by a woman tells the story of an observer in Shelbyville, Kentucky, who eschewed The HCO/Coast Survey station at Shelby College and instead chose to stand alone on a hill six miles away. The byline is anonymous.

At that moment flashed a light of inexpressible richness—pure, brilliant, glittering—seemingly a star of tenfold magnitude—a great diamond; or, it seemed, a lamp of purest fire, swung by an unseen Hand, in the depths of eternal space. Every beam it shot forth was full of joy and lighted up the soul with the brightest happiness. It seemed like Goodness, and Truth, and Bliss laughing, and dancing, and leaping forth out of the infinite heavens. It twinkled suddenly into extinction, the landscape fell into darkness, the whole sun was obscured, and there, in the violet sky, was the round, dark Moon encircled by the corona, which looked more like a wreath than like a halo.

It would be an easy narrative for me to write: the contrast between nineteenth-century bias and twenty-first enlightenment. However, history is more nuanced than that:

> All this equipment, scores of experts and the national press were impressive, but what really got the attention of Burlington was the arrival of Miss Marie Mitchell, a Vassar College astronomer, who brought with her eight attractive female students to view the eclipse.
> Sun shadows were impressive, but not as exciting as the college lovelie [*sic*], dressed in the height of East Coast fashion, featuring wide hoop skirts and tiny wasp waists, set off by showy parasols.

That passage was written and published in 2021.

* * * * *

In 1869, roughly 200,000 persons of Hispanic descent lived in the United States proper. They ranged from those whose families had lived in the country (or what would become the USA for many generations to more recent immigrants.

Yet *"eclipse total de sol"* does not appear in contemporary narrative sources. (It is typeset in Spanish language almanacs published outside the United States.) Nor do any traditionally Hispanic names. If members of this American culture gazed up at the total eclipse, as they surely must have, these they were invisible to those chronicling the event while they did so.

* * * * *

Benjamin Banneker (1793–1806) was a free African-American whose father had been a slave. He became known for his astronomical knowledge as demonstrated by his production of useful almanacs in the eighteenth century, including superior eclipse predictions. Banneker assisted with the initial survey of the USA federal district, the District of Columbia, and was a correspondent of science enthusiast Thomas Jefferson.

Difference generations created their own Benjamin Banneker. In the nineteenth century he was put forth as an exception to a rule. A historical curiosity. In the twentieth, he was a token role model to science-minded children, meant to exemplify a fictional diverse discipline of astronomy. The accomplishments of contemporary African-American astronomers were overlooked.

Banneker was pointing out an error in others' solar-eclipse predictions when he wrote, "It appears to me that the wisest men may at times be in error ..." Or was he writing about more than an eclipse?

The scientists who traveled to observe the 1869 total eclipse of the Sun were virtually all male and white (and—to complete the 'WASP' acronym—at least nominally Christian[4]). There were no African-American astronomers in 1869.[5] Harvard would not appoint an African-American faculty member for another hundred years.

The principal astronomers of the 1869 total eclipse of the Sun were men who belonged to the side that had won the Civil War. Nevertheless, because their positions often exempted them from combat service, they were little affected personally by the conflict over slavery.[6] (John Eastman was an exception; Charles Young served as a captain in a state regiment, but only for four months.[7])

Even if they had been, only a few (statistically) would have recognized abolition as a cause that was being fought over. Indeed, most were from the northeast, an industrial region of the country that profited heavily from raw products produced by slave labor. (Mitchell acknowledged this by refusing to wear cotton, even after the War.)

These astronomically minded individuals adverted little opinion on, specifically, the major historical event of their lives. More generally, most kept their attitude toward race to themselves. Benjamin Peirce openly sympathized with the South, though with the excuse (which many used) that it was only because a dissolution of the United States was too terrible to contemplate. More bluntly (but only second-hand) we have, "Asaph Hall, who was outside the pale of any religious sect, disbelieved in woman-suffrage, wasted little sympathy on negroes, and played cards!" (Interestingly, his wife Angeline Stickney was a Unitarian, suffragette, and abolitionist.) Conversely, we do know that the pseudo-professional Gilman was an "original, though not outspoken, abolitionist."

My limited research indicates that only Stephen Alexander, the oldest of the government-eclipse-expeditions' membership, even employed or associated with an African-American colleague. In 1866 Alexander joined the American Colonization Society. The racist sentiment of this organization was ambiguous: Offering former slaves passage to Africa (even though their families may have spent generations in North America) nominally was viewed as some sort of reparation, but its result in the extreme would have been the establishment of a completely white United States. The relationship between the first and last sentence in this paragraph is unclear.

---

[4] While most kept their religion close to their vest, we can say that, as a group, they were nominally Mainline Protestants. Examples included Cleveland Abbe's Episcopal Church and Thomas Stafford's Lutheranism, along with a disproportionate number of Unitarians, *e.g.*, Benjamin Gould, Benjamin and Charles Peirce, James Clarke, and Maria Mitchell. (Mitchell, Samuel Gummere, and Robert Warder were raised Quakers.)

[5] Edward Bouchet received a PhD, in Physics from Yale University in 1876.

[6] Even Joseph Zentmayer was exempt, because he made all the military's microscopes.

[7] Edward Goodfellow suffered sunstroke before his first battle and was discharged from the Army.

Fig. 17.2 Stephen Alexander's assistant, Alfred Scutter (dates unknown). Between 1860s and 1870s. {Princeton University Archives, Department of Rare Books and Special Collections}

Why is all this important to the subject at hand? The trip west and (particularly) south to the total eclipse of the Sun may not only have been for these individuals a rare experience in the rural USA, it also may have been a first significant experience with a diverse population. Alas, they were not the lot to document their thoughts and feelings, at least for public consumption.

For instance, Bristol and its contiguous town of Goodson straddle the Tennessee-Virginia border. Both had once been part of a single, huge plantation. The USCS astronomers established their observatory on land cleared by a "Rev. King's slaves." Did they find anything odd about clergy owning a large number of slaves? (Many Americans of the 1860s would not have.)

What did the astronomers temporarily relocated to Shelbyville, Kentucky, think of the local rumor that the eclipse was part of President Grant's Reconstruction policy? It was "gotten up for the purpose of making every body [*sic*] look as much like a negro as possible?" This was a political joke, of course. Did they find anything out-of-the-ordinary about it?

History has left us little regarding the reaction of any members of minority populations in the United States *themselves* to the total solar eclipse. At the time of the eclipse, formerly enslaved people had been considered citizens of the United States for just over a year.

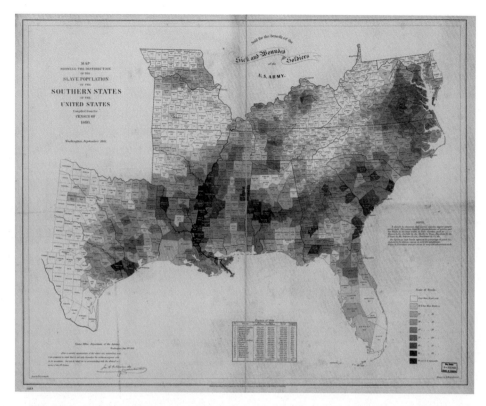

Fig. 17.3 In this contemporary United States Coast Survey map (based on census data from the year 1860), we see that Shelby County, Kentucky, likely had a high population of former slaves at the time of the total solar eclipse. 1861. {Library of Congress Geography and Map Division}

As for the 14,000 inhabitants of Lexington, Kentucky, one third were classified African-American in 1869. News from that point on the path of totality provides us—ever so faintly—with the closest thing to an African-American insight into the 1869 total eclipse of the Sun. "The crowds, which commenced to gather on the streets at an early hour in the afternoon, grew larger and larger as the momentous period approached, and by the time the great celestial wonder began its sublime, visible work, it seemed that all living Lexington had abandoned shelter and emptied itself out into the main street." Of course, this passage could refer to the citizens of any city in totality.

Be that as it may, the Lexington *Evening Express* chose to publish:

An old colored woman in the West End took a position in the middle of Jefferson Street, and for some time steadily and intently gazed at the great performance going on above. Then "totality" occurred, the woman took to her heals and ran into her house, muttering, "I golly; if dis yere wurl's comin' to an end, I'se gwine to look after dem yere children I is."

There are other examples of a similar tone.

The difficulty with assessing the African-American experience of the total eclipse of the Sun is the racism that gets in the way within specific references to it. When we read that during totality, "Men's faces looked like the faces of negroes, and some amusing mistakes occurred on the streets, by persons thinking others they met were 'American citizens of African descent,'"[8] it tells us nothing about African Americans in the path of totality and begs many questions about the author.

(These include whether or not those belonging to people of "African decent" are the faces of "men." There is also the writer's use of quotation marks. We can guess the bent.)

Fig. 17.4 Another staged, less formal, photograph of the United States Coast Survey/ Harvard College Obervatory temporary eclipse observatory, including staff, at Shelbyville. Pictured: probably Francis Blake, jr. (seated, looking through the telescope toward the left); George Dean (seating on crate); George Clark (standing in line, at the telescope); Arthur Seale (standing in line, at the telescope); Possibly John Prendergast (standing in line, at the telescope); John Whipple (standing at the small telescope); Joseph Winlock (standing at the large telescope); and Alvan G. Clark (seated, looking through telescope toward the right). {National Portrait Gallery, Smithsonian Institution}

---

[8] The language use could be worse. This passage is from a Quaker publication.

Yet look carefully at the photograph of the USCS/HCO astronomers *et al.* in Shelbyville, standing and sitting in front of their make-shift observatory. To the right of the observers, assistants, and family members stand several likely African-Americans.

Who are they? They must be local citizens. Did they participate in the transport of equipment? Help build the modest observing station? Provide other assistance? The astronomers were not obliged to include them in their photograph of record for the occasion. The presence of these individuals is the only connection of African-Americans to the scientific study of the 1869 total solar eclipse.

(In the lithographic version of this scene that appeared on the front page of *Harper's Weekly*, most of the black adults are cropped out. Further editing would interfere with the view of the telescope; the remaining black man and boy appear white.)

To reiterate, I have found extremely little historically useful written about African Americans and the 1869 total solar eclipse Sun. In what was written, racism is inescapable.

There is *one* first-person account, from Nancy Whalleh of 924 Pearl Street, Albany, Indiana. She was at an estimated age of 81 when interviewed. This would make her nine years old at the time of the total solar eclipse. She was interviewed as part of a Depression-era jobs program for writers. Her words were collected by white interviewers and written down by white writers. These lenses are apparent.

> The old woman remembers the Big Eclipse of the sun or the "Day of Dark" as she called it... the darkies all thought the end of the world had come... and everyone was scared to death.

We are told that Whalleh, "... lived down in Kentucky [further location unknown] after the [Civil] War until she was quite a young woman and then came to Indiana where she has lived ever since." In 1869, she was a recently emancipated slave.

As well-known cultural astronomer Jarita Holbrook concludes from these passages,

> African Americans are seen as childlike, ignorant, speaking in the vernacular, and superstitious—though the women could have been echoing Christian end-of-the-world sentiments. The nineteenth-century goal was always to show that my people deserved to be one-time slaves, needed to be managed by whites, and thus were incapable of exercising real American citizenship.

Now let us consider the writing of Samuel Langley. Langley is famous in both astronomy and aeronautics and epitomized in Langley Air Force Base, various school names, and the Langley, a unit of solar radiation. The National Center for Atmospheric Research describes Langley's book this way: "In 1888 Langley published a beautifully written, popular science book entitled *The New Astronomy* with a strong emphasis on recent developments in Solar Physics, which gained a very broad readership and did much to popularize the rising science of astrophysics." It was essentially the textbook of its time.

Fig. 17.5   Samuel Langley. 1887. {Smithsonian Institution Archives}

The New Astronomy has not much to say about the total solar eclipse of 1869 itself, nearly two decades in the past. What, then, is Langley's purpose in including this passage, rather suddenly, in the middle of his account?[9]

> In this part of Kentucky the colored population was large, and (in those days) ignorant of everything outside the life of the plantation, from which they had only lately been emancipated. On that eventful 8th [7th] of August they came in great numbers to view the enclosure and the tents of the observing party, and to inquire the price of the show. On learning that they might see it without charge from the outside, a most unfavorable opinion was created among them as to the probable merits of so cheap a spectacle, and they crowded the trees about the camp, shouting to each other sarcastic comments on the inferior interest of the entertainment. "Those trees there," said one of the observers to me the next day, "were black with them, and they kept up their noise till near the last, when they suddenly stopped, and all at once, and as 'total-

---

[9] Remember that Langley was not there. He was observing a several kilometers to the south at the time.

ity' came, we heard a wail and a noise of tumbling, as though the trees had been shaken of their fruit, and then the boldest did not feel safe till he was under his own bed in his own cabin."

Langley then returns, without transition, to describing the astronomy and physics of the Sun with a mere change of paragraph.

The *same* awkward phenomenon occurs in the travelogue of Edward Ashe and suggests a literary gimmick. In the midst of carefully describing his passage through the western United States *en route* to the path of totality, Ashe interrupts his narrative, with no attempt to disguise his words as anything other than cultural assumption, to write:

> ... the African negro, and no matter what you do to him he thrives under the treatment; whether free or in slavery he multiplies and is happy. Strange that rum which kills the Indian, only makes him fat.

Again, even—as hard as it to do—disregarding the explicit content, this seems to us today to be dramatically out-of-place from the pen of a would-be naturalist. It was not in 1869. The intention of such asides can be construed as insertions to maintain the attention of a voyeuristic white audience.

The purpose of observing, experience of, and physics behind a total eclipse of the Sun may have stressed the reading audience's intellectual horizon in order to expand it. These lurid anecdotes appear meant to give the reader momentary rest by reinforcing comfortable and familiar world views. I conclude that the most serious critique to be acknowledged here is that *little thought* appears to have been given these 'throw aways.' There was no editorial push back evident today; they were habitual.

That this material was published comes without surprise. The upshot here is that these archetypical quotations are from the professional *scientific* literature.

Whiggish history assumes, without justification, some sort of cultural assent, from 'bad' times to 'better' times. It is that written solely from the skewed point of view that is the here and now. It disregards the context of the time it is trying to recreate and is professionally considered poor history. Thus, it is important to point out that everything in their biographies suggests that Langley, Ashe, and other eclipse observers who we have encountered were typical northern intellectuals of their times.

Robert Warder, who 'cut his scientific teeth' on the Cincinnati Observatory eclipse expedition, went on to teach science at the most prestigious of historically African American universities, Howard University. We do not know if he taught astrophysics. If he did, the best resource available to him would have Langley's *New Astronomy*.

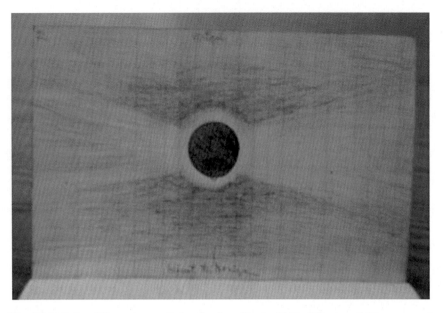

Fig. 17.6 Robert Warner's pencil sketch of totality in 1869. {Cincinnati Observatory}

18

Fire Cloud *

Lakȟóta Sun/Earth symbol

Another large swath of the Earth experienced the total solar eclipse of 1869. However, the record in all but extreme-eastern Asia is silent and dark. Searches in Arabic and Cyrillic yield nothing. Countless people must have awoken to the sight that 8 August 1869. We may never know their story.

From the Old World to the New. The most famous moment in history involving Native Americans and a total eclipse of the Sun is not a happy one. More than half a century earlier, Tucumseh was a well-known warrior among the Shawnee. His brother, Tenskwatawa, was a shaman. Together, they imagined a nation of united Native Americans that could stop the westward encroachment by people of European decent.

As the pair's influence grew, the governor of the then Indiana Territory became alarmed. He had little way of challenging Tecumseh's political acumen; instead, he decided to take on Tenskwatawa's claim to existential support for the brothers' plan. So, William Harrison challenged Tenskwatawa to perform a biblical-sounding miracle: influence the Sun and the Moon.

Bad timing on Harrison's part. Tenskwatawa agreed to the test and prophesized a total eclipse of the Sun, which duly came to pass on 16 June 1806.

---

* Maȟpíya Yapȟéta, a total eclipse of the Sun in Lakȟóta

This was likely no divination. Tenskwatawa could have learned of the upcoming celestial conjuncture from an almanac. Others have postulated that he heard about it from a British spy. It did not matter.

Nor did it seem to matter that the total-eclipse path passed slightly to the north of where the Prophet was living: He only could have seen a deep partial eclipse himself. The point was made. Recruiting escalated at Tecumseh's and Tenskwatawa's new village called Tippecanoe.

Words failed to squash the nativist movement. Now came raw military strength. A Harrison instigated battle took place with casualties on both sides. The next day, Harrison burned Tippecanoe, along with the villagers' winter food supply, and desecrated their buried dead. The living escaped, though the idea of resistance did not. Harrison was rewarded for his war crime by election to the office of President of the United States.

How does this tie in with our story? Historians later in the century were so appalled that (white) readers might think that Tenskwatawa actually had calculated the circumstances of a solar eclipse that they put forth a scenario in which he was alerted to the matter by astronomers visiting Springfield, Illinois, and Des Moines, Iowa, for the eclipse. But there *was* no Springfield, and certainly no Des Moines, in Tenskwatawa's time. Nor, for that matter, were there astronomers trekking to the (by then) Midwest. Those authors, intentionally or not, convolved 'our' 1869 total eclipse of the Sun with that of Tenskwatawa (1806).

The idea of a Native-american confederation occupied only the past by 1869. The Little Big Horn still lay in the future, but so did Wounded Knee. By the time of the trans-american eclipse, Native-american independence existed only on lands that the United States government had not yet bothered to reapportion for other purposes. The Great Sioux Reservation was created in 1868.

Serena Washburn

I would like to be able to tell you of an 1869 first-person, total-eclipse report from the Montana Territory. Alas, I cannot. I *might* have been able to, though:

Serena Washburn was well aware of the upcoming solar eclipse as she and her husband, Henry (a former congressman disabled in the Civil War) traveled to take up residence at Fort Benton. (He had been appointed Surveyor General for Montana.) But their Missouri River steamboat sank, and the Washburns had to turn around and make their way back east. Both ways they encountered Native Americans, some of whom may well have seen the total eclipse of the Sun.

Serena Washburn a partial eclipse from Leavenworth City, Kansas. It was a mostly cloudy day; "Still, it is wonderful the beautiful shadow and the small silver rim." Henry set out again (alone, this time, to Montana by another route) but grew sick, returned home, and died shortly thereafter.

Fig. 18.1 Washburn Family. {Montana State University Archives and Special Collections}

The 'elephant in the room,' can be seen on the 1869 eclipse-path map of North America: *Most* of totality occurred over unsettle lands inhabited by autochthonous[1] peoples. Their experience with the solar eclipse is largely self-undocumented and, hence, lost. We have only those reports of white men who interacted with Native Americans/aboriginals and Alaskan Natives at the time. These are colored by the stereotypes prevalent in the nineteenth century.

---

[1] The term "Native-American" is not used in Canada, while the term "aboriginal" is not used in the United States.

And even then: Among those who watched Cleveland Abbe at work in the Dakota Territory were local Dakȟóta-speaking Yanktonai. Yet Abbe tells us *nothing* about their reaction to a total eclipse of the Sun.

(It is perhaps worth mentioning that, after Dakota, Robert Abbe became a life-long 'collector' of Native-american artifacts. His museum still exists today.)

Here is an example of skewed reporting regarding Alaskan Natives and the total solar eclipse: Many secondary-source accounts of George Davidson's United States Coast Survey expedition include words along the lines of, 'The lives of the party likely were spared by the Indians, because of the scientists' prediction of the approaching eclipse.' Yet I can find no original, authoritative antecedent for this belief.

It is true that there would have been potential sources of friction. Americans from the United States were beginning to infringe upon Tlingit lands, using up local resources, and had been attacked.

Resentment went the other way, too: Members of the Lingít-speaking Chilkaht who were invited to Sitka, Alaska, to host Davidson were arrested briefly. They were charged with "some petty offense." One was killed trying to escape. Those incarcerated included a local chief, Kohklóx (meaning "hard to kill"). He and the others were promised their freedom if they would guide Davidson to his total-eclipse observing site. They agreed.

The Army considered the Chilkaht to be the most hostile people of the Alaskan Coast. Still, Davidson and assistants chose to travel without soldiers for protection.

According to one observer, Davidson appeared "less anxious about the [Chilkaht] than about the weather," The war-canoe pilots were five (armed) Tlingit. The destination was the village of Kulkwan [Tlákwaan], home to about 400 people. Guns fired as they approached, but Davidson assured others that it was only an act of greeting.

Of chief Kohklóx, Davidson recounted:

> The chief was a man of commanding presence, nearly six feet high, broad chest, and a well-formed head that measured twenty-four inches in circumference. He carried a bullet-hole in his cheek. He was held to be the greatest warrior and diplomat of all the tribes north and west of the Stak-heen [River]. In our future relations we found him truthful and absolutely honest. With all the instruments, tools, camp equipage, stores, carried and handled by his people, we never lost a single article during our day at his strong village … he fulfilled in spirit and letter every promise, and our every wish was attended to.

Kohklóx also fed the white men "fish and fowl" and lent them his house. This partly freed up a tent that could be used instead to protect the telescope.

The Chilkats did not generally know of total solar eclipses. Sympathized Davidson,

About the time the Sun was half obscured the chief Kohklóx and all the Indians had disappeared from around the observing tent; they left off fishing on the river-banks; all employments were discontinued; and every soul disappeared; nor was a sound heard throughout the village of fifty-three houses… The natives had been warned of what would take place, but doubted the prediction. When it did occur they looked upon me as the cause of the Sun's being "very sick and going to bed." They were thoroughly alarmed, and overwhelmed with undefinable dread.

After first contact, work in the village ceased. Some say locals drew closer to the astronomers, others that they retreated their houses. Some fled to the woods. Sheet'ká Tlingit visiting from Sitka knelt and recited the *Lord's Prayer* as instructed by Orthodox priests. Many assumed that the total eclipse was caused by Davidson's instruments, a source of great magic.

The Chief exhibited a different reaction. Kohklóx " … delighted with the great trick of his friend {Davidson], made a serious offer of all his canoes, blankets, and wives, if the astronomer would tell him 'how he did it' …"

All was well after the total solar eclipse. It led to a ceremonial exchange of gifts between Davidson and his Chilkat hosts. Davidson painted for Kohklóx and his two wives the total eclipse as seen through his telescope. This depiction is said to exist still among the Chilkats.

The family, in turn, presented the geographer with a map of their surroundings. It proved so accurate that it became the basis for future published charts of the region. The map survives in the hands of the USCS's modern equivalent to this day.

William Seward's experience with Alaskan eclipse-viewers was equally benign, though it delayed his voyage. Journalist Eliza Scidmore (1856-1928; National Geographic Society), inspired by her friend naturalist John Muir, wrote the first book about travel in Alaska. In it, we read:

Mr. Seward and his son, and General Davis, with two staff officers, and others of the party, left the ship in three [cedar] canoes early on the morning of the day of the [eclipse]… the shadow began to cross the Sun, and the weird, unearthly light fell upon the land. The Indians in the canoe said the Sun "was very sick and wanted to go to sleep," and they refused to paddle any further.

Once they reached Kulkwan, all were more relaxed. Kohklóx and Seward got along well. Perhaps they admired each other's scars.[2] Quoting Ted Hinkley, an award-winning historian of Alaska,

---

[2] Seward was attacked as part of the plot on the night of President Lincoln's assassination.

The chiefs wanted to know more about the eclipse. Seward "using as illustrations the cabin lamp to represent the Sun, and an orange and an apple to represent the Earth and Moon," described the phenomenon. He asked if they comprehended. "The chiefs have understood much," they replied, though not all, great Tyee has told them. They understood him as saying the eclipse was produced by the Great Spirit and not by man. Since he says so, they will believe it. They have noticed [,] however, that the Great Spirit generally does whatever the "Boston Men" want him to.

Now contrast the above with reports from Asaph Hall's United States Naval Observatory adventure on the Bering coast. Hall himself makes only a passing attributable comment on the local population: " … the Indians huddled together in awe." Just half a sentence. And the fact that he uses the term "Indian," which was not only literally incorrect for the indigenous Sireniki[3] but was not applied across the Straight, makes Hall a less-than-credible anthropologist. (Karl Neiman merely says that the Chukchi [lyg'oravetl'a] hid in their yurts.)

An unsigned article about the same episode, on the other hand, goes on at length about those native to the expedition's station. Here is that source—apparently a member of the expedition—on the total eclipse of the Sun itself:

The Indians exhibited a good deal of alarm; but were very incredulous when we told them what was going to happen. Some of them, who had intercourse with fur-traders and whalers, and had learned little English, would say, "Me think the Sun no break to-day," but when it did break, they were filled with consternation.

The anonymous scribe does not stop there. He also includes 'facts' such as that members of this Siberian population were, dirty, drunkards, wife beaters, easily amused by trinkets, unable to smoke, and (as something of an afterthought) small footed. (A scene described in which the author talks a woman into removing her moccasin sounds very uncomfortable to a modern ear.) A concession of sorts is made that:

The Esquimaux [sic] seemed to be perfectly inoffensive, unwarlike people, good-natured, and readily pleased; they are not much given to begging, nor are they possessed of the thievery propensity so peculiar to Indian races. I was not able to discover that there was any form of government among them, except a sort of parochial one, though they do not seem to be subject to the will of any Chief.

Between Davidson's patronization and the Hall crew's straightforward racism, we are left with no objective historiographic source on reception of the total solar eclipse in the far North.

---

[3] The last Serenik language speaker died in 1997.

British North America[4] is the great *terra incognita* of the 1869 total solar eclipse. In Manitoba, members of the Dakȟóta-speaking Sisseton told anthropologists long after the fact, that they, too, viewed the solar eclipse with alarm. It had served as a warning to prepare for disaster. Notwithstanding, the informants' forebearers likely had seen only a partial eclipse on 7 August 1869.

By coincidence, the 1869 path of totality followed the Missouri River for 1,600 kilometers. Reactions of Native Americans who lived in its drainage differed.

Fig. 18.2 Two Bears (dates unknown). {Oklahoma History Center}

Many Dakȟóta-speaking Yanktonai had been driven out of Iowa and Minnesota onto reservations by 1869. One group that still roamed the plains hunting buffalo was led by Two Bears, and other survivors of the 1862 Whetstone Massacre. It was camped outside Fort Rice (Dakota Territory) on the occasion of the total eclipse of the Sun. Throughout the summer, the officers, soldiers and, especially, post surgeon (and future ethnographer) Washington Matthews (1843-1905; United States Army), explained what was to happen. On eclipse day, Two Bears (who spoke English) and his followers were invited to watch the spectacle with the soldiers.

---

[4] The Dominion of Canada had just been constituted (1867), but consisted at first of only Nova Scotia, New Brunswick, Quebec, and Ontario.

Fig. 18.3 Fort Rice. {State Historical Society of North Dakota}

There was great anticipation among the Yanktonai. Nevertheless, on the day of the total solar eclipse, they were quiet. Pipes were lit. Sage, sweetgrass, and cedar were burned as incense. After the eclipse, the Yanktonai left the fort.

One total-eclipse record that does not include Native American/white visitor interaction also involves the Yanktonai. We know of the following *bon mot* because another group of these people happened to be camped that August near Fort Peck[5] (eastern Montana Territory). Unfortunately, there is no detail.

We learn second hand that a contingent of Lakȟóta-speaking Sihásapa on the Great Plains happened to be in the Moon's umbra on 7 August 1869.

> So everyone … yelled very loudly, and wailed and cried and prayed. Finally, the Sun began to get brighter and finally came to life again.

Such reactions are totally logical considering that a total solar eclipse was regarded as a marvel in which a sky creature swallows the Sun. Eventually, the Sun burns or cuts through and is revealed again.

Dakȟóta- and Lakȟóta-speaking people count years (*waníyetu*) in terms of winters that have passed. A Winter Count is a pictographic history, usually recorded on animal hide but also perhaps on paper. A new symbol was added annually that reflected a significant incident occurring during the previous year. The symbols begin in the center and spiral outward counterclockwise.

---

[5] not identical with the modern town of Fort Peck, Montana

Fig. 18.4 Lone Dog (dates unknown). {National Museum of the American Indian Cultural Resources Center}

One Yanktonai company's Winter Count was painted on muslin. We know that the most recent addition represents the year corresponding to 1871. Thus, simple subtraction shows that this Winter Count represents a remarkable 71 years. In 1870, the Keeper of the group's Winter Count was Lone Dog. It was up to him to decide what occurrence best symbolized the past year. He chose a black circle and two red 'stars.' The circle is the appearance of the Sun (*Wí*) during total solar eclipse (August 1869). The 'stars' might well be the planets Venus and Mars.

Colonel Garrick Mallery published a copy of this Winter Count in 1877. We do not know if he had permission.

A member of the Lakȟóta-speaking Oglála, chief Black Bear (dates unknown), stood side-by-side with soldiers at Fort Laramie (Wyoming Territory) during the 1869 eclipse of the Sun. He recorded it on his Year Count, too, as a black circle surrounded by stars, even though the eclipse he saw was not total. (The Oglála said that a total eclipse results from a dragon devouring the Sun.)

There are examples of potential solar eclipses on other Winter Counts, too. Here is a Winter Count by The Swan (dates unknown), a Lakȟóta-speaking Mnikȟówožu.

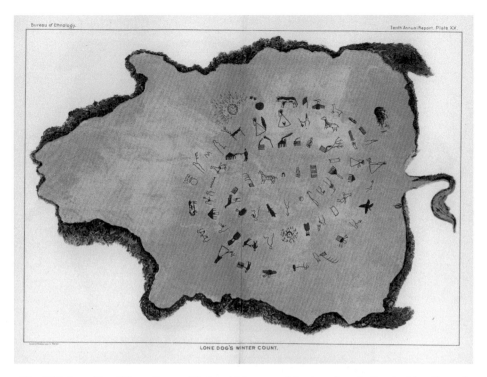

Fig. 18.5 Lone Dog Winter Count. {National Museum of the American Indian Archive Center}

Fig. 18.6 Fort Laramie. 1868. Pictured: Washington Matthews (1843-1905; United States Government Peace Commission), Mountain Tail, Pounded Meat, Black Foot, Winking Eye, White Fawn, White Horse, Poor Elk, Shot in the Jaw, Crane, The Pretty, and Young Bull. {Minnesota Historical Society Library}

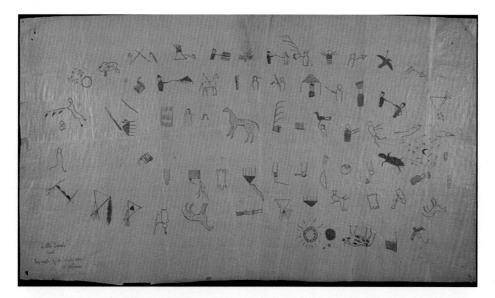

Fig. 18.7 Swan Winter Count. Covers 1800 to 1870. {National Anthropological Archives, Smithsonian Institution}

Here is a story with a twist. After service in the Civil War, Captain DeWitt Poole (1828–1917; USA) continued a military career. In 1869, he was ordered to assume the duties of Indian Agent "for Indians in the Sioux District, located upon a reservation at Whetstone Creek, Dakota Territory." He kept a diary of his subsequent experiences, which was published at book length.

Indian Agents were notoriously corrupt. Poole may have been an exception. The Oglála and associated Lakȟóta-speaking Brulé, for which he supposedly was responsible, only recently had been forced into reservation life so as to make easier the construction of the Transcontinental Railroad.

Reading the following narrative, recall that Mark Twain's A *Connecticut Yankee* in *King Arthur's Court*, which includes a parallel scene, had not yet been published. However, Christopher Columbus's lunar-eclipse extortion of native Jamaicans was in the history books. Poole tells of a similar situation that ends differently.

Some days before the great eclipse of August, 1869, Dr. C_____, physician for the Indians at the agency, concluded to try his skill as a magician, and impress the Indians with his magic art, inseparably connected in their minds with the healing art.

Fig. 18.8 DeWitt Poole (back, right; others unidentified). 1870. {USA}

The doctor announced to some of the principal chiefs and warriors the coming event, telling them the precise time (taken from an almanac) when the sun would be obscured and darkness follow, until he saw fit to have it pass away. When the day and the hour arrived, the doctor had his audience in readiness, duly armed with smoked glass. Being within the line of totality, and having a cloudless sky and the clear, delightful atmosphere of the plains, the phenomenon was observed under the most favorable circumstances. There was no mistake as to time; the Moon gradually crossed the disc of the Sun, a black, spherical mass, surely putting out its light.

The Indians were impassive lookers on, until, as the eclipse reached its culmination, leaving only a narrow, bright rim around the outer edge of the Sun, the deepening steel-gray shadows attracted their attention, as well as that of beasts and birds. Then, concluding that the exhibition had gone far enough, and that they must drive away the evil spirits, they commenced discharging their rifles in the air.[6] The light of the sun gradually returning, they were thoroughly convinced that it was the result of their efforts, and that the Indians' medicine was better than the white man's.

The doctor could predict the eclipse, but they could drive it away ...

---

[6] As recently as 1994 in South America, I heard gunshots during a total eclipse of the Sun.

19

What Did It All Mean?

"Scientific men all over the line of country in which the eclipse was visible have made extensive observations, and will publish them for the benefit of the world."

"The observatories must have been left undirected; the mathematical chairs of colleges must have been emptied," reads a contemporary mussing on the 1869 total eclipse. The United States Naval Observatory, for instance, was staffed by only one thirty-year veteran astronomer in August of 1869. Professor Mordecai Yarnell (1816–1879; USNO) had to stay behind to tend the clocks and coordinate the longitude-establishing telegraph signals! It is true: American astronomy had gone all out for the total eclipse of the Sun.

And the lion's share of traveling astronomers did see the total solar eclipse: from more than two dozen expedition stations. That the 1869 solar eclipse happened to take place in August vouchsafed mostly clear skies in central North America, where the great majority ventured to observe it.

Once back home, a tremendous amount of time was taken reducing the records of time and location from the total solar eclipse. For all that, the details of this tedious work are of little interest to us today. My GPS sampling suggests that the 1869 measurements of latitude and longitude made for the total-eclipse expeditions deserved at least one fewer significant digit than that which was published. (The uncertainty in measurement was at least ten-fold greater than that recorded.)

Almost immediately after the total eclipse of the Sun, and with little consideration of it, Simon Newcomb set about honing the parameters of the Solar System within the laws of Newtonian gravity. This meant refining the constants in the equations for the Solar System, particularly those for the Moon's orbit. For instance, what is the size of the Moon's orbit? What is its shape? (How elliptical is it?) How is this orbit inclined (tipped) with respect to the ecliptic? At what points does it cross the ecliptic? (Where are the nodes?) How is it oriented? (In what direction is its major axis—its apside—pointed?) In addition, the Moon's location at any one given time is necessary to establish it position at any other time.

By the time of his retirement, Newcomb had produced the best model for the motions in the known Solar System ever. Newcomb even challenged whether there was much more to be learned from classical astronomy. (The forthcoming study of dynamics outside our Solar System gave almost immediate lie to this contention.) It appears that Newcomb's monumental *cause célèbre* owes little to the total solar eclipse of 1869. He did not collect second and third contact times at future total eclipses.

The View from the Bottom

It was time to write the reports: step-by-step accounting of the planning and execution of the total-eclipse expeditions. Such communiques are written by the academics in charge. But what did those at the base of the scientific hierarchy have to say? Here is a rare insight, that of one such member. He was an undergraduate/writer for his student newspaper, who jumped aboard the Nautical Almanac Office enterprise at the last minute.

The writer claims that he,

suggested the name Observatory Hill, which was adopted both by the party and the people of Ottumwa [, Iowa], and will likely always cling to it.[1] It commands, I think, one of the finest views we saw on the whole trip. Below lies the city, in almost bird's-eye view. Beyond is a prairie, bounded by a range of hills, distant about six miles, all dotted with houses and farms, and through it runs the Des Moines River.

Official statements gave perfunctory acknowledgment to municipal leaders who often provided unquestioning support to an expedition, in a tangible manner. The authors then write as if the 180° of reality that is not above the horizon did not exist. Our junior correspondent took more interest in the new surroundings in which he found himself.

All eyes were fixed upon our proceedings, and some agog with wonder...[A] verdant youth .... was around prying into every corner and under every box. The writer, after answering many questions, was urging him to try and rise to a higher station, citing the example of Mr. Lincoln, late President of the United States, when the boy addressed to him a vacant question about that country that showed he did not know

---

[1] The quoted author is proud to have named Observatory Hill; however, I can find no use of the name after 1899.

he lived in it …A man in town reported that there was a lot of fools up on the hill to see an eclipse. He asserted that there never had been such a thing and never would be.

Official summaries also minimize failure. Nowhere in the NAO report submitted to the Government Printing Office do we learn that:[2]

An accident occurred that caused very much additional labor in determining the meridian. The transit instrument had been very carefully fixed on a table and its bearings ascertained. In the meanwhile someone came along and turned the instrument and its telescope toward the High School, had a view, and then very nicely put it back; and though very neatly replaced, all the previous calculations had to be also neatly replaced . . .

…Two of us were on guard on Thursday night. Early on Friday morning a heavy rain and wind set in. Our roof was not made to endure such treatment, and soon began to leak very badly. The chronometers, & c., were put under an umbrella we fortunately had. Then one corner of the roof blew off entirely.

The telegraphic wire hung over the corner of the screen, so that persons might pass under the wire. As the screen [to protect an instrument until totality] was taken away this fell to the ground, and in the darkness … some person caught his foot in the wire and broke it, so all the record after that time was lost. The person also at the chronometer misunderstood a sign of the Professor's, and did not record the time several seconds before the totality.

And, finally, a very personal vision of totality:

The fever heat of excitement had now been reached; for myself I scarcely remember any three more exciting minutes in my life than those of the total eclipse, if I except several minutes in which I was once held in a drowning condition at the bottom of a stream and the review of my life passed before me . . .

Of most use out of what corporally remained after the astronomers vacated their temporary observatories on the late solar-eclipse path were not timings but the descriptions of their actual observations, drawings, and photographs. The latter were considered the most precious. Glass copies were made for astronomers who did not attend the total eclipse of the Sun. Paper prints were sold to the public!

---

[2] The writer uses a pseudonym.

(Stereophotographs were a popular fad of the day; when viewed this way, two eclipse images taken half a minute or so apart result in a view in which a darkened Moon eerily hangs in front the Sun. The 1869 total eclipse of the Sun was the first after which images of it were mass-marketed.)

Fig. 19.1 Commercial stereophotograph of the 1969 solar eclipse. It is titled, "Views of the Great Eclipse" and is based on a image produced by Henry Morton's Philadelphia Photographic Corp. The user would have to have the appropriate viewer. {Hallmark Cards, Incorporated}

Foremost was the appearance of the total solar eclipse frozen into them for all time. Astronomers, once home, poured over the images. They used elaborate measuring machines in an attempt to extract from them the shape and apparent size of the Sun. Nothing resulted from this tedious business, though, that could not be measured by other means.

Still, 1869 made a difference. Writing about the Sun in 1865, influential astronomer and best-selling author/popularizer Camille Flammarion (1842–1925; French Bureau des Longitudes) did not mention the layers of the solar atmosphere distinguishable during a total eclipse of the Sun. (Flammarion even puts the term "photosphere" in quotes.) After 1869, this would never again be the case.

The chromosphere was first mentioned in 1851. In the 1860s, it was discussed as a feature of the 'solar atmosphere.'

Astronomers now unquestionably distinguished between the photosphere and chromosphere. It is clear what they are describing because of common reference to a layer of pink, appearing and then quickly disappearing after second contact and reappearing and re-disappearing briefly before third contact.

It is true that, in the immediate post-1869-eclipse literature, little was said in *print* about the chromosphere. An exception was, "The body of the Moon fully obscured the Sun, and around its edge was a crimson circle from which rays shot forth."

During the 1869 total eclipse of the Sun, the Moon covered the photosphere by definition, but, as always, temporarily left exposed the chromosphere alone as the Moon ingressed and then egressed the solar disk. The time during which the chromosphere was visible, but not 'drowned out' by the glare of the much brighter photosphere, admittedly was brief. All the same, it should not have been that difficult to see based upon telescopic brightness. Some observers admitted after the fact that their contact times were suspect due to misidentifying the photosphere with the emerging and disappearing chromosphere.

The chromosphere's extent was judged to be $1/60^{th}$ of the solar diameter, a reasonable value. Yet only Joseph Winlock referred specifically to spectroscopy of the chromosphere in 1869, Edward Curtis much later to photography of the chromosphere. Lewis Swift complained that he inexplicably failed to see it altogether.

Perhaps it is not surprising that little attention was given the chromosphere considering the magnitude of this total eclipse. Even if one stood on the centerline and saw an eclipse in which the geometric center of the Moon passed exactly over the geometric center of the Sun, it was a 'tight fit.'

The total solar eclipse of concern had a magnitude of 1.02 at most for the majority of observers. (This also is why the path width was not particularly broad.) Solar eclipse magnitudes can approach 1.1. In 1869, there were admittedly only seconds available during which to see the chromosphere without photospheric competition nor obstruction by the Moon.

More attention was given to the prominences, known of since at least 1706 but now considered to be extensions of the chromosphere. (Fortuitously, the 1869 solar-eclipse magnitude was such that they could be seen all around the cloaked photosphere near maximum eclipse.) Yet only starting with the 1842 eclipse were they well-documented.

At the 1851 total solar eclipse, the phenomenon was first ascribed to the Sun. The 1860 total-eclipse photographs proved that the chromosphere and prominences are real, and a solar feature, because sequential views showed the Moon covering them on one solar limb while exposing them on the opposite and *vice versa.*

Astronomers of 1869 attempted to give prominence lengths in miles; however, their values were dependent upon the physical size of the Sun, still not known to many significant digits. More useful were estimates of the prominences' apparent angular sizes. Typical values were $1/10^{th}$ of a solar diameter. This is reasonable, though much longer prominences have since been observed since.

Total-eclipse expedition members concluded that prominences were confined largely to the same low solar latitudes as sunspots. (Curtis went so far as to say that the prominences visible during this particular total eclipse of the Sun and the sunspots seen on surrounding days happened to be more plentiful in the Sun's Southern Hemisphere.) This was used as early evidence that both were related to the same phenomenon, which we now know to be local areas of intense solar magnetism on the Sun.

Prominences were not simple flames shooting radially away from the Sun. The 1869 recognition of *detached* prominences was significant. From their complex forms, astronomers eventually would realize that these features were plasma condensing and following local magnetic field lines.

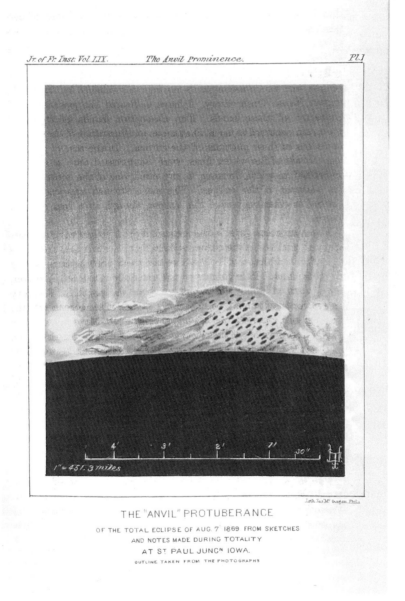

THE "ANVIL" PROTUBERANCE
OF THE TOTAL ECLIPSE OF AUG. 7 1869 FROM SKETCHES
AND NOTES MADE DURING TOTALITY
AT ST PAUL JUNCᴺ IOWA.
OUTLINE TAKEN FROM THE PHOTOGRAPHS

Fig. 19.2 Winthrop Gilman's "colored sketch" of what he called the "Anvil" protuberance, made immediately after the 1869 total solar eclipse. {Photograph by the author}

Much ink was spent describing the varied appearances of prominences. The reader may have noticed how often observers compared prominence morphologies to animals—real and imagined (*e.g.*, "stag's antlers" or "…the semblance of a horse's tail . . ."). This was done by individuals not heretofore known for great feats of imagination. Yet given the ill-defined borders of these features, explaining the shape of prominences as seen through the telescope is reminiscent of children picking out familiar figures in the clouds.

William Sheehan is well known as a historian and author in astronomy. He is also a licensed psychiatrist who has written about perception at the eyepiece. We agree that recognition of animals may be a survival mechanism dating back to when nomadic hunters first studied them carefully as a food source. This was the case over the hundreds of thousands of years in which we evolved and during which our brains were hardwired.

Ultimately, total-eclipse astronomers had little to show for their effort to find an apt visual metaphor for prominences. Shape not only varied from prominence to prominence but ultimately was qualified by the mind's filter of the eyewitness.

The detailed (and fanciful) morphologies reported in 1869 did allow prominences seen by one group to be matched and compared with those seen by another far away. It was learned that prominences do not vary—Edward Ashe's claim notwithstanding—over the time it takes the Moon to trace out a total-eclipse path. They are long-lived on the timescale of a total solar eclipse.

(The 1869 photographs of the prominences include one oddity: The base of the prominence sometimes seems to extend within the limb of the Moon. Some thought that this was due to their reflection upon the lunar surface. Instead, it quickly was realized that the phenomenon was illusory and caused by the movement of the limb relative to the Sun during the exposure.)

Depending upon who you asked, the prominences were:

"red"
"pinkish red"
"bright pink"
"pinkish"
"rose"
"pale rosy"
"ruby"
"brilliant deep crimson"
"carmine"
"copper-colored"
"fire-orange"
"yellow"
"yellowish white"
"greenish"

"reddish purple"

"gold"

Maria Mitchell provided a partial explanation:

> Any correct observation of color is, however, impossible. Beside the differ-
> ent perception of the eye, in its normal state, the retina cannot instantly lose
> the effect of the colored glass. I had just left an orange glass, and was quite
> insensible to that color; while one of our party who had been using a green
> glass declares the protuberances to be orange-red.

Astronomers had time (just barely) to make spectroscopic observations of the
chromosphere and prominences, confirming the presence of hydrogen in them.
This had been ascertained just the year before, during the first application of spectroscopy to a total eclipse of the Sun. However, the weather in 1868 had made
results from that year's eclipse expeditions suspect.

A trip to a chemist's laboratory shows that pure hydrogen, in a Bunsen burner,
glows some tint of red. (We do not know which 1869 total-eclipse astronomers
made such a trip.) Yet the devil is in the details: Minor contaminants can cause that
color to vary.

Were there any other chemical elements present in the prominences? The 1869
eclipse spectroscopists confirmed the answer to this question by detecting the
spectral signature of various metals. It seemed odd to them that heavy elements
rose high above the photosphere.

On the other hand, it did not take much of such trace elements: These species
have many spectral lines. Their spectra produced some of their brightest lines in
the spectral wavelength region astronomical spectroscopists examined during the
1869 total eclipse of the Sun.

While the spectra of multiple prominences revealed the same composition, sug-
gesting uniformity, it continued to be difficult to prove that every single one was
chemically the same and in the same proportion. Still, in the proud mind of
Winlock, "It may almost be said that nothing definitive was known of [the solar
chromosphere/prominences] until 1869."

What about the corona?

The corona was said to be "…perhaps the most thrilling effect in nature." Here
is where the most effort was focused. If 1868 was the total solar eclipse of the
chromosphere/prominences, 1869 was the total solar eclipse of the corona. Indeed,
1869 solar-eclipse observers in the United States were spared spending more time
with the prominences because of investigations made from India in 1868. Arguably,
they could have spent even less time, as Pierre-Jules-César Janssen (1824–1907;
École Speciale d'Architecture) and Norman Lockyer had since both discovered
how to observe prominences (spectroscopically) outside of total eclipse. This
news does not appear to have reached the USA yet.

Fig. 19.3 French visitor Étienne Trouvelot's (1827–1895; illustrator) lithograph based on a Joseph Winlock/John Whipple photograph of the corona. The original photograph was made at Shelbyville, Kentucky. 1876. {Astronomical Laboratory, Normal School of Science (Science Museum of London)}

1869 estimates of the corona's extent ranged from a minimum of 1/10th of a solar diameter to 2/3rds or more. For the first time, the distinction between an inner and outer corona was made consistent. John Eastman put the inner corona (called the "leucosphere," briefly) at about 1/30th. (We now know that the full corona extends for many solar diameters.) The variety of impressions is understandable given variables such as telescope aperture and dark adaptation.

Color descriptions for the corona included "soft white," "pearly white," "silvery white," "silver gray," "not bluish," "fine violet," "greenish-violet" and even "reddish. (Edward Pickering suggested this last one was due to contrast with trees and grass within eyeshot.) It was as if astronomers wished to avoid the conclusion that, to the human eye, the corona is simply, ghostly white.

This was the perfect total eclipse of the Sun at which to ask questions about the *nature* of the corona. It shown with great display, unlike at some other recorded total eclipses when it was reported to be even less extensive and distinctively shaped. And while earlier solar-eclipse photographs may have technically recorded the corona, this was the first time that photographs made during total eclipse included the never-before-imaged outer corona. For the first time, photographs of the 1869 corona did justice to it.

To begin with, the corona was expected to be a simple aureole. Attention was called to its sometime acircular nature only in 1851. In 1860, it was pointed out that coronal streamers (now known to follow the Sun's magnetic field lines) could be curved. Nonetheless, during the 1869 total eclipse of the Sun, the radially-asymmetric streamers of the faint, outer corona were so dramatic and complex that eclipse-observing astronomers felt that they had been misled by their predecessors.

Eventually, it was realized that both sets of observers were right: Sometimes the corona was round, at other times it was more pronounced at the solar equator and less so at the poles. Like so many other things associated with the Sun, this changing appearance of the corona follows the 11-year, solar magnetic cycle exemplified by sunspot number.

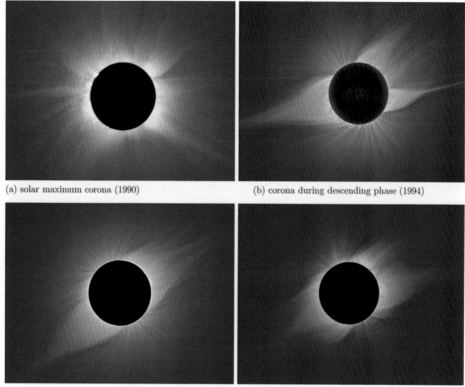

(a) solar maximum corona (1990)   (b) corona during descending phase (1994)

(c) solar minimum corona (1995)   (d) corona during ascending phase (1998)

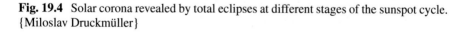

Fig. 19.4 Solar corona revealed by total eclipses at different stages of the sunspot cycle. {Miloslav Druckmüller}

Yes, streamers definitely were real: While any bright light seen against a dark background might produce an iridescence beyond its source, that the coronal streamers were unaffected by Simon Newcomb's screens proved that they were not the product of the eye and brain. Indeed, anybody who thought to rotate their head during totality, and note the unchanged corona, could demonstrate this to be true.

For centuries, people had longed to make the Earth's satellite inhabitable. One of the last to postulate air on the Moon was Stephen Alexander! Regardless, anybody who still held hope that the corona was, in fact, a lunar atmosphere, suffered the final blow in 1869. Total-eclipse photographs proved what visual observers already had come to realize, "that as the moon advanced, the corona was progressively covered," just as with prominences. It did not track with the Moon. Curtis assured those who thought that they saw in the photographs an 'atmospheric' glow over the Moon's silhouette that they were seeing a photographic artifact; it was not real. Charles Young swore that there was absolutely no evidence of a lunar atmosphere to be seen through his spectroscope, either.

Into the vacuum caused by the elimination of the lunar-atmosphere stepped Benjamin Gould. He resurrected an old idea that the solar corona was something of an *illusion*—merely an optical effect caused by the Earth's atmosphere.

Gould did not dwell on the *source* of the corona's light. A minority of astronomers who observed the 1868 total eclipse of the Sun and agreed with Gould ventured that the corona was photospheric light that somehow made its way around the obstruction of the Moon.

Sunlight scattered in the Earth's atmosphere is polarized. If the coronal light was not, that would serve to distinguish the two. At the 1868 total eclipse of the Sun, it was decided that the corona was polarized. There must be some particulate component to it.

The 1869 polarimetric observations, involving a greater extent of the corona, swung back and forth on the presence of polarization. So, this physical technique failed to decide the matter, after all.

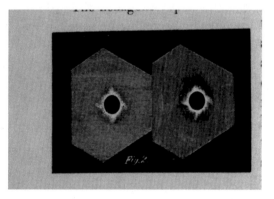

Fig. 19.5 Edward Pickering's drawing of coronal polarization patterns. {United States Naval Observatory}

The atmospheric theory was incredibly tempting. Weather phenomena produced various fuzzy circles about celestial orbs all the time. The idea was comfortable. That the corona of 1869 was *not* circular was what froze observers dead in their tracks. (And vexed more than a few.) What they were seeing flaunted two-and-a-half millennia of Pythagorean-aesthetic symmetry.

William Harkness entered the debate. He was the atmospheric theory's major detractor. He complained that the Moon's umbra is big: If the Earth's atmospheric brightness was affected by the eclipsed photosphere, without the existence of a corona, the sky should be exclusively *darkened*. Moreover, no terrestrial atmospheric physics accounted for the "trapezoidal" (now often called "butterfly") shape of the corona that all observers agreed upon.

The Iowa-based astronomers noted the difference in the appearance of the corona as seen from the relative lowlands of that state and that described as seen from a mountain peak in Virginia. Albert Myer, stationed at an elevation at which more air was below the observer and less above, reported a view of the corona that was seemingly superior to that of those on the plain. The weather was similar at both locations. Rather than inducing the corona, the atmosphere actually was a hindrance that interfered with viewing the outermost solar layer. The major cause of coronal light was decided to be a physical corona itself.

(While observers differed in describing the appearance of the corona, nobody went so far as to suggest that he or she and any other observer were not looking at the same thing. Better, photographs, taken at widely separate spots along the total-eclipse path, confirmed this. The biggest accomplishment of 1869 eclipse photography was *proof* that the corona was not a cis-lunar or local phenomenon; it was far away. The corona and its light were intrinsic to the Sun.

In the name of complete honesty, results of *spectroscopy* were a little more nuanced, though.

Any illumination originating in the photosphere would, it was thought, exhibit its absorption lines. None of the 1869 total-eclipse spectroscopists saw such lines. As a major contributor to coronal light, this mechanism was indeed a 'dead horse.'

These observations did not mean that they were not present—scattering of *some* photospheric light does take place in the corona—they just were not observed. (Since the 1930s, we know that this is because scattering particles—electrons—in the inner corona are moving so fast that these spectral lines are smeared out by the Doppler Effect.[3]) At first glance, a primarily continuous spectrum was a real impediment to understanding the corona.

Saying where most of the light from the corona does *not* come from (Moon, solar photosphere) is not exactly the same as saying that it is self-luminous. There remained resistance to a corona that was some sort of extended solar atmosphere. Had not the Comet of 1843 passed through it unscathed? (There really is much empty space within the corona.) This resistance extended to the proposition that it was self-luminous.

Despite such inertia—once all the data were collated—after the 1869 total eclipse of the Sun, the corona could be associated confidently with the physical Sun. It was

---

[3] For any wave source, its measured wavelength is lengthened when the source is moving radially away from the observer and shortened when moving radially toward the observer.

an outermost solar atmosphere, both illuminated by the photosphere *and* (surprisingly) self-luminous. This was established on two, independent grounds.

First, of course, there was the trapezoidal/butterfly shape of the corona. There was some attempt to correlate coronal streamers with different elevations on the lunar limb or with prominences, but these did not get very far. Streamers had nothing to do with either the Moon or (in any morphological way) the photosphere. This shape was observed because it was 'baked' into the corona (or, as we now know, the solar magnetosphere).

Second, and most convincingly, there was Young's and Harkness's—some historians list Harkness first, others Young—initial, high-resolution spectroscopy of the corona. It showed the presence of *emission* lines, too. Such lines should never be produced in the Earth's comparatively cold atmosphere: They always appear exclusively in the spectrum of a hot, low-pressure, glowing gas. In other words, spectroscopy pointed directly to a corona that was, for the most part, a rarefied, gaseous[4] 'atmosphere.'

A concession to a lesser, scattering part of the corona (outer) was explainable, too. A long-standing debate about the zodiacal light was coming to an end in the 1860s. (Zodiacal light is sunlight reflected from interplanetary dust, which fills the ecliptic plane of the Solar System.) The Solar System is a dusty place. Why not the vicinity of the Sun?

So, the corona was the Sun's outermost atmosphere. This was still a tough 'pitch' because of the difficulty in believing that an 'atmosphere' could extend so far from the body it enveloped. Also, planetary atmospheres cool with upward distance. Why was the corona hot enough to emit light?

Was the corona behaving like an aurora ("northern lights") on the Earth? For a brief time, a popular theory suggested that it was. The aurora was a fashionable scientific topic at the time, due to dramatic descriptions provided by seafarers on recent voyages of discovery to the far North.

At one time, such an analogy even was supported by Young. Later he disavowed it. On closer examination, the induced aurora spectrum was very different from that of the corona. Despite this, the theory made headlines; for a while, it overshadowed a more important conclusion about the Sun.

Any planetary analogy just did not work. Not only was the corona extremely thin, for some reason it achieved great *altitude*. At the same time, it reached great temperature. And its appearance could change! This was not how atmospheres were supposed to work.

Did the corona change over the time scale of a total solar eclipse? The coronal "streamers" certainly appeared dynamic.

---

[4] plasma

Fig. 19.6 *Aurora Borealis* by Frederic Church. 1865. "During the Civil War, the auroras [*sic*]—usually visible only in the North—were widely interpreted as signs of God's displeasure with the Confederacy for advocating slavery …" {Smithsonian American Art Museum} (the most famous American painter of the 1860s)

Here is an example of photography's value. While they do not print well, a few original plates show these features in the outer corona. Comparison of drawings made during the 1869 total eclipse suggested change in them over the time it took totality to move along the umbral path. The objective photographs show no such change. The apparent changes that appeared when comparing sketches were the result of the human element inherent in subjective artwork.

However, the corona does transition over a longer interval of time. The most recent total eclipse before that of 1869, which was well-observed by astronomers, was that of 1860. The year 1860 was a sunspot maximum year. The Sun was half-way through its eleven-year cycle. The corona reacts to the sunspot cycle such that it is most symmetrical at sunspot maximum and least at sunspot minimum.

During the next cycle, the year 1869 occurred after sunspot minimum had just passed, in March 1867. The result was a brighter but more asymmetric and less broad corona than astronomers had understood that they would encounter. Hence, the missed guesses at sky brightness and photographic exposure times. Hence, the trapezoid.

Even by late 1870, the corona already had changed in appearance once again: Simon Newcomb was disappointed at the total eclipse of the Sun that year, saying, "Instead of the gorgeous spectacle I witnessed in 1869, I saw only the most insignificant corona. . ."[5] The correlation of number of sunspots and length of coronal streamers would be used by later astronomers in the formulating the theory that most solar activity responds to the sunspot cycle.

Remember that the fact that the Sun had an extended outer atmosphere at all continued to distress some. Expected gaseous layering of increasingly less density that far from the center of the Sun was hydrogen was expected to have been far too thin to be seen, much less photographed in 1869. The fact that the glowing corona *is* visible so far from the rest of the Sun should be true only if, in the words of Simon Newcomb, its temperature is "many times 100,000°." This was certainly an impossible number in his view. (As we shall see, Newcomb's statement ["100,000°"] is ironically low.)

It took a while for science to realize it, but the spectroscopy of the solar corona in 1869 ultimately led to what solar astronomers Leon Golub and Jay Pasachoff refer to as "the single most outstanding problem in astrophysics for fifty years." Recall that those who first turned a spectroscope toward a total eclipse of the Sun (354 days earlier) identified emission spectra coming from the chromosphere and prominences. Most obvious was that produced by the element hydrogen, but there were those produced by other, heavier elements, too. No readily identified lines of any kind had appeared to them in the spectrum of the corona—admittedly, not their principal target. (Descriptions of absorption lines were spurious.)

In Burlington, Iowa, Young was able see three emission lines in the spectrum of the corona; each were later identified as coming from a different element. Two, including hydrogen, were the same as those seen in the Sun's lower layers. But one other line was unidentifiable! Harkness saw it, too, from Des Moines. (As we know, others tried to do spectroscopy at the 1869 total eclipse of the Sun, but with small instruments inadequate to the task.)

What caused the unidentified, comparatively bright, green, coronal line— seen no where else—superimposed upon the faint, continuous spectrum, all observed by both well-equipped eclipse spectroscopists? It was eventually decided that it belonged to common iron! (This was the only emission line that Harkness saw, but unlike Young, he was pointing his spectroscope as far out from the center of the eclipsed solar disk as he could, into the fainter and fainter corona, so as to avoid light contamination from other parts of the Sun.)

Young was unusual in that he had the physical training to interpret what he saw through his spectroscope. That does not mean that his interpretation was always right, though. The iron identification was wrong.

---

[5] The fact that he was looking through far from transparent skies in 1871 (quite different than in 1869) must bear part of the blame.

Shortly after the 1869 total eclipse of the Sun, Young conceded that, "…the absence of hundreds of other and more important iron lines from the coronal spectrum, and the difficulty of supposing the density this metal less than that of hydrogen, or of accounting for its presence in such quantities in the upper portions of the solar atmosphere, made it from the first highly improbable that this line could be due to iron." He had miscorrelated the line with that of normal, gaseous iron.

In Young's vision of the corona, more-and-more of the *lighter* elements should be prevalent farther from the center of the Sun (and less-and-less of the heavier ones). Yes, a few metals were identified in the photosphere, chromosphere, and even the base of prominences. That was not alarming: How conspicuous a line is and the abundance of the element that produces it do not necessarily go hand-in-hand. Even so, a really heavy element such as iron, so high above the bottom most layer of the Sun visible, bordered upon the absurd to Young. He was probably glad to see it go. Nevertheless, at least the iron theory did not require any new chemistry.

So, the real mystery began back home, after the solar eclipse. On Young's closer examination of his coronal spectrum, he could now see that there was really a *pair* of emission lines, instead of just a single line in the green. One did show the presence of a bit of iron, but he was now positive that the other, abutting it, was produced by *no known element*.

As an aside, it is interesting that Young might have seen the unidentified line appear before his very eyes earlier in the total eclipse than it did, and have longer to study it, but at the time he was looking for another phenomenon: He hoped to see the so-called reversing spectrum, predicted on theoretical grounds to occur when absorption lines suddenly appear to convert to emission lines, as the photosphere is eclipsed and the chromosphere becomes visible. (He did not see it in 1869 but did so at the total eclipse of the Sun visible in Spain the next year.[6])

Shortly after the total solar eclipse of 1868, Jules Janssen and Norman Lockyer independently identified a new line in the spectrum of the Sun. It, too, did not seem to correspond to any element. Boldly, they decided that they had discovered a new element, perhaps one not present on the Earth. Lockyer named it helium, after the Greek god of the Sun. Much later (1895), helium was discovered escaping from a mineral on the Earth. We now know that helium is the second-most common element in the Universe, after hydrogen.

Therefore, back in 1869, it was not out of bounds for Young to suspect that he, too, had discovered a new chemical element. This supposed element was named coronium.

Problems began almost immediately. Coincidentally in Spring 1869, Russian Chemist Dmitri Mendeleev[7] (1834–1907 Saint Petersburg University) had organized the chemical elements into the famous Periodic Table based upon common

---

[6] I have seen the reversing spectrum; it is very brief.

[7] In 1887, Mendeleev would see his own total eclipse of the Sun—from the altitude provided by a hot-air balloon.

Fig. 19.7 Newspaper headline touting (imaginary) applications of coronium. {*Times-Dispatch*}

properties. There was no place available for coronium! Yet Young and Harkness confirmed the presence of the coronium line in the coronal spectrum that they observed during the total solar eclipse of 1870. Meanwhile, geologists and chemists found no examples of coronium anywhere on or in the Earth.

Because coronium supposedly rose high into the solar atmosphere, it was thought to be very chemically light—lighter than hydrogen, already known to be diaphanous. (Recall the supposed problem when the prominent line was thought to belong to iron.)

But then New Zealand physicist Ernest Rutherford (1871–1937; Victoria University[8]) formulated his watershed model of the atom: a nucleus of protons and neutrons surrounded by 'orbiting' electrons at different orbital radii, with the number of protons indicating the type of atom. Possessing only a single proton in its nucleus, there could be no element 'lighter' than hydrogen.

The enigma of coronium was solved only between 1939 and 1943 by German astronomer Walter Grotrian (1890–1954; University of Potsdam), Swedish engineer Hannes Alfvén (1908–1995; University of Uppsala), and Swedish physicist Bengt Edlén (1906–1993; Lund University). After Rutherford, it was recognized that atoms could exist in different varieties—atoms with the same number of protons, and therefor chemically similar, but with different numbers of electrons. Atoms in which the number of protons and electrons do not match are called ions, and have a positive or negative electrical charge, depending upon whether they have a surplus or (more commonly) deficit of negatively charged electrons. Elements are ionized by, among other things, high temperature. The spectra of ions are slightly different than the spectra of their neutral relatives.

As astrochemists explored characteristics of ions raised to higher and higher temperatures, they observed the spectra produced by elements more-and-more ionized. With so many electrons (26 when neutral), iron could be terrifically ionized when very hot. There, in iron's spectrum—when ionized a whopping 13 times (thirteen missing electrons)—appeared the celebrity green line of 'coronium.' But only at incredible temperatures—millions of degrees!

Yet such an ion could remain with the corona for a long time, if the corona is even more rarefied than was realized by the likes of Young and Harkness (pressures equivalent to a laboratory vacuum), and the high-speed atomic collisions that might help neutralize these shredded atoms happen rarely.

Young was wrong that he was wrong. Coronium *is* merely iron, but not the iron with which he was familiar. Its coronal form is too exotic to be found naturally

---

[8] today the University of Manchester

anywhere on the Earth. The inevitable conclusion: The solar corona must be much, much hotter than any other visible layer of the Sun.

We are used to objects becoming cooler with distance from the center. Why is the corona to such a degree warmer compared to the Sun immediately below it? Thus, began a scientific puzzle of the twentieth century, which in the twenty-first still has not been solved to everyone's satisfaction.

* * * * *

In 1869, astronomers began to quantify the Sun. For instance, the solar corona was not merely fainter than the photosphere, it was much fainter—in proportion to the length of time required to record it on a photographic plate.

Another example, this time from the total-eclipse environment: Solar radiation was not just cooler during a total eclipse of the Sun, the change in temperature of the local air was measured on a thermometer. Perhaps most importantly, the 1869 eclipse provided a baseline of objective data that could be compared to that gathered at all future solar eclipses.

There would be more similarly studied total eclipses of the Sun to come. Nonetheless, 1869's provided the template for how total-eclipse expeditions would be conducted in the future.

I end this portion of the chapter with an individual who heretofore has been a figuratively parenthetical participant in our story. He was Homer Lane.[9] Recall that Lane was a career, District of Colombia, civil servant, who nonetheless floated between jobs after the Civil War before joining the United States Bureau of Weights and Measures in 1869. He had time for a vacation to observe a total eclipse of the Sun. To my knowledge, it was his only significant travel.

Simon Newcomb knew Lane as a member of a Washington scientific club. He described him as an:

> odd looking and odd mannered little man, rather intellectual in appearance, who listened attentively to what others said, but who, as far as I noticed, never said a word himself. Up until the time of which I am speaking, I did not even know his name.

But it was none other than Lane who figured out what the Sun really *was*. We do not know to what extent his epiphany came from witnessing the total solar eclipse; he had a modest record as an astronomer. Perhaps it was the result of a much longer intellectual campaign. Regardless, after his return to Washington, Lane published a scientific paper in the prestigious *American Journal of Science* that

---

[9] He is not to be confused with his contemporary, the famous professional wrestler Homer Lane (1828–). It is unlikely that someone from their time would have done so.

J. HOMER LANE

AUGUST 9, 1819—MAY 3, 1880.

Fig. 19.8 Homer Lane. {Yale University}

secured his name for posterity *vis à vis* the Lane-Emden Equation. (There is no "Newcomb Equation.")

As recently as the 1850s, it still was thought possible that the Sun is a cool, dense body with a luminous atmosphere. After the formulation of the rules of spectroscopy, it was acknowledged that the unseen bulk of the Sun could be hot and luminous throughout. It could be made of the three known states of matter: solid, liquid, or (dense) gas. Liquid was an easy 'sell,' because sunspots rotate around the Sun at different rates indicated a fluid photosphere. Gas was a greater problem for the imagination.

When we consider a sphere of gas, we mostly think of that in, for example, a low-density balloon. Such a toy is always shrinking or expanding. The Sun does not do this. Even Alan Fiala failed reliably to find its diameter change measurably over any length of time kept by humankind.

Lane showed how the Sun might vary greatly in density at different radii, and at the same time be entirely a gas. Such a Sun could be stable. No balloon. Each infinitesimal layer radially outward from the Sun's midpoint is balanced in place by 1) the pull of gravity (inward), at that distance, and 2) pressure as a function of temperature and density (outward), at that same distance.

Lane initiated solar modeling, in which a wholly gaseous Sun is very dense and hottest at its center. It becomes less dense and cooler in a rigorously prescribed way through the photosphere, so as to remain in (formally) hydrostatic equilibrium over intervals of deep time. Lane further described how the interior of the solar sphere would change as it radiated heat.

The conceptual road that Lane started out on ultimately led toward the answer to the great question asked about the Sun since time immemorial: What is its *source* of energy? It was to be found in the Sun's core, under extreme conditions that yielded direct conversion of mass to energy, the most efficient energy source known.

This mechanism was, of course, elucidated by Albert Einstein, born the year before Lane died. But it could be said that Lane, arguably the seemingly least likely figure in the story of the 1869 total solar eclipse to do so, was the person who made the greatest contribution, of those in attendance, to our understanding of the Sun.[10]

Charles Himes tended to exaggerate its scientific importance when he recalled the total solar eclipse of 1869. Still, as it incorporated aspects of astronomy, physics, and chemistry—at once practical, observational, and theoretical—he had a point when he reflected on the newly *interdisciplinary* nature of eclipse science. It was a vanguard of how science would be undertaken in the future:

> Always impressive, this [total eclipse of the Sun] had been thrust, by very recent progress in science, to an unusually prominent position. Never before had such an event promised so much of a revelation. No phenomenon shows more beautifully, in the means of its observance, the startling progress of science in the last decade, and the growing interdependence of its various branches. The seeming lines of cleavage, upon which the study of nature has been divided for the sake of convenience, grow fainter with every advance.

---

[10] In an 1869 lecture, Irish chemist Thomas Andrews (1813–1885; Queen's University of Belfast) presented work that supported Lane's theory.

20

… And What Happened after That?

"Providence has made of the pastoral State of Iowa one of the most important locations in the history of solar physics."

Sir Norman Lockyer, doyen of British physical astronomy, later wrote: "… never before was an eclipsed Sun so thoroughly tortured with all the instruments of Science." Also,

> The Government, the Railway and other companies, and private persons threw themselves into the work with marvelous earnestness and skill; and the result was that the line of totality was almost one continuous observatory from the Pacific to the Atlantic … 'there seems to have been scarcely a town of any considerable magnitude along the entire line, which was not garrisoned by observers, having some special astronomical problem in view.' This was as it should have been, and the American Government and men of science must be congratulated on the noble example they have shown to us, and the food for future thought and work they have accumulated.

What was it all about?

Significance for Astronomers: In 1869, astronomy was moving out into the galaxy: observing distant binary stars, variable stars, and nebulae. All the same, the nearby Sun still was largely a mystery.

The change was that new instruments finally had matured by 1869 to the point where they could be taken 'on the road' and yield useful measurements. This is particularly true of the new art of photography. The photograph was coveted because, in the words of frequent commentator Charles Himes, "Its unblanching

© The Author(s), under exclusive license to Springer Nature Switzerland AG 2023

T. Hockey, *America's First Eclipse Chasers*, Springer Praxis Books, https://doi.org/10.1007/978-3-031-24124-6_20

eye, with its highly sensitive and indelibly impressible retina, sees everything, and leaves no chance of slips of memory."

Over 240 images were made of the solar eclipse. By far, most of these were of the partial phases and of limited scientific value. Despite this, enough were of totality to prove that doing so would become routine in the future. It is noteworthy that eclipse-bound astronomers convinced some of the nation's best commercial photographers to flip their "Open" signs to "Closed" and accompany the observers (often near to clueless about photography themselves).

Photography was the natural culmination of description or drawing at the eye piece—Galileo would have understood its purpose. Meanwhile, though, it was recognized that science undertaken during a total eclipse of the Sun ought to include spectroscopy and other instruments of light analysis. This was the case even though no one in 1869 was trained in such techniques. All who undertook them were self-taught.

It is true that the attempt to comply often was done naively. Astronomers pointed handheld, stand-alone spectroscopes at the Sun and hoped to solve its riddles. Often their instruments were not constructed specifically for astronomical use but were meant for the laboratory bench. Many of these people appear to never have done spectroscopy before. Even among those better prepared, it is surprising that anybody except Charles Young (and William Harkness) managed to get useful spectroscopic data at the 1869 total eclipse.[1]

Ironically, it was classical astronomy—in this case, measurement of time and position by use of total solar eclipses—that faltered. Or at least it had reached its limit in 1869. Samuel Langley's "New Astronomy," on the other hand, began to flourish.

As a bottom line, all the discoveries about the Sun made in previous total eclipses of the Sun were verified in 1869. None of the discoveries made during that total eclipse were contraindicated in future eclipses.

One significant difference was that the 1869 observations were made under nearly ideal conditions: Remarkably, starting with Cleveland Abbe and working eastward, almost every expedition experienced clear and dry skies. Often at respectable elevation. References to 1869 added verisimilitude to observations made during other eclipses, in the near past or future, many of which were compromised, at least to some degree, by weather.[2]

---

[1] Upon his return from the total solar eclipse, Young commission George Clark to make for him a *thirteen*-prism astronomical spectroscope.

[2] William Harkness experienced *hail* during totality in 1870!

Fig. 20.1 (**a–d**) Typical cloud cover along the total-eclipse path in August. {National Oceanographic and Atmospheric Administration}

Fig. 20.1 (continued)

At the Margin of the Map

By 1869, eclipses transcended their original interest to government astronomers—that of a tool for navigation. Yet even then there was a notable exception.

Famed explorer of the West, Professor John (Wesley) Powell (1834–1902; Illinois Wesleyan University) and troop were attempting to navigate the Colorado River in the summer of 1869, the first white men to do so. But how far west was West? The white-water journey was perilous for men in an open boat, never mind a chronometer. They were fifteen-hundred kilometers from a telegraph pole.

Fig. 20.2 John Powell (on horseback) at the Colorado River. Other man unidentified. {United States Geological Survey Library}

The total eclipse of the Sun could supply Powell with this vital information. The moment of an eclipse plus the journeyman product of John Coffin, Simon Newcomb, William Harkness and the rest (the *Nautical Almanac*) could yield the correct time.

Powell probably cared little about the exotic observations other men were making on the day of the total solar eclipse. It was their routine day-to-day work in Washington that counted. The moment of first (or last) contact, compared to that predicted by the premiere *Almanac*, would provide a sufficiently correct time difference with which to pin down longitude.

But on Eclipse Day, the Powell Expedition happened to be in a canyon! The barranca walls mostly blocked view of the partially eclipsed Sun. So, Powell and his brother climbed the side of the chasm to the plateau above for a better look. Students of western American history recall that Powell did this without one arm, it having been lost in the Civil War. Surely some sort of record for shear physical effort to observe an eclipse was set that day.

The Powells mostly saw clouds and rain. But it was enough. The 1869 solar eclipse helped fill in the map of the one yet uncharted region in what was to become the United States of today. The Powell Expedition started into the Grand Canyon four days after the eclipse.

The outcome of the various total-eclipse expeditions in 1869, ignoring for the moment scientific productivity, demonstrated that American astronomers could organize multiple, advanced ventures into unknown country. Doing so was acknowledged at the time to be practice for the much more-ambitious, global transit-of-Venus expeditions that would be launched in a few short years.

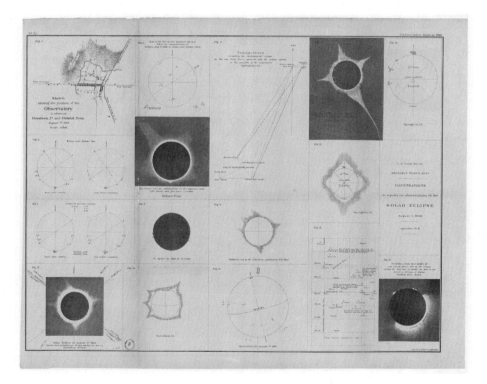

Fig. 20.3 Benjamin Peirce's summary of United States Coast Survey activity along the 1869 total-eclipse path. {National Oceanographic and Atmospheric Administration, Historical Map and Chart Collection}

A transit of a planet in some ways is similar to a total solar eclipse. In this upcoming and much-rarer event, the planet Venus was to pass over the disk of the Sun. Such a phenomenon can be viewed from greatly disparate locations on the Earth. The difference seen from various points of view (in where the planetary body of much-smaller apparent size transits the Sun and at what exact time) is used to calculate the distance between the Earth and Sun through the geometric optical property of parallax.

The Earth-Sun distance was the 'grail' of astronomers since the seventeenth century. It allowed calculation of the distances to all of the other planets—the scale of our Solar System—and even nearby stars. (Joseph Winlock explicitly stated that his 1869 total-eclipse images were a dry-run test for using photography at the transits of Venus.)

American astronomers now felt that they were up to the task of joining their European brethren in this 'noble' task. Simon Newcomb was rewarded for his 1869 total-eclipse expedition's success by appointment as secretary of the international Transit of Venus Commission, which shortly thereafter began to make its ambitious plans for the 1874 event. But that, as they say, is another story.

Table 20.1 1869 Total Solar Eclipse Observers Who Participated in Transit-of-Venus Expeditions. (Cross-referenced with Dick 2003, Tables 7.3 & 7.5.)

1874	1882
Davidson	Davidson
Hall	Eastman
Harkness	Hall
Peters	Harkness
Ranger	Newcomb
Watson	Rock
Young	Very

Nevertheless, there is a scientific connection between the transits of Venus and the 1869 total eclipse of the Sun, too. Once the distance to the Sun was known, the angular size of the Sun yielded its physical size through simple trigonometry. In anticipation, Charles Peirce, much later, back at Harvard, attempted to measure the apparent size of the Sun from the photographic plates taken at Shelbyville, Kentucky.

Charles Schott did the same with the Springfield, Illinois, images. (Schott has been described as, "the Survey's greatest computer …"). Neither Charles Peirce nor Scott could be certain that the plates were oriented exactly orthogonal to the direction of the Sun at the time of exposure. In other words, the solar image may have been foreshortened slightly.

Moreover, both of the telescope-plus-cameras that had produced the images under study were chromatically dependent. It was the old problem of different colors brought to slightly different focuses. This left the solar limb less sharp than it needed to be for exact measurement. Neither were very successful at their task.

Significance for Technology: I previously described the importance of the railroad in making modern total-eclipse expeditions—now with portable laboratories and laboratory technicians accompanying them—possible. Here, I discuss three other technologies that affected observation of the total eclipse itself.

Telegraphy. In 1869, a judge oversaw an entire criminal trial—in a 'virtual courtroom'—conducted over the telegraph. It was the original Zoom. Clearly, by that year, the telegraph as a communications tool was here to stay[3] (until superseded by still more advanced electrical technology). On the other hand, its use in science, specifically the observation of a total solar eclipse, was new.

The telegraph proved useful during the 1869 total eclipse of the Sun in several ways. The timing of a total eclipse, undertaken far from an observatory, was made more exact using telegraph signals. Geographic coordinates were established to the required level of accuracy by the exchange of such signals. Telegraphy was used to set the fine chronometers lugged from Washington (which amazingly worked in spite of midwestern dust, heat, and weather).

In fact, developments in electrical communications and high-tech clocks now had surpassed the ability of astronomers to make the most use of them. The precision of contact timing became limited by human ability to decide when the Moon first formed a 'nick' in the extremely bright solar disk (first contact) or finally left it (fourth contact). The fact that the limb of the Sun is darker than its mid-disk, and that the limb of the Moon is not smooth, make this a less-than-obvious act within the brain. Similarly, when the last glimmer of photosphere (second contact), and the first (third contact), occurred was equally a judgment call and a function of personal equation.

About half an hour before totality in 1869, Young realized that in the future he could use the spectroscope's revelation of the appearance and disappearance of spectral lines to establish contact times, instead of estimating those times with just an optical telescope.[4] The method also could be used to accurately measure the thickness of the chromosphere. He proposed to try the technique at the Transit of Venus, too.

Nobody in, perhaps, Bloomington, Illinois, thought to telegraph someone in, perhaps, Bristol, Tennessee, and warn them what to expect. Or, maybe jury-rigged wires were insufficient for complicated messages. Or western astronomers were too busy and lacked enough time post-totality to alert eastern astronomers. Or such a deed was thought to compromise scientific objectivity. Or they simply did

---

[3] By 1869, recent mergers and acquisitions had resulted in the entire technology becoming synonymous with the Western Union Telegraph Company.

[4] He did not know that this technique had been proposed by French astronomer Hervé Faye (1814–1902; École Polytechnique) earlier in the year.

Fig. 20.4 Charles Peirce. {Houghton Library, Harvard University}

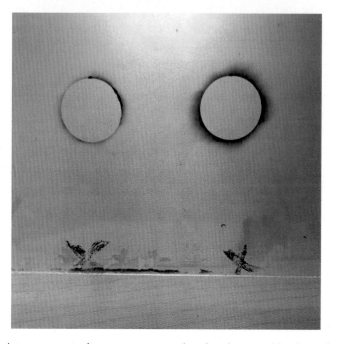

Fig. 20.5 Astronomers took measurements using the photographic plates themselves, which were in negative. For example, these images were made by Joseph Winlock/John Whipple at Shelbyville, Kentucky, and analyzed by Charles Peirce in Washington. {Harvard College Observatory}

not know what to say. I do not know; the idea is not mentioned even as a possibility in any of the literature or archived documents. This strategy awaited the 1878
total eclipse of the Sun.

Another eclipse application of the telegraph was in the rapid dissemination of
journalistic information. Americans in the populous cites of the east coast could
read in their newspapers about the total solar eclipse as seen in, for instance,
Vincennes, Indiana, or Greensboro, North Carolina, the day after the event.
Historians-of-astronomy Stella Cottam and Wayne Orchiston suggest that this
solar eclipse helped restore the public's faith in astronomers, who had predicted a
great meteor shower for 1866—which did not appear.

Photography. The factoid about the 1869 total eclipse of the Sun that often
appears in writing, if there is anything at all, is that this eclipse was when Baily's
Beads originally were photographed (by Charles Himes). No matter—the bigger
point is this: For the first time, photographs of the Sun were made of a quality such
that they might be scientifically useful, and not just a chemical stunt. Proof of this
is that they were referenced by astronomers who did not witness the total eclipse
themselves. This was made possible by those observers willing to travel with a
chemical laboratory and put up with the juggling act that was the first such use of
wet collodion photography.

To be honest, *most* of the photographs taken of the 1869 total solar eclipse were
not that great. Despite all efforts to calculate and experiment before the eclipse,
there remained the ambiguity of correct exposure time. One only could bracket
exposures in the hope of hitting the correct one for a given circumstance during
the eclipse. These include, for instance, 1) partial phases, which incorporate the
profile of the lunar limb and sunspots before second contact, 2) Baily's Beads at
second contact and the much fainter prominences moments later, and 3) the corona
during totality. No one could time the beginning of a photograph to capture a
single desired moment in the motion picture show that is an eclipse. Exposure
requirements were too long. For example, one dare not start an exposure of the
corona late enough such that it drifted past third contact. Edward Curtis complained that there was no suitable time to take a picture of the chromosphere. That
he could make both prominences and the corona appear at least once in the same
photograph was through great serendipity.

One could argue that specialization should have taken place. One photographic
expedition might have ignored the partial phases so as to be fully prepared for
totality. One might have placed 'all its money' on first and third contact. Another
might have been 'all in' for prominences. Or the corona. This took place to some
extent. But how do you convince persons, some longtime professional photographers, to submit themselves and their cherished equipment to the vicissitudes of
long-distance odysseys and then hold back on a probable once-in-a-lifetime
opportunity? That opportunity was the chimera of a photograph that somehow
captures the full effect of the phenomenon. It was not in their blood.

HARPER'S WEEKLY

JOURNAL OF CIVILIZATION.

VOL. XIII.—No. 661.]   NEW YORK, SATURDAY, AUGUST 28, 1869.   [SINGLE COPIES, TEN CENTS. $4.00 PER YEAR IN ADVANCE.

Entered according to Act of Congress, in the Year 1869, by Harper & Brothers, in the Clerk's Office of the District Court of the United States, for the Southern District of New York.

SOLAR ECLIPSE, 1869.

TOTAL eclipses of the sun are very rare. HALLEY computed in 1715 that up to that date not one had occurred in London for a period of 575 years. And since that date London has not been favored with this singular phenomenon. In this country since 1834 no total eclipse of the sun, until the recent one, has occurred, and there will not be another during this century.

The total eclipse of this year was visible along a track about 140 miles wide, and more than 6600 miles long. When this track is laid down on a map, throughout its whole extent, it looks like a narrow ribbon, stretching across North America and a portion of Asia. It begins in Siberia, where it takes a northeasterly course, till it crosses a little south of Bering Straits, after which it turns its course southeasterly, traversing portions of the Territory of Alaska, thence into British America, and through Montana, Dakota, Nebraska, Iowa, Missouri, Illinois, Indiana, Kentucky, Tennessee, and North Carolina. It ends in the Atlantic Ocean, off the coast of the last mentioned State. On the central line of this track the total obscuration of the sun lasted for a period of from two and one-half to nearly four minutes.

To various points along the line of total eclipse scientific parties went to make observations—some sent by the Government, and some by private enterprise. But the principal point to which the attention of these parties was directed was the corona, observed in previous eclipses, and about which there is much curiosity among astronomers. We now know that the sun is surrounded by an atmosphere of flaming gas, which is the principal source of the light he sends us. This incandescent atmosphere is called by astronomers the sun's photosphere. During a total eclipse of the sun, his whole sphere, including the photosphere, is covered and hid from view by the interposed body of the moon. Nevertheless, at the very crisis of the central eclipse, there is seen a brilliant white corona or halo around the moon's dark body, like that which painters place around the heads of saints. In this corona, moreover, are frequently seen red or rose-colored projections, of irregular form and position, around the disk. The explanation usually given of the white corona is, that it betokens a transparent, non-luminous atmosphere, extending beyond the photosphere or luminous atmosphere, analogous to our own atmosphere. The white light of the corona is accounted for by the reflection of that of the photosphere, very much as our own evening and morning twilight is produced by the reflection of the solar rays in our atmosphere. The irregular red masses seen projected into the white corona may prove to be immense volumes of thin cloudy smoke, or solid vaporous particles precipitated from the hot gaseous atmosphere in which forms the corona.

Another question to be settled, if possible, was whether there is a planet between Mercury and the sun, as LE VERRIER supposed there was, on account of extraordinary perturbations of Mercury not otherwise explained.

The corona, or luminous ring around the moon in a total eclipse, was observed at Montpellier, in France, 1706, and at London, by HALLEY, in 1715. In the latter case the red protuberances were also observed. In 1724, MARALDI observed, for the first time, that the luminous corona was not concentric with the moon. At the beginning of the eclipse it appeared larger on the eastern than on the western side; at the end, on the contrary, it was larger on the west than on the east. The northern border was also somewhat larger than the opposite. The importance of these observations was that they proved the corona to be concentric with the sun instead of with the moon; that it is a phenomenon closely connected with the sun's physical constitution. Important additions were made to these observations during the total eclipses of 1778, 1806, and particularly in 1842. On the 8th of July in the latter year, ARAGO and other astronomers observed the luminous corona in all its splendor. The distinguished French astrono-

THE SOLAR ECLIPSE, AUGUST 7, 1869.—HARVARD ASTRONOMICAL EXPEDITION MAKING OBSERVATIONS AT SHELBYVILLE, KENTUCKY.
[PHOTOGRAPHED BY J. A. WHIPPLE.]

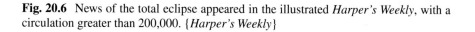

FIGURE 1.   FIGURE 2.   FIGURE 3.   FIGURE 4.

FIGURE 5.   FIGURE 6.   FIGURE 7.   FIGURE 8.

SOLAR ECLIPSE, AUGUST 7, 1869—PHASES OF THE ECLIPSE, AS SEEN AT SHELBYVILLE, KENTUCKY, FROM THE BEGINNING TO THE POINT OF TOTALITY.
[PHOTOGRAPHED BY J. A. WHIPPLE.]

Fig. 20.6 News of the total eclipse appeared in the illustrated *Harper's Weekly*, with a circulation greater than 200,000. {*Harper's Weekly*}

Still, there were so *many* photographs taken, one or more were *bound* to be satisfactory. And, indeed, photography surpassed the eye in one respect: The corona in Whipple's images extended even farther than in any previous documentation by naked-eye observers.

Telescopes. I have one more note about technology. No fewer than 36 telescopes, of 3-inch aperture or greater, were pointed at the 1869 total solar eclipse. (All but two had to be transported for the occasion.) At least ten of these were made by Alvan Clark. The solar eclipse was the first of any such astronomical event for which Clark telescopes were used extensively.

The business of Alvan Clark & Sons was already in ascendancy. Alvan Clark had received the endorsement of one of England's foremost amateur astronomers (Royal Astronomical Society gold medalist, Reverend William Dawes [1799–1868; Nonconformist]). Clark had won awards and successfully ground a commissioned 18-1/2 lens. The company supplied optics for the North in the recent war; a modicum of political payback was called for afterward.

Fig. 20.7 Alvan Clark (center) and sons, George and Alvan G. *Circa* 1870. {Lick Observatory}

But sales really took off after 1869. Clark & Sons became the USA's foremost telescope maker, and the quality of its instruments resulted in their becoming collector's items today.

The publicity of so many Clarks in use at the 1869 total eclipse of the Sun surely helped. Astronomers of the day usually did not spend much time discussing their telescopes with one another, but they liked to extol their Clark-made instruments. The total eclipse also revealed a shortage of portable telescopes, which Clark & Sons was happy to help eliminate.

The act that may have sealed the company's reputation was undertaken by Newcomb. Upon his return to Washington, he was given the perk of ordering a new signature telescope for the Naval Observatory. Congress had provided overdue funding. Newcomb had seen a lot of Clark & Sons achromats at work during the total eclipse of the Sun. As described previously, he was no virtuoso of the telescope. Nonetheless, he knew a good buy when he saw one. The Clarks' talent was under-valued he felt. Why put up with the price, tariffs, and logistical complications of ordering from abroad? Explained Newcomb,

> … [Clark's] genius in this direction had not been recognized outside of a limited scientific circle. The Civil War had commenced just as he had completed the largest refracting telescope ever made, [but] the excitement of the contest … did not leave our public men much time to think about the making of telescopes.

Alvan Clark's place in history was assured when, for $50,000, he produced the 26-inch refractor of the USNO. It was the largest telescope in the world for a decade.

For Science: The American total-eclipse expeditions of 1869 did not go unnoticed in Europe, the 'VIP club' for world science. German chemist Hermann Vogel (1834–1898; Technische Hochschule of Berlin), who spent his career improving photography, had this to say:

> The report of the American expedition for observing the recent solar eclipse, is being discussed in scientific and photographic circles with the liveliest interest.
>
> You have carried out a great scientific undertaking on the grandest scale, and the results are correspondingly great. The combined great powers of England, France, North Germany, and Austria succeeded in arranging two telescopes for the observation of the eclipse of 1868. One instrument was furnished by England, the other by North Germany. The other countries ignored photography as a means of observation entirely. Your expeditions, on the contrary, were furnished with about twenty of the finest instruments arranged for photographic observation, and hundreds of photographers and astronomers were engaged on them simultaneously. It really was a photographic battle crowned by a complete victory.

This imposing demonstration in the realm of science, surprises and grati-
fies us the more, as, generally, your countrymen are accused of caring only
for material matters. Your successes will, undoubtedly, give to celestial photography a new impulse, and refute many of the objections that are now
advanced against it.

Hundreds of people *observed* the 1869 solar eclipse. Countless others beheld it, of
course. Thus, it was the biggest simultaneous scientific enterprise in the United
States up to that time.

They did this even though the 1868 total eclipse of the Sun had yielded few
definitive answers. Nearly all who observed that total eclipse had invested less
time and effort in doing so than those attending the 1869 eclipse.

Scholar Alex Soojung-Kim Pang, in his otherwise excellent *Empire and Sun:
Victorian Solar Eclipse Expeditions*,[5] contrasts British and American scientific
approaches to total eclipses of the Sun: "American expeditions, in contrast, were
local affairs, planed, funded[,] and outfitted by individual observatories and colleges, or by astronomers borrowing instruments and traveling on small grants
from various institutions or agencies." This was not true.

The herculean effort of 1869 was made possible by the first-ever federal government investment in a strictly scientific undertaking. More famous expeditions
proceeded it, but there always was the hope of economic benefit. British astronomers were amazed: To them, government funding meant supply of a ship. That
was all. This American precedent was the major political outcome of this total
eclipse of the Sun.

Funding solar-eclipse expeditions was ultimately sold to Congress on patriotic
grounds: Britain was able to send an expedition all the way to India for the previous year's total eclipse of the Sun and would expect their American counterparts
to go to similar lengths for a total eclipse on their own soil.

There also was the vital assistance of private companies. The railroads and
telegraph were supported by the government by, for instance, the provision of
right of ways. It would be easy to conclude that these up-and-coming industries'
support for civil eclipse activities was *quid pro quo* for favorable legislation.

Historian of public science Marc Rothenberg reminds me that this conclusion
would be based upon the more adversarial relationship between the fed/state and
private sectors that exists today. In 1869, the political, military, and corporate
leaders of the USA all knew each other personally. They were neighbors. They
had gone to school together. They had fought the Civil War together. Favors passed
routinely between them. The publicity generated by making possible the showy
total eclipse expeditions was in the best interest of all involved. It was an easy deal
to make. American science would not forget this.

---

[5] Stanford, California: Stanford University Press. (2002)

Why *three* separate government expeditions? (It did not happen again.) I have failed to answer this question satisfactorily. That both the USNO and USCS ended up in Des Moines, Iowa, suggests lack of coordination. Simple competition? More data is better data?

Ultimately, the motivation is not of *scientific* importance. After it was all over, everybody concerned got one of several, thick, well-written reports on the total eclipse of the Sun for the bookshelf. The downside of much of the science coming out of the 1869 total eclipse appearing in government reports is that said reports take a long time to publish. The immediate influence of, and long-term historical recognition of, this eclipse might have been greater if its scientific dissemination had been different.

The reader no doubt noticed something about the previous chapter. There were notable exceptions; however, even while contemplating what they saw while pre-paring words to put into print, most astronomers who observed the 1869 total eclipse of the Sun did not include a great many *interpretations* with descriptions of their research. The Baconian tradition was affirmed.

Interpretation was left, in many cases, to others. Or, at least, to a later time. Richard Cutts explained: "In the preceding report I have confined myself to a mere statement of facts and description of the phenomena as they appeared to me, deeming any discussion of theories or conclusions by any separate observer to be both inappropriate and unsafe until all the proof and evidence collected … shall be brought together, criticized, and compared."

Still, the work of the 1869 eclipse-expedition astronomers verified a model for the Sun still in use today: a dense, self-luminous body surrounded by a hot atmo-sphere. It is one that follows all the physical laws acting on the Earth yet to which all life on our world is tied.

Finally, if one is to pick a single year, 1869 marked the full professionalization of astronomy in America. And, perhaps, the beginning of cutting-edge science, as a professional endeavor, from America in general.

This 'licensing' came at a nostalgic cost. The year's total eclipse of the Sun was also the 'last hurrah' for the Grand Amateur in astronomy, at least in the USA. Thereafter, such observers could not successfully compete with the profes-sionals in terms of equipment and access to often mountaintop, permanent obser-vatories. No longer could the Paines, Gilmans, and Blickensderfers assume that their reports would be published side-by-side with those of the professionals.

The big takeaway? For me, it is the reminder (if such a reminder is needed) that science—like everything else—evolves. Yet the pace is not a steady one. I have given a great deal of attention to state-sponsored eclipse expeditions, those sent by the kinds of institutions stereotypically slow to change. But they do change.

Astronomy was about measuring position and time for millennia. The invention of the telescope augmented that task but did not immediately change it fundamen-tally. Then, suddenly (on the time scale to which I am referring), astronomers were willing to replace their eye and attach instruments to the ends of their telescopes in order to collect data outside of the normal four dimensions. Spectroscopy. Polarimetry. The rudiments of photometry (which heretofore had not made much progress since the days of Hipparchus of Rhodes).

What did those willing to make the transition think? What were their conversations like with those who preferred traditional measurements, or, perhaps, pictures? Was there a continuum of opinion on how one should spend the brief minutes or seconds of a total solar eclipse? Ah, to be a fly on the wall at the Brazelton House Saloon!

* * * * *

It cannot be simple coincidence that a total eclipse of the Sun often coincided with a turning point in the métier of those who observed it. Of course, *post hoc, ergo propter hoc* does not always apply: Sometimes the total eclipse precipitated a change in situation, for others a change in circumstances allowed participation in observation of the eclipse.

So, what happened to those who observed the 1869 total eclipse of the Sun? The age following each name below is that of the individual at the time of the total eclipse (with potential for one year's error due to significant digits). Note how many were early in their careers at the time.

- Two years after the 1869 total solar eclipse, Cleveland Abbe (age 30; Cincinnati Observatory Expedition) became chief scientist at the fledgling United States Weather Service under Myer.
- Robert Abbe (age 18; Cincinnati Observatory Expedition), who accompanied the longest eclipse expedition by land, was the youngest member of any such expedition. His peregrinations, which started with a prairie schooner, culminated with an airplane flight across the English Channel late in life.
- Stephen Alexander (age 62; NAO/Franklin Institute Expedition) went back to university teaching. He was at Princeton for nearly fifty years.
- Edward Ashe (age 55; Canadian Expedition) continued his mid-career interest in the Sun. His 1867 theory of sunspots proved to be incorrect. After retirement, Ashe was succeeded at the Quebec Observatory by his son.
- Frederick Bardwell (age 37; USNO Expedition) joined the University of Kansas faculty the month after the 1869 total eclipse of the Sun. He headed to Colorado in order to observe the next American total eclipse but became terminally ill and had to turn back. He was not yet 50.
- Jacob Blickensderfer (age 53; independent) became chief engineer for several railroads. He retired to a grand house he had built for his wife, but she died the night before they were to move in.
- The rest of John Coffin's (age 53; NAO/Franklin Institute Expedition) career as superintendent of the Nautical Almanac Office was unremarkable: Upon his retirement in 1877, he was replaced by Newcomb.

- Alfred Compton (age 34; Cincinnati Observatory Expedition) was from New York. Therefore, he sets an eclipse record for traveling the longest in order to claim his time in the umbra—without the use of a ship.
- Edward Curtis (age 31; USNO Expedition) turned his interest in photography from the telescope to the microscope, developing techniques for use in medicine. Through one of those wonderful coincidences that scholars love, Edward Curtis [1868–1952] was one of the great photographers of endangered Native-American culture in the Old West.)
- George Davidson (age 44; USCS/HCO Expedition) discovered a vast deposit of iron ore in Alaska. He became the most famous scientist on the United States's West Coast. The highest of San Francisco's fabled Seven Hills is named Davidson.
- John Eastman's (age 33; USNO Expedition) lifetime preoccupation was preparation of the USNO's star catalog.
- Winthrop Gilman (age 61; independent) liked what he saw in Sioux City, Iowa: He developed property there in a neighborhood still called Gilman Terrace.
- Edward Goodfellow's (age 41, USCS/HCO Expedition) eyesight was slow to recover. A modern reading of his obituary points to possible Major Depressive Disorder. He died of "accidental asphyxiation."
- Benjamin Gould's (age 44; NAO/Franklin Institute Expedition) astronomical career was in doubt before the 1869 total eclipse of the Sun. Notwithstanding, he successfully 'jump-started' it in 1870, when he left the USA for Argentina. There he made the observations for and created the—at the time—definitive catalogue of South Celestial Hemisphere stars. (His assistant was Miles Rock.)
- Asaph Hall (age 39; USNO Expedition) would be in charge of the world's largest refracting telescope for sixteen years; he used it famously for his observations of Mars. After retirement from the USNO, Hall III taught at Harvard. His son, Asaph Hall IV (1859–1930; Detroit Observatory), held the observatory directorship once belonging to James Watson.
- William Harkness (age 31; USNO Expedition) championed the use of photography for the 1882 transit-of-Venus measurements. Based on the resulting images obtained from around the world, he calculated a value of the Sun-Earth distance equal to 149,300,000 kilometers—the distance currently in use is 149,600,000 kilometers. Harkness succeeded Newcomb as head of the Nautical Almanac Office, while
- Julius Hilgard (age 44, HCO/USCS Expedition) became superintendent of the United States Coast Survey.
- Charles Himes (age 31; NAO/Franklin Institute Expedition), enamored of the camera, used photography as an instructional tool during his long, uneventful, college-teaching tenure.

- On the other hand, things got so bad between Gustavus Hinrichs (age 33; University of Iowa) and his colleagues—they described him as "volatile, abrasive, and sometimes indictive"—that the state legislature had to intervene. Afterward, with an assignment including less teaching and more time for research, Hinrichs identified and named the meteorological phenomenon that devasted Iowa in 2020, the derecho.[6]
- George Hough (age 32; Dudley Observatory Expedition) left the Dudley directorship to become Director of the (easily confused) Dearborn Observatory.
- Charles Irish (age 35; University of Iowa), was appointed head of the United States Bureau of Irrigation and Inquiry in the second Grover Cleveland administration.
- Homer Lane (age 49; HCO/USCS Expedition) was elected to the National Academy of Sciences, along with other 1869 total-eclipse participants, but never became part of the Washington scientific elite.
- Samuel Langley (age 34; USCS/HCO Expedition) invented the bolometer. He became Secretary of the Smithsonian Institution. He also narrowly lost to the Wright Brother in the race to achieve heavier-than-air flight.
- Lone Dog (independent) was the last keeper of a Yanktonai Winter Count.
- Charles Marsh (age 34; independent) served in the Illinois State Senate. James M'Clune (age 50; NAO/Franklin Institute Expedition) returned to Haverford, Pennsylvania, following the total eclipse of the Sun, and slips from sight.
- Demonstrating the importance of networking, Henry Morton offered his solar-eclipse companion Alfred Mayer (age 32; NAO/Franklin Institute Expedition) a professorship at the new Stevens Institute of Technology. Mayer accepted. Sadly, his eyesight failed, and he did not pursue astronomy.
- Maria Mitchell (age 51; Vassar College Expedition) was not invited to join any of the Newcomb transit-of-Venus expeditions. Mitchell's students during her long teaching career included such luminaries as Antonia Maury (1866–1952; HCO), Caroline Furness (1869–1936; Vassar College), and Margaretta Palmer (1862–1924; Yale University).
- Henry Morton (age 32; NAO/Franklin Institute Expedition) was known for teaching astronomy with a set of lantern slides depicting every aspect of a total solar eclipse. He eventually became president of the Stevens Institute of Technology. In the nineteenth century, one could not underestimate the prestige of his appointment to the United States Lighthouse Board.
- Albert Myer (age 40; independent) came down from his eclipse-watching mountain to found the Army Signal School at what is now named Fort Myer.

---

[6] Perhaps the lyric "only the good die young" applies: Hinrich lived to tell of the 1869 total solar eclipse for 53 more years.

Under his guidance, meteorology became a significant part of the Signal Corps's mission. (Myer's nickname was "Old Probabilities.")

- Karl Neiman (age 40s; Chukotka Expedition) continued to explore Siberia and only returned to Estonia after two decades in the far East. Back home, he married his fiancé; she had waited twenty years for him.
- As Superintendent of the Nautical Almanac Office, Simon Newcomb (age 33; USNO Expedition) continued to quantify the motions of the Solar System. If Newcomb indeed wished to immortalize himself with the discovery of an astronomical object—two 'quick and dirty' searches for Vulcan not withstanding—he never really made the effort to look for one. Newcomb became first president of the fledgling American Astronomical Society in 1899.
- Cursed with a famous family name, Robert Paine (age 34; independent) spent his life in another kind of shadow: that of his famous forbearers.
- Most mathematicians do their best work while young. Benjamin Peirce (age 60; USCS/HCO Expedition) published his greatest contribution to mathematics, linear associative algebra, not long after the eclipse—at age 63.
- Charles Peirce (age 29; USCS/HCO Expedition) worked on and off at the USCS for over thirty years. Regardless, Peirce primarily devoted his mind to philosophy. Called the Father of Pragmatism, today he is considered one of the most important among American philosophers.
- Christian Peters's (age 55; Litchfield Observatory Expedition) behavior at the 1869 total eclipse of the Sun might seem to indicate that he was not a fan of new physical techniques in astronomy. But Peters was no luddite. For instance, he was one of only three Americans to participate in the 1887 Paris Conference that led to the audacious, international plan for photographing the entire Celestial Sphere: the *Carte du Ciel*.[7]
- The young Edward Pickering (age 23; NAO/Franklin Institute Expedition) may have appeared a bit indolent in his up-scale hotel room; the older Pickering was far from it. He replaced Joseph Winlock upon the latter's death. As HCO Director, he organized astrophysics on an industrial scale. Indeed, as the influence of the United States Naval Observatory waned, under Pickering the Harvard College Observatory became the preeminent astronomical institution in the USA.
- Truman Safford (age 33; USNO Expedition?) went broke. He landed well, though: prestigious Williams College, with a smaller telescope but for more money.
- Charles Schott (age 43; USCS/HCO Expedition) never became a star in the history of the USCS. He was adroit with compasses and did shine briefly as the United States' representative to the International Conference on Terrestrial Magnetism.

---

[7] It was never completed.

- Arthur Searle (age 31; USCS/HCO Expedition) was acting Director of the HCO after Winlock died. He joined the Harvard faculty and became an expert on zodiacal light.
- Even though he had no undergraduate degree, Lewis Swift (age 49; Dudley Observatory Expedition) was awarded a PhD. by the University of Rochester. Besides comets, his discoveries eventually included more than a thousand nebulae, star clusters, and galaxies.
- John Van Vleck (age 36, NAO/Franklin Institute Expedition) served as Acting President of Wesleyan University. Both his son and grandson (Nobel Laureate John H. Van Vleck) became famous academics.
- Robert Warder (age 21; Cincinnati Observatory Expedition) graduated from Harvard University and became a university professor. He does not appear to have ever strayed so far from home again.
- James Watson (age 31; NAO/Franklin Institute Expedition) left Michigan and moved to the Washburn Observatory of the University of Wisconsin. They paid twice as much. Watson designed and started to build a new kind of telescope, which his colleagues found ludicrous, but died before he could use it.
- Mary Whitney (age 21; Vassar College Expedition) proved that the world of the 1869 total eclipse of the Sun was a small one. She became a special student of Benjamin Peirce at Harvard: He had to escort her personally on and off the male-only campus. Whitney succeeded Mitchell at Vassar College.
- Joseph Winlock (age 43; USCS/HCO Expedition) took to heart the criticism of his colleague John Whipple's tiny-but-elegant solar-eclipse images: To produce larger ones of similar quality, he invented the instrument called the photoheliograph for his use during the 1870 total eclipse of the Sun. This specialized telescope/camera became the standard for producing solar images in and out of eclipse. However, Winlock did not live to see the next American total eclipse.
- Charles Young (age 34; NAO/Franklin Institute Expedition) was ahead of his time. He an outburst from the Sun, now called a solar flare, with his spectroscope in 1872 and noted its coincidence with a geomagnetic storm on the Earth. In 1877, Young moved to Princeton University and became Observatory Director. His final scientific paper speculated that "atomic" energy might explain the Sun's enormous output.
- Joseph Zentmayer (age 43; NAO/Franklin Institute Expedition) was inspired; he left Iowa for home where he constructed an optical-mechanical eclipse simulator. While no scientist, he was awarded a medal from the Franklin Institute for his life's work: turning the imaginings of experimental scientists into physical reality.

* * * * *

Seven years after the 1869 total eclipse, the Sun and the Moon danced again. They conspired to produce another trans-American eclipse in 1878. In fact, the two paths crossed at the Alaska-Yukon border. An imaginary fur trapper there would have witnessed a total eclipse of the Sun twice within a decade—without ever leaving the clearing surrounding his cabin.

The successes of 1869 emboldened the 1878 total-eclipse expeditions. They headed to the now-further-expanded American West, dwarfing those to 'our' solar eclipse in size and complexity.

Young led a Princeton expedition to Denver, Colorado. Mitchell once again took a crew of women with Vassar affiliations to observe the eclipse, also at Denver. She had with her the trusty comet finder.

Hall, typically, located himself at remote Fort Lyon, Colorado.

Myer (inspired by mountaintop observing on his Virginia total-eclipse expedition of 1869?), Cleveland Abbe, and Norman Lockyer (visiting from England) climbed the 4,302 meter Pike's Peak for the highest-ever recorded observations of a total eclipse of the Sun up to that time. Abbe suffered acute altitude sickness and almost died before he could be carried down. He only was able to watch the total eclipse from a cot at a much lower elevation—without his instruments.

Though he was not technically an astronomer, Morton made the trip to Rawlins, Wyoming, for another total eclipse of the Sun. Harkness led an expedition to Creston, Wyoming. Among those with him was Alvan G. Clarke. Not really a town, Creston was a watering stop for the railroad. It was one of the most destitute spots, for looking at a solar eclipse, imaginable.

In Chapter 15, I mentioned Newcomb and Watson observing side-by-side at Separation, Wyoming. Newcomb repeated the use of his screen to cover the inner corona; it helped but was countered by the fact that the sky brightness was greater than it was in August 1869. (Remember that July 1878 was near sunspot maximum, five-months hence; the appearance of the corona differed from that of 1869, too.) Busy with his own work during the 1878 total eclipse, Newcomb was not able to confirm Watson's putative (and specious) 'discovery' of Vulcan.

Another total eclipse of the Sun passed over the upper-middle United States in 1889 (on 1 January!). This twenty-year cluster of American total eclipses, during the time when science was first coming to grips with the structure and inner workings of our star, is an appropriate end to our story.

The long-term personal effect of a total solar eclipse is hard to guess ahead of time. Young clearly enjoyed the experience: It marked the formal beginning of a successful career as a solar astronomer. After 1869, Young led total-eclipse expeditions in 1878, 1887, and 1900 (the last when he was 66 years old). In between, he was a popular speaker about eclipses on the lecture-for-entertainment circuit.

For Young, total eclipses were the normally mum Sun's occasional whispers about its own essence. He was the preeminent American expert on the subject of solar astronomy during the latter third of the nineteenth century.

For Newcomb, solar eclipses became the source of fond, after-dinner anecdotes to tell, during a lengthy life in the public eye. Be that as it may, the coincidence of three such eclipses in his 'backyard' ultimately influenced his career very little.

Besides figuring out the orbits of the planetary system, Newcomb became the Carl Sagan (or, for a younger generation, Neil de Grasse Tyson) for the turn of the twentieth century. His writings and lectures established him in the public mind as *the* voice of American astronomy.

For Alexander, 1869 must have been bittersweet. The eclipsed Sun had set for him. He was no longer America's expert on the subject. Solar science had passed him by. Alexander never again would witness a total eclipse of the Sun.

It was a philosopher, Charles Peirce, who enunciated most clearly: If you know something is there, you will see it. For Watson, obsession came with what he thought he 'saw' during total eclipse. He tried to build a pseudo-telescope designed to recapture Vulcan *in the daytime*. Between observing at night and constructing an instrument that undoubtedly would fail, Watson exhausted himself and died. He was only 42.

Would Watson have lived had he not imagined a planet within the weird scenery that appears only during those minutes when day becomes night? Death by total eclipse of the Sun?

Fig. 20.8 The influence of the eclipse lingers today. Of all the solar eclipses that have since taken place, the musicians who recorded this album chose as cover art an illustration from the 1869 total eclipse of the Sun. {from the collection of the author}

Postlog

I began with one of the most prominent names in the history of American astronomy. I end with one of the most obscure.

Sometimes, a total eclipse of the Sun will take hold of a person. And not let go. It is commonly stated (admittedly, without much attribution) that only one out of ten thousand people have seen just one. That is now. What about the mid-nineteenth century? What about *more* than just one?

Robert Paine appears never to have had more than perfunctory connections with professional astronomers. As a lawyer, his everyday life had less to do with total eclipses of the Sun than nearly every other character who has appeared in this memoir of totality.

With his portable telescope small and designed only for direct observation, Paine could not hope to make much of a contribution to solar science. He rarely committed his astronomical observations to publication. (Sadly, what unpublished notes he may have made were lost shortly after his death.)

Perhaps few knew the whole story: While still a young man, Paine saw the annular eclipse of 1831 from a lighthouse. To do so, he "was obliged to make an inclement passage by sea of seven miles in an open rowboat."

He witnessed his first total solar eclipse from a War-of-1812-era arsenal in South Carolina (1834). Annulars observed followed in 1838, 1854 and 1865. (For the last, Paine returned to South Carolina; Charleston was still burnt from the siege of War.) That of 1875 was watched from his home in Brookline, Massachusetts. Post 1869, the otherwise urban Paine would shutter his practice for the second time and again travel from the Atlantic to west of the Mississippi, in order to observe the 1878 total eclipse of the Sun. By then he was, by his own admission, "sick and nearly blind."

T. Hockey, *America's First Eclipse Chasers*, Springer Praxis Books, https://doi.org/10.1007/978-3-031-24124-6

In 1880, though definitely "infirm," he traveled from Massachusetts to *California* alone. "On reaching his destination, he was left by the train on a tree-less prairie, with no human being in sight or within twenty miles." While we may not be able to say that he 'saw' the eclipse in the traditional sense, he still was satisfied that he had experienced its totality.

At the time of his death, Paine was planning a trip to the wilds of Montana for the 1885 annular eclipse in his 81st year.

Paine led such an uneventful, even sheltered, life that it is difficult to find biographical information on him. Uneventful, that is, except for when a solar eclipse was to occur. All told, he witnessed *nine* central eclipses (four total) as well as fifteen partial eclipses of the Sun.

Sometimes it never lets go. No other natural phenomenon arouses such horror and sublimity at the same time as does a total eclipse of the Sun.

It did so in 1869. It will do so again in 2024.

Appendix A: Simon Newcomb's Instructions to Amateur Astronomers Along the Edge of the Total-Eclipse Path

This was the first use of crowd-sourcing in astronomy.

The following information is desired by the United States Naval Observatory, namely: *The duration of total eclipse at various places along the line of totality, situated between one and ten miles from its limits.*

To obtain this information the observatory invites the co-operation of intelligent citizens residing near the limits, and the following instructions are prepared for the use of those who will co-operate:

Instruments.—The only indispensable instrument is a good watch, provided with a second hand, and having a white face. The minute hand should be carefully set, so as to be on an exact minute when the second hand is at 60^s. This being done it is no matter how far wrong the watch may be. A good auxiliary will be a common spy-glass [*sic*] lashed to a round post, so as to be steady enough to give an easy view of the Sun. To lessen the brilliancy of the Sun, cover the object glass with a cap having a round hole, three-fourths of an inch in diameter, cut in its center. The spy-glass will be worse than useless unless one is accustomed to its use, and has it fastened so as to be steady. A smoked glass should also be prepared, but a part of the glass should be very lightly smoked.

*Arrangements for observ*ation.—Each observation should be made by a party of three persons. Only one instrument of each kind, watch, glass, &c., is needed by a party. A station should be selected where they will be free from all interruption, either in the open air or at an open window facing west. One, at least, of the party must have a pencil and note-book in hand to record the time.

T. Hockey, *America's First Eclipse Chasers*, Springer Praxis Books, https://doi.org/10.1007/978-3-031-24124-6

The observation.—When the visible part of the Sun is reduced to the narrowest crescent, the holder of the watch, keeping his eye on the face, will begin to count the seconds aloud;[1] the holder of the smoked glass, with or without the spy-glass, will watch for the last ray of true sunlight, being careful to look through the brightest part of the glass the eye will bear without inconvenience; and the third observer, if there be one, will look for the disappearance of sunlight with the naked eye, and stand ready with pencil and paper to record the time. When the last speck of the Sun has disappeared, the observer with the glass will call "*time*," and the exact second at which the call was given must be immediately written down. The minute, also, must be carefully noted and recorded.

The observers will then await the return of sunlight, the count of the seconds being kept up, if the face of the watch can be seen, which it probably can if held so that the light of the "corona" shall fall upon it. The duration of total eclipse will generally fall between half a minute and a minute and a half, depending on the position of the observer. The first flash of true sunlight will seem to burst out suddenly, and the minute and second of its appearance must be recorded with the same care as the time of disappearance. The difference of the two times gives the duration of totality.

Special precautions.—In judging the beginning of totality, there is danger of error from two sources. The first is that the Sun's crescent may become so narrow as to become invisible through the smoked glass, if it be too dark, several seconds before it is really all covered, and thus the observer may call "time" too soon. Such a mistake may be detected and corrected by the third observer looking on with the naked eye, if the following circumstance be attended to:

The beginning of total eclipse is marked by a very rapid increase in the darkness, caused by the advent of the moon's shadow. If, then, the darkness increases more rapidly after *time* is called than it did before, time was called too soon, and must be repeated.

The other danger is of the opposite kind, and should be equally avoided. It is that the light of the brilliant rose-colored protuberances which surround the dark body of the Moon during the total eclipse may be mistaken for sunlight, and thus the critical moment be suffered to pass. In this case each observer must determine separately as to the exact second at which it ceased to grow darker, and if they agree within one or two seconds, the time thus judged may be supposed correct, and each one's estimate may be written down separately.

The observer with the smoked glass will be most liable to the first of these mistakes; the naked-eye observer to the last.

---

[1] It would be well for the observers of each party to practice beforehand the counting, calling, and recording.

The return of sunlight will also be preceded by a reddish glow on the border of the dark Moon, which must not be mistaken for the Sun. Indeed, if the observers be near the edge of the shadow, it is probable that this red glow, which comes from the hydrogen atmosphere of the Sun, may be visible during the whole time of totality.

All the recorded times, with an estimate of the uncertainties to which the observers think they were liable, and a statement of the place where made, giving distance in miles and direction from the court-house [*sic*], if it be a county town, and from the railroad station, if a railroad pass through, should be immediately certified by the signatures of all three observers, and forwarded to the Naval Observatory, Washington.

It is particularly requested that each party send off its report before comparing notes with any other party; also, that the original pencil record, however imperfect, accompany the report. All will be carefully preserved in the archives of the observatory for the use of astronomers.

Appendix B: Really Bad Advice

"Bits of glass of every description were at a premium, and grandfather's spectacles, mother's court-house [*sic*], if sister's lorgnette, all underwent the process usually confined to hams, beef and sausages . . ."

A second edition of Thomas Webb's *Celestial Objects for Common Telescopes* was published in 1868[1]. This very popular guide to practical observing contains advice on 'safe' solar observation that appears to parallel the practices used by most who observed the 1869 total eclipse of the Sun. It is alarming, then, that almost *NONE* OF THE ADVICE THAT APPEARS BELOW WILL RENDER THE SUN HARMLESS TO VIEW THROUGH A TELESCOPE.

> With due precaution, there is no danger; but the eye and hand had better first acquire experience elsewhere. Much depends on the dark glass of the solar cap, which is to be screwed on the eye-piece [*sic*]; red is often used, and *may* be dark enough—it is not so always—but it transmits too much heat; green is cooler, but sufficiently thick. The Germans have employed deep yellow, probably to save the brightest and most central rays of the spectrum. Herschel I adopted, with great success, a trough containing a filtered mixture of ink and water. The late Mr. Cooper (of Markree Castle, Ireland) used a "drum" of alum water and dark spectacles, and could thus endure the whole 13-3/10th inches, of his great 25-foot achromatic. . . . In screen-glasses combinations of color are good. Red succeeds perfectly with green. Herschel II used green and cobalt blue. If there is to be only one solar cap, deep bluish

---

[1] London: Spottiswoode and Company

T. Hockey, *America's First Eclipse Chasers*, Springer Praxis Books, https://doi.org/10.1007/978-3-031-24124-6

grey, or neutral tint, will be quite satisfactory; if several, it would be worth while [*sic*] to have different colors, Secchi's observations at Rome seeming to show that the visibility of very delicate details may depend on the tint. In the absence of a proper screen, smoked glass may be used: it is said to intercept heat very perfectly, by Mr. Prince, who places it within the eye-piece, close to the "stop," or circular opening, which bounds the field; but thus it can have only one degree of depth, and must be taken out to view other objects. A strip of glass may be smoked to different in different parts, and held between the eye and the eye-piece; but it should be protected from rubbing by a similar strip of glass placed over it, and kept from touching by bits of card at the corners, the edges of two strips being bound round with gummed slips of paper, or tape. Where expense is not regarded, an optician will provide a delightful graduated screen with two wedges of glass, plain and colored.

NO METHOD DESCRIBED ABOVE GUARANTEES THE SAFETY OF THE HUMAN EYE. The danger lies in the Sun's invisible infrared rays: Damage to the retina from heat causes no alerting pain. It is, perhaps, fortunate that telescopes were not common in the hands of the general population *circa* the 1860s. Dilettantes may have been protected by the small apertures available to them. Professionals might have experienced a reluctance to observe the eclipse at length prior to second contact (and instead used the time to prepare for totality) or after third contact (in order to use the time to record and collate their observations of totality). The latter group was wise enough to stop down their large-aperture instruments; an eight-inch telescope otherwise could quickly and unexpectedly impair vision. Naked-eye eclipse watchers were at less risk—theirs likely were mere glances at the bright photosphere.

Appendix C: A Rare Eclipse Experiment

The popular picture of a scientist is that of one performing experiments in a laboratory. In the field sciences, such as geology, astronomy, and even branches of biology, this normally is not the case. Observations of nature are made under as controlled circumstances as possible, and then interpretations are made. The sequence of hypothesis-test-confirm/falsify is all the rarer during the limited time of solar-eclipse totality.

One exception is the Tennessee man who in 1869 noted the time between the disappearance of sunlight on nearby mountains and darkening sat his own location. He knew how far away the mountains were, so was able to measure the speed of the eclipse umbra across the Earth.

Thus, we might hope that the story below, of an actual (if whimsical) experiment performed during the 1869 total eclipse of the Sun, is not altogether fictional.

A Scientific Experiment

A[n anonymous] correspondent of the Chicago *Evening Journal* visited Springfield, Ill., on the occasion of the recent eclipse, to ascertain, from actual observation, the truth or falsity of the statement that chickens go to roost during a total eclipse. To test the matter, he procured for thirty-five cents, the use of a game-cock—fine bird, thorough-bred [*sic*], head erect, beautiful plumage, the victor of many a hard-fought battle, and the loser of none. The rooster

T. Hockey, *America's First Eclipse Chasers*, Springer Praxis Books, https://doi.org/10.1007/978-3-031-24124-6

was fastened to the corner of a fence, with a string and a comfortable roost arranged for his use. The scientific correspondent reports as follows:

"Three, minutes before 5 o'clock, not more than one eighth of the sun's face is visible, and my rooster has certainly stopped eating, preparatory to going to roost."

"How wonderfully grand as the total eclipse approaches. The situation is peculiarly impressive, and my rooster is surely getting ready to go to roost."

"At ten minutes after 5 o'clock, the Sun's face is entirely obscured; three stars appear, and for two minutes exactly by my watch, the scene is awfully grand and impressive. My rooster retires to the fence corner, shuts his eyes, but seems to have no idea of flying on the roost. For the moment I am entirely oblivious to my rooster and watch with intense interest the scene around me. Silence reigns supreme; the thousands around me are quiet as the grave, and speak only in whispers. To attempt any description of the scene would be useless. No pen could picture the sensation that fills the thousands of people who are impressed to perfect stillness. Mercury and Mars appear, in rear of the Sun. The gentlemen at the telescopes and photographing apparatus are busy, and all is breathless quietness. The cause of science is advanced by a cloudless sky and favorable weather for making observations."

"The photographers are getting some executed impressions, but my observations and are not to be successful. I approach my rooster and hope to induce him to go to roost. Surely he will not decline now. He is master of the situation, however, and although evidently drowsy, and thinks it bedtime, yet stubbornly refuses to go to roost. One minute passes, and only a single moment of darkness remains according to astronomical observations. You must be busy, 'General,' if you get to roost before morning. I count the seconds as they pass, and pray for a chicken tongue to ask him what he is waiting for. Thirty seconds more, and still he remains quiet in on the fence corner. Thirty seconds more are ticked into the past and then the other any side of the sun's face is visible, and day light appears. My rooster picks up his head and I am waiting to hear him crow, but he is evidently not in crowing mood, and soon goes to pick up food from the grass around me."

"My observations are a failure, and the world is not yet enlightened on this important fact. I am told that about the city the chickens actually did go to roost, but of this cannot say with certainty. My rooster did not, neither did he crow, and I returned him to his owner. He may be a be success as game cock, but for the cause of science is not. In fact I incline to the opinion that spring chickens are better both for domestic and scientific purposes, and at the next eclipse I shall not try a game cock. They are too intelligent, and cannot be fooled by any such device. Sadly I returned the 'General' to his owner and coop. The boy asked me how he answered my purpose. 'Not at all,' said I, 'he cannot be coaxed to roost,' 'To roost!' exclaimed the boy; 'he don't go to roost—*too old*.' 'How does he sleep?' said I. 'Why, he squats. It would take more'n forty eclipses to get that rooster to fly on a roost.'"

"Whether this was spoken in a jest or not, I any unable to say; but I am certain that he 'squatted' all the time during the total eclipse of August 7th, 1869, and when next I attempt an observation of this kind, shall select a less gamey bird."

Appendix D: Persons Who Observed the 1869 Solar Eclipse in Totality

Using a most conservative number, tens of thousands of people watched the 1869 total eclipse of the Sun. In addition, I have collected the names of 515 individuals who observed it with some purpose in mind. The total eclipse likely resulted in the astronomical event with the most intentional (if passive) participants ever, up to that time.

Eighty-two locations are listed in order of increasing (decreasing west) longitude, thus in time order of the umbra passing over the observer.

−173.4°   Mys Gaydamak, Siberia[1]   Elevation = 161 meters

Baker, Richard   USA
Franklin, Samuel Rhoads   USN (Captain, USS *Mohican*)
Hall, Asaph III   USNO
Rogers, Joseph A.   United States Hydrographic Office
Very, S. W.   USN

−135.9°   Klukwan, Alaska   Elevation = 83 meters

Davidson, George   USCS
Davis, Jefferson Columbus   USA
Douglas, James   British nobleman
Seward, Frances Adeline Miller   (Mrs. William Seward) abolitionist and suffragette

---

[1] Modern place names are used.

T. Hockey, *America's First Eclipse Chasers*, Springer Praxis Books, https://doi.org/10.1007/978-3-031-24124-6

Seward, Frederick William   (son of William and Adeline Henry)
Seward, William Henry   former United States Secretary of State
Throckmorton, S. R., Jr.

−96.7°   Sioux Falls, South Dakota   Elevation = 448 meters

Abbe, Cleveland   Cincinnati Observatory
Abbe, Robert   physician (brother of C. Abbe)
Compton, Arthur G.   professor
Haines, James H.   Cincinnati Observatory
Longstreth, S. N.   photographer
Taylor, W. O.   photographer
Warder, Robert Bowne   professor

−96.2°   Le Mars, Iowa   Elevation = 376 meters

Ball, Seymour
Farrell, N. E.   farmer
Farrell   (Mrs. N. E. Farrell)
Gilman, Daniel T.   banker
Gilman, Winthrop Sargeant, Jr.   banker
Locklin, Eugene
Phelps, Lucius C.   civil engineer
Vincent, Leon Henry   civil engineer
Wood, Vincent

−96.1°   Onawa, Iowa   Elevation = 320 meters

Geddes, Addie

−95.6°   Cherokee, Iowa   Elevation = 364 meters

Blickensderfer, Jacob   railroad engineer
Hinkley, Myron   farmer

−95.2°   Red Oak, Iowa   Elevation = 320 meters

Corker, John F.
Cousins, E. B.
Humes, John E.
McClelland, G. F.
Osborn, Marcus B.

−94.4°   Jefferson, Iowa   Elevation = 325 meters

Ashe, Edward Davis   Royal Navy
Douglas, James   physician
Falconer, Alexander Pytts
Heizer, A.   minister
Stanton
Vail, Hugh D.   professor

−94.0°   Adel, Iowa   Elevation = 272 meters

Williams, Ora   youth

−93.9°   Boone, Iowa   Elevation = 346 meters

Goldthwait, Nathan E.
Goldthwait, M.
Holcombe, Benjamin Byron   judge
Holt, H. C.   minister
Paine, Robert Treat   attorney

−93.6°   Des Moines, Iowa   Elevation = 291 meters

Armstrong   college student
Beeman, M. V.
Brennan, M. S.   USA
Cecil, Sackville Arthur   British Lord
Colbert, Elias   journalist
Curtis, Edward   USA
Eastman, John Robie   USNO
Eastman, Mary Jane Ambrose   (Mrs. John Eastman)
Faulkner, Robert
Fraiser, John   professor
Goodfellow, Ogden Edward   USCS
Hall, Isaac H.
Harkness, William   USNO
Hilgard, Julius Erasmus   USCS
Hilgard, T. C.   physician
Hubbard, Frederick
Jassman, J. R.   USN
Lane, Jonathan Homer   formerly, United States Patent Office
Le Merle, A. E.   USA
Marryatt, D. P., Jr.   college student
Meade, A. E.   USA

Newcomb, Simon   USNO
Norton, William A.   professor
Peters, Christian Heinrich Friedrich   professor
Rogers, William A.   professor
Safford, Truman Henry   professor
Stone, Ormond   professor
Ward, E. J   photographer
White, S. V.   journalist

−92.7°   Oskaloosa, Iowa   Elevation = 256 meters

Gummere, S. G.   professor
M'Clune, James H.   teacher

−92.6°   Toledo, Iowa   Elevation = 276 meters

Moore, J. S.   photographer

−92.5°   Cedar Falls, Iowa   Elevation = 268 meters

Anderson, W. J.   county surveyor
Anderson, W. S.   civil engineer
Brown, E.
Dean, George M.
Horr, Asa   physician
Horr, E. W.
Overman, Dempsey C.   miller
Overman, John Milton   miller
Stanley, J. H.   jeweler
Toketon, G. O.
Tollerton, J. J.   judge
Wormwood, W. J.
Wormwood, W. W.

−92.5°   Waterloo, Iowa[2]   Elevation = 269 meters

Lackty, John A.   farmer

−92.5°   Waverly, Iowa   Elevation = 278 meters

Hoover, Henry S.   surveyor
McKinney, J. W.

---

[2] outskirts of

−92.4°   Ottumwa, Iowa   Elevation = 205 meters

Alexander, Stephen   professor
Baker, W. J.
Browne, John C.   photographer
Graham, Thomas L.
Halsted, Nathaniel Norris   USA
Himes, Charles Francis   professor
Hooper, George H.   college student
Moelling, E.
Moore, Charles H.   college student
Morton, M. P.
Peters, John E.   college student
Wilson, Edward Livingston   photographer
Yeisley, George C.   college student
Zentmayer, Joseph   optician

−91.8°   Bonaparte, Iowa   Elevation = 167 meters

Barber, G. W. clothier

−91.7° Monroe, Missouri Elevation = 227 meters

Weissner, John   NAO
Weissner (son of John Weissner)
Weissner (son of John Weissner)

−91.6°   Iowa City, Iowa   Elevation = 204 meters

Hershire, A. J.   newspaper editor
Hinrichs, Gustavus Detlef   professor
Irish, Charles Wood   mining engineer
Irish, John Powell   newspaper editor/state legislator
Irish, Thomas Myrick
Koegler, J. H.
Marquart, O.
Parvin, Theodore Sutton   professor
Startsman, J. P.   jeweler

−91.6°   Marion, Iowa   Elevation = 259 meters

Christie, John Thomas
Clark
Durham, Samuel W.   county surveyor
Gifford   (Mrs.)
Graves, James B.
Gray, G. A.   county surveyor
Hayzlett, John G.   sheriff
Holmes, George W.   physician
McClellan, J. W.   school principal
McEllhenny, J. W.
Ristine, Henry M.   physician
Samson, E. L.   jeweler
Simmons, Susan
Smith, E. M.
Tennison   (Mrs.)

−91.6°   Mount Pleasant, Iowa   Elevation = 218 meters

Alexander, B. C.
Carbutt, John   photographer
Clifford, H. M.   photographer
Cremer, James   photographer
Hoover, James V.   photographer
Johnston, William A.   civil engineer
Krause, Frank
Leisenring, James B.   photographer
Leisenring, J. R.   photographer
Mansfield, Arabella Babb   attorney
Mansfield, J. M.   professor
McIntyre, D.   university treasurer
Merriman, George B.   professor
Morton, Henry Jackson   professor
Pearson   physician
Pickering, Edward Charles   professor
Ranger, W. V.   photographer
Van Vleck, John Monroe   professor
Watson, Annette Waite   (Mrs. James Watson)
Watson, James Craig   professor
Whiting, J. H.   banker
Wilson, Edward Livingston   photographer

−91.4°   Hannibal, Missouri   Elevation = 153 meters

Bayse, J. W.
Bayse, L. E.   (Mrs. J. W. Bayse)

−91.4° Keokuk, Iowa Elevation = 184 meters

Libby, E. P.   photographer

−91.2°   Bowling Green, Missouri   Elevation = 271 meters

Cluster, Albert
Corthell, E. L.   civil engineer
Fiero, W. B.
Lawry, W. P.   school principal
Leighton, John   minister
Lockling, F. R.   civil engineer
Sheldon, Augustus C.
Sheldon, Nellie H.
Shields, J. H.
Smith, Melville
Sumner, H. M.   civil engineer
Williams, G. H.

−91.2°   Mechanicsville, Iowa   Elevation = 281 meters

Bradshaw, Homer S.   school principal
Bradshaw, Peter Randall
Golding, John F.

−91.1°   Burlington, Iowa   Elevation = 212 meters

Baldwin, A. C.
Blatchley, Sarah Louise   poet
Bonsal, J.
Carter, Isabella (Mrs. James Carter)   teacher
Clarke, James Freeman   minister
Coffin, Elizabeth R.   (daughter of John Coffin)
Coffin, John Huntington Craine   NAO
Eaton, D. G.   professor
Ely, Achsah M.   professor
Emmerson
Greeley, Samuel S.   municipal surveyor
Glazier, Sarah   teacher

Gould, Benjamin Althorp   astronomer
Kendall, Otis E.
Mahoney, J.
Mayer, Alfred Marshall   professor
McLean
Mitchell, Maria   professor
Montfort, Albert W.   photographer
Morton, G. C.   railroad agent
Phillips, H. C.
Reybold, Mary Rhoades   student
Roach   judge
Roach   (Mrs.)
Rock, Miles   civil engineer
Rorer, Delia M. Viele   abolitionist
Stark
Stockwell, John N.   NAO
Stockwell   (Mrs. John Stockwell)
Stoddard, Charles   college student
Thielson, Henry B.   railroad surveyor
Way   minister
Whitney, Mary Watson   student
Wild, Edward A.   USA
Willard, O. H.   photographer
Young, Charles Augustus   professor

−90.6°   Davenport, Iowa   Elevation = 177 meters

Baldwin, E.   jeweler
Barler, A. W.   geologist
Bowman, W. E.
Cressy, R. M.
Gayford, A. B.   photographer
Hazen, Edward Hamlin   physician
Jones, Paul R.   photographer
Lighton, Thomas   professor
Lynch, E. P.   manufacturing company president
Murray, Thomas   municipal engineer
Newbury
Parry, Charles Christopher   physician/botanist
Pratt, W. H.   publisher
Roundly, D. C.   physician
Sheldon, David S.   professor

−90.6°   Rock Island, Illinois   Elevation = 171 meters

Butler, W. P.   USA
Comly, Clifton   USA
Metcalf, Henry   USA
Oland, M. S.   USA
Parker, F. H.   USA
Schaff, Morris   USA
Simmons, M. L.
Yule, Samuel   farmer

−90.4°   Colona, Illinois   Elevation = 175 meters

Bell, James   merchant
Bell   (Mrs. James Bell)
Weed, S. H.   minister
Weed   (Mrs. S. H. Weed)

−90.4°   Henderson, Kentucky   Elevation = 124 meters

Blacknell, P. L.   professor
Crosby, T. H.   civil engineer

−90.2°   Genesco, Illinois   Elevation = 196 meters

Brown, W. C.   physician
Campbell, L. C.   justice of the peace
Dresser, J. F.   banker
Moderwell, E. C.   attorney
Parker, C. E.   merchant
Reter, John   dentist
Smith, John
Stein, F.   jeweler

−90.2°   Saint Louis, Missouri[3]   Elevation = 142 meters

Burgas
Cobb
Eimbeck, William   professor
Golden
McKown

_____

[3] outskirts of

McMath
Pitzman, Julius   county engineer
Rogers, W. H.   clock maker
Schmidt
Soldan

−90.1°   Alton, Illinois   Elevation = 151 meters

Long, T. M.   county surveyor

−89.9°   Gridley, Illinois   Elevation = 229 meters

Boris, W. H.

−89.9°   Kewanee, Illinois   Elevation = 244 meters

Austin, E. P.   NAO
Covert, A. L.

−89.7°   Springfield, Illinois   Elevation = 170 meters

Aycrigg, John Benjamin   USCS
Benjamin   clothier
Black, John Wallace   photographer
Carter, George T.
Dudley, Timothy
Dupuis, Nathan Fellowes   chemist/astronomer
Evans, Alexander Sr.   former Maryland congressman
Fay, C. N.   college student
Fitzgerald, Richard   photographer
Gorman,   photographer
McLeod, Robert A.   college student
Meek, Fielding Bradford   Geological Survey of Illinois
Montague, W. P.   Harvard College alumnus
Pourtalis, L. N.   USCS
Schott, Charles Anthony   USCS
Seaver, E. P.   college trustee
Sexton, T.   photographer
Shaler, Nathan   professor
Twining, Alexander Caitlin   civil engineer for the railroad
Warner, J. W.
Wines, Fred H.   minister

−89.4°   Carlyle, Illinois   Elevation = 143 meters

Barkley, James   newspaper editor
Case, Zophar   clerk of USA circuit court
Robinson, William A.

−89.4°   Lacon, Illinois   Elevation = 143 meters

Davis   Emily K. Sheaff   (Mrs. George Davis)
Davis, George   physician

−89.0°   Bloomington, Illinois   Elevation = 243 meters

Frink, W. R.
Jackman, John A.
Peirce, James Mills   professor
Roe, Edward R.   newspaper editor
Warner, Joseph B.   HCO

−89.0°   El Paso, Illinois   Elevation = 228 meters

Gibson, George L.   mayor
Moore, Joseph H.   postmaster
Torrey, W. R.

−88.7°   Chenoa, Illinois   Elevation = 215 meters

Elder, C. S.   physician

−88.4°   Mattoon, Illinois   Elevation = 223 meters

Bostwick, C. B.   newspaper editor
Easterday, L. M. F.   professor
Hill, Thomas   college president
Hough, George Washington   astronomer
House, John C.   professor
Keifer, Joseph Warren   USA
Marshall, C. H.   minister
Murphy, David   professor
Simmonds, Thomas   astronomer
Smith, F. H.   professor
Swift, Lewis A.   hardwareman

−88.1°   Paxton, Illinois   Elevation = 243 meters

Corwin, A. R.
Glasener, E. T.   postmaster
Hanley, John M.   school principal
Harkey, J. S.   minister
Hasselquist, Tuve Nilsson   professor
Howe, H. J.   county surveyor
Jones, Lewis K.
Pells, W. H.   real estate developer

−87.9°   Mount Vernon, Indiana   Elevation = 121 meters

Pearse, S. H.   physician
Willoughby, E. C. H.   minister

−87.4°   Terre Haute, Indiana   Elevation = 152 meters

Bosworth, R. F.   professor
Halford, E. W.   journalist
Hobbs, Barnabas Coffin   school superintendent
Holliday   journalist
Hufman   photographer
Moore   professor
Olcott, J. M.   professor
Potter   USA
Tingley   professor

−87.2°   Washington, Indiana   Elevation = 153 meters

Lovitt, T. D.

−87.1°   Danville, Illinois   Elevation = 182 meters

Chandler, William P.
Cunningham, W. T.
Dunbar, Warren
English, J. G.   banker
Lemon, Theodore E.   physician
Winslow, J. C.   mayor/paleontologist

−86.8°   Greencastle, Indiana   Elevation = 257 meters

Langsdale, George J.   newspaper editor

−86.5°   Bedford, Indiana   Elevation = 29 meters

Joseph Gardner   physician

−86.5°   Bloomington, Indiana   Elevation = 235 meters

Allison, James Benjamin   photographer
Ballantine, Elisha   professor
Foster   USA
Gabe, William A.   newspaper editor
Kirkwood, Daniel   professor
Lee
Wylie, Samuel Brown   photographer
Wylie, Theophilus Adam   photographer

−86.4°   Oakland Station, Kentucky   Elevation = 178 meters

De Brie, N.
Donaldson, J. L.
Edwards, C. R.
Hilburn, John J.
Langley, Samuel Pierpont   astronomer
Reise, W. I.   USA
Strange, Robet F.   attorney
Timmons, J. A.
Younglove, John Edwards   pharmacist
Younglove, Timothy Meigs   pharmacist
Wilder, Graham

−86.1°   Franklin, Indiana   Elevation = 220 meters

Brown, F. W.   professor
Hougham, John Scherer   professor
Morey, A. B.   minister
Scott, W. T.

−86.0°   Edinburgh, Indiana   Elevation = 205 meters

Baker, Aaron   civil engineer
Boynton, Charles
Fritts, W. W.   physician
Logan, John B.   minister
Milner, S. J.   minister

−85.9°   Columbus, Indiana   Elevation = 192 meters

Beck, Albert T. attorney
Comstock, L. W.
Dickey, N. S.   minister
Graham, A. H.   school superintendent
Rabb, George G.   physician
Smith, J. Lawrence

−85.7°   New Albany, Indiana   Elevation = 137 meters

Ayerigg, Benjamin   civil engineer
Bradford, Charles H.   journalist
Bradford, Herman Gleason   jeweler
Campbell, John L.   professor
Clapp, W. A.
Crosier, Edward Stokes   physician
Daily, Reuben   journalist
Fales, J. C.   professor
Haskins, F. C.
Heimberger, Christian   photographer
Hoffeld, Alfred   pharmacist
Jocelyn, George B.   college president
Laselle, Charles B.
Llewellyn, Russell   merchant
Matthews, Lucien Gustavus   newspaper editor
Mitchell, F. A
Morse, F. L.
Pell, F. A.   college student
Pratt, William Moody   minister
Reid, A. M.   school superintendent
Reynolds, J. B.
Smith, George M.   municipal engineer
Smith, L. S.   youth
Vance, Hart Vanderpool   civil engineer
Williams, A. A.   journalist
Wilson, James A.   photographer
Wilson, John M.   postmaster

−85.5°   Hanover, Indiana   Elevation = 239 meters

Allison, L. H.
Archibald, G. D.   college president
Bean, G. W.

Eastman, John C.   minister
Gill, Heber   minister
Mulvey, Oliver   photographer
Thomson, J. H.   professor
Thomson, S. H.   professor
Wiley, Harvey Washington   college student

−85.5°   Bardstown, Kentucky   Elevation = 209 meters

Peirce, Charles Sanders   USCS
Shaler, Nathan Southgate   professor
Smith, John Lawrence   retired professor

−85.4°   Burksville, Kentucky   Elevation = 177 meters

Alexander, R. M.   physician
Hunter, W. G.   physician
Waggener, L. A.   clerk of county court
Walker, Scott   attorney
Williams, J. W.   clerk of circuit court

−85.3°   Shelbyville, Kentucky   Elevation = 232 meters

Agnew, F. H.   USCS
Blake, A. F., Jr.,   USCS
Bowditch, Jonathan Ingersoll   professor
Clark, Alvan Graham   optician
Clark, George Bassett   optician
Dean, George Washington   USCS
Dixwell, J. J.
Frankenstein, John Peter   artist
Hannaman, T.   blacksmith
Lewis, Robert
Pendergast, John W.   photographer
Searle, George Mary   professor
Seymour, C. B.   professor
Sharrod, R. E.
Sitter   photographer
Stevens, Albert
Tevis, Robert C.   college trustee
Whipple, John Adam   photographer
Williams, John W.   photographer
Winlock, Anna   youth (daughter of Joseph Winlock)
Winlock, Joseph   HCO

−85.2°   Albany, Indiana   Elevation = 277 meters

McLeod, Robert A. minister

−85.1° Vevay, Indiana Elevation = 147 meters

Boerner, Charles G.   clock maker

−84.7°   Midway, Kentucky   Elevation = 252 meters

Bell, David N.   USN

−84.6°   Nicholasville, Kentucky   Elevation = 286 meters

Locke, Robert J.   professor

−84.3°   Catawba, Kentucky   Elevation = 177 meters

Crozer, D., Sr.
Crozer, D., Jr.

−84.3°   Falmouth, Kentucky   Elevation = 168 meters

Arnold, W. E.   civil engineer
Arnold (Mrs. W. E. Arnold)
Barbour, James H.   physician
Clarke, A. R.   attorney
Cooper, W. G.
Crozer, W. C.   civil engineer
Grant, R. W.
Hudnall, James I.   judge
Ireland, William W.   Judge
Johnson
Johnson, B.
Knight, J.
Lee, Joseph   judge
Murphy   (Mrs.)
Scott
Searle, Arthur   HCO
Wandelohr, Carston A.   clerk of the court
West   telegraph operator
Woodson
Yelton, D.

−84.3°   Mount Vernon, Kentucky   Elevation = 360 meters

Christy. C.   professor
Hutchison, John Calvin   professor
McFarland, George A.   professor
McFarland (daughter of George McFarland)
McFarland (daughter of George McFarland)
Oldfather   student
Scott, E. S.   student
Stoddard, William Osborn   professor

−84.0°   Carlisle, Kentucky   Elevation = 262 meters

Fritts, W. W.   physician
Logan, John B.   minister
Milner, S.

−82.5°   Eden Ridge, Tennessee[4]   Elevation = 399 meters

Smith, F. H.   professor

−82.2°   Bristol, Tennessee   Elevation = 511 meters

Allen, Charles J.
Bardwell, Frederick W.   USNO
Brown, W. Leroy   professor
Cutts, Richard Dominicus   USCS
Davidson, Thomas   USN
Hayward, T. W.
Longstreth, Miers Fisher   physician
Marshall, Thomas D.   "intelligent gentleman"
Mosman, A. T.   USCS
Perkins, Frank Walley   USCS
Walthall, Thomas Jefferson

−82.0°   Whitetop, Virginia   Elevation = 1,682 meters
Myer, Albert James   USA
Winthrop, W.   USA

−81.6°   Abington, Virginia   Elevation = 636 meters
Coale, Charles B.   newspaper editor

---

[4]The town no longer exists.

−81.3°   Hickory, North Carolina   Elevation = 362 meters
Crozier, D., Jr.
Crozier, William   minister

−81.2°   Taylorsville, North Carolina   Elevation = 377 meters
Boyd, A. C

−80.9°   Cornelius[5], North Carolina   Elevation = 245 meters
McDonald, J. L.

−80.8°   Davidson, North Carolina   Elevation = 256 Meters
Phillips, Charles   professor

−80.5°   Salisbury, North Carolina   Elevation = 241 meters
Ramsey, John Andrew   surveyor
Robbins, William McKendree   attorney/state senator

−80.3°   Lexington, North Carolina   Elevation = 247 meters
Robbins, F. C.

−79.0°   Chapel Hill, North Carolina   Elevation = 148 meters
Mason, James Bruce   attorney

−78.8°   Greensboro, North Carolina   Elevation = 272 meters
Caldwell, S. C.   professor

−78.6°   Raleigh, North Carolina   Elevation = 96 meters
Primrose, William Stuart   banker

−78.5°   Franklinton, North Carolina   Elevation = 125 meters
Conway, C. W.   pharmacist
Neal, George W.   school principal
Pritchard, Calvin   justice of the peace
Winston, Robert N.

---

[5] There is no "Coulord," North Carolina. Because of its longitude and recorded eclipse duration, this location is taken to be a misprint for "Cornelius."

−78.5°   River View, North Carolina   Elevation = 12 meters
Baldwin, M.

−77.9°   Wilson, North Carolina   Elevation = 33 meters

Benton, A. A.   minister
Dunham, John W.
Fugna, J. A.
Husk, Robert S.
McBride, John

−77.8°   Rocky Mount, North Carolina   Elevation = 30 meters
Fountain, S.   postmaster[6]

_____

[6] last person I know of by name to witness a total solar eclipse until 1870

Sources

Roman numerals are provided as a finding help in Citations.

I.   ARCHIVES[1]

<u>Congress, Library of</u>
Simon Newcomb Papers

   Diary 26 July 1869 through 14 September 1869, Box 1

   Eclipse, Box 50

   Letters, Box 4

United States Nautical Almanac Office incoming correspondence, Box 20

United States Nautical Almanac Office outgoing correspondence, Box 15

United States Naval Observatory Records, Box 13

<u>Harvard University Archives</u>
Eclipse of 1869[:] Reports, 1869

KG11365-6, phaedra0361, Box: 9

Measurement of Photographs of Eclipse, 1869/1870

Harvard College Observatory: observations, logs, instrument readings, and calculations

---

[1] Direct quotes are referenced in Citations.

© The Editor(s) (if applicable) and The Author(s), under exclusive license
to Springer Nature Switzerland AG 2023
T. Hockey, *America's First Eclipse Chasers*, Springer Praxis Books,
https://doi.org/10.1007/978-3-031-24124-6

KG11365-6, phaedra0325, Box: 8

Illinois State Capitol
Jenkins, Troy. Dedicated Trees, Plaques & Stone Markers at Capitol Complex in Springfield Illinois as of 2009. Three pages. Unpublished document. (2009)

National Oceanographic and Atmospheric Administration
Cloud, J. *Benjamin Peirce and The Science of Necessary Conclusions (1867–1874)*. Unpublished manuscript. (circa 2010)

Putnam Museum
Davenport Academy of Natural Sciences, Eclipse Committee, correspondence, 1869

State Historical Society of Iowa
Vertical file: "Eclipses."

Thomas Hockey, Collection of
Fiala, A. D. IowaSites.txt. One page. Unpublished document.

Fiala, A. D. Report: Visit to Cedar Falls, IA, June 28, 2004. Three pages. Unpublished document.

United States Army Medical Museum
Photographs of the Total Eclipse of the Sun, August 7, 1869. Surgeon General's Office, Washington, D. C., October 9, 1869. Two pages. Unattributable document.

Gardner, J. The Eclipse: Success in Getting Photographs. One page. Unattributable document. Curtis dated 4 September 1869.

United States Naval Observatory Archives
Bound Notebook: Letters Received 1869 (copies of incoming correspondence)

Bound Notebook: Letters Sent 1869 (copies of outgoing correspondence

Envelope: Expeditions/Solar Eclipse/Des Moines, Iowa, 1869.

Personnel Records

Schedule of United States Naval Observatory Instruments Stored and Not in Use 1898

II.   REPORTS OF ECLIPSE EXPEDITIONS[2]

Government Publications
Coffin, J. H. C. *Reports of Observations of the Total Eclipse of the Sun, August 7, 1869, Made by Parties Under the General Direction of Professor J. H. C. Coffin, U. S. N.* Washington: United States Government Printing Office. (1877a)

---

[2] Direct quotes are referenced in Citations.

Includes:

Alexander, S. "Report of Prof. Stephen Alexander, Ottumwa, Iowa." P101. (1877)

Austin, E. P. "Report of Mr. E. P. Austin, Kewanee, Illinois." P73. (1877)

Coffin, J. H. C. "Report of Professor John H. C. Coffin, U. S. N., Burlington, Iowa." P7. (1877b)

Coffin, J. H. C. "Supplemental Report by Prof. J. H. C. Coffin." P111. (1877c)

Eaton, D. G. "Report of Prof. D. G. Eaton, Burlington, Iowa." P63. (1877)

Gould, B. A. "Report of Dr. B. A. Gould, Burlington, Iowa." P29. (1877)

Himes, C. F. "Report of Prof. C. F. Himes, PhD, of Dickinson College, Pennsylvania." P51. (1877)[3]

Meyer, A. M. "Report of Alfred M. Mayer, PhD." P129. (1877)[4]

Mitchell, M. "Report of Miss Maria Mitchell, Burlington, Iowa." P53. (1877)

Morton, H. "Reports of the Philadelphia Photographic Expedition." P115. (1877)[3]

Pickering, E. C. "Report of Prof. Edward C. Pickering, Mount Pleasant, Iowa." P91. (1877a)

Pickering, E. C. "Notes on Photographing the Corona." P156. (1877b)

Stockwell, J. N. "Report of Mr. John Stockwell, Burlington, Iowa." P59. (1877)

Van Vleck, J. M. "Report of Prof. J. M. Van Vleck, Mount Pleasant, Iowa." P85. (1877)

Watson, J. C. "Report of Prof. James C. Watson, Mount Pleasant, Iowa." P77. (1877)

Weissner, J. "Report of Mr. John Wiessner." P69. (1877)

Young, C. A. "Report of Professor Charles A. Young, Burlington, Iowa." P39. (1877)

Peirce, B. *Report to the Superintendent of the United States Coast Survey, Showing the progress of the Survey During the Year 1869*. Appendix No. 8. Washington: United States Government Printing Office. (1872a)

Includes:

Blake, F., Jr. "Report by F. Blake, Jr., United States Coast Survey." P141. (1872)

Blickensderfer, J., Jr. "Observations of the Eclipse at Cherokee, Iowa." P176. (1872)

[3] also published in the *Journal of the Franklin Institute*
[4] also published in the *Journal of the Franklin Institute*

Cevil, S. A. "Report of Mr. S. A. Cevil." P169. (1872)

Cutts, R. D. & Mosman, A. T. "Observations from Bristol, Tennessee." P117. (1872)

Davidson, G. "Observations at Kohklux, Chilkaht River, Alaska." P177. (1872)

Dean, G. W. "Report by G. W. Dean, Assistant United States Coast Survey." P137. (1872)

Eimbeck, W. "Report by Professor William Eimbeck to Julius Pitzman, Esq." P174. (1872)

Fay, C. N. "Report of C. N. Fay, Aid in Observing Eclipse of August 7, 1869, at Springfield, Illinois." P160. (1872)

Goodfellow, E. "Reports of Observers Attached to the Des Moines Party." P165. (1872)

Hilgard, J. E. P175. (1872)

Horr, E. W.; Anderson, W. I.; & Wormwood, W. W. P172. (1872)

Lane, J. H. "Report of J. H. Lane, Esq." P167. (1872)

Langley, S. P. "Report by Professor S. P. Langley of Observations at Oakland, Kentucky." P134. (1872)

McLeod, R. A. "Report of Mr. R. A. McLeod." P157. (1872)

Peirce, B. "Observations at Hanover College, Indiana." P161. (1872b)

Peirce, B. "Observations at Des Moines, Iowa, Near Cedar Falls, Iowa, and Near Saint Louis, Missouri." P163. (1872c)

Peirce, C. S. "Report on the Results of the Reduction of the Measures of the Photographs of the Partial Phases of the Eclipse of August 7, 1869, Taken at Shelbyville, Kentucky, under the Direction of Professor Winlock." P181. (1872)

Peirce, C. S. & Smith, J. L. "Observations at Falmouth." P126. (1872)

Peirce, J. M. "Reports of Gentlemen Attached to the Springfield Party." P153 & P159. (1872)

Pitzman, J. "Observations Near Saint Louis to Determine the Southern Limit of Totality." P173. (1872)

Schott, C. A. "Observations at Springfield, Illinois." P145. (1872)

Searle, A. "Memorandum by Arthur Searle of Work Done at Falmouth, Kentucky, on August 6 and 7, 1869, with Reference to Observations of the Solar Eclipse." P128. (1872)

Searle, G. M. P135. (1872)

Seaver, E. P. "Report of Mr. E. P. Seaver." P156. (1872)

Seymour, C. B. P144. (1872)

Twining, A. C.; Carter, G. T.; & Ayerigg, B. "Reports of Professor A. C. Twining, Mr. George T. Carter; & Mr. Benjamin Ayerigg." P161. (1872)

Warner, J. B. "Report of Mr. J. B. Warner." P155. (1872)

Sands, B. F. *Reports of Observations of the Total Eclipse of the Sun, August 7, 1869*. Washington: United States Government Printing Office. (1869a)
Includes:

Bardwell, F. W. "Report of Mr. F. W. Bardwell." P187. (1869)

Curtis, E. "Report of Dr. Edward Curtis, U. S. A." P121. (1869a)

Eastman, J. R. "Report of Professor John R. Eastman, U. S. N." P97. (1869)

Gilman, W. S., Jr. "Report of Mr. W. S. Gilman, Jr." P171. (1869)

Hall, A. "Report of Professor Asaph Hall, U. S. N." P197.

Harkness, W. "Report of Professor William Harkness, U. S. N." P23. (1869)

Myer, A. J. "Report of Brevet Brig. Gen. Albert J. Myer, Chief Signal Officer U. S. A." P191. (1869) [Includes a letter from Charles B. Coale].

Newcomb, S. "Report of Professor Simon Newcomb, U. S. N." P5. (1869b)
Sands, B. F. "Report of Commodore B. F. Sands, U. S. N." P3. (1869b)

Non-government
Abbe, C. Observations on the Total Eclipse of the Sun of 1869." *American Journal of Science*. #16. P264. (1872) [Cincinnati Observatory Expedition]

Ashe, E. D. *Proceedings of the Canadian Eclipse Party. Quebec*: Middleton & Dawson. (1870a) {Canadian Expedition]

Bean, J. "The Eclipse." *Friends' Review*. xxxiii, #1. P3. (1870) [Earlham College Expedition]

Blogs.IU.edu (accessed 25 February 2022) [Indiana University Expedition]

Bond, W. C. & Winlock, J. "Work of the Observatory." *Annals of the Astronomical Observatory of Harvard College*. viii. #1. P47. (1876) [Harvard College Observatory Expedition]

[Cherokee, Iowa] *Daily Times*. P1. (20 June 1989) [Blickensderfer Expedition]

Crozier, E. S. *An Account of the Observations Made at New Albany, Indiana, During the Total Eclipse of the Sun, August 7th, 1869*. New Albany (Indiana): New Albany Society for Natural History. (1869a) [New Albany Society for Natural History Expedition]

Includes:

"Report of Col. Benj. Aycrigg, Civil Engineer, of Passaic, N. J." P7. (1869)
Pell, F. A. "Report of Mr. Pell of Princeton College." P8. (1869)

Hawk Eye. Burlington (Iowa). (2021) [Pilger Expedition]

Hazen, E. H. *Photographs of the Eclipse of the Sun, August 7, 1869. Davenport* (Iowa): Griggs, Watson & Day. (1869) [Davenport Academy of Natural Sciences Expedition]

Irish, C. E. *Observations of the Total Eclipse of the Sun of August 7, 1869, at Iowa City, Iowa.* Iowa City (Iowa): Iowa Academy of Science. (1869) [University of Iowa Expedition]

M'Clune, J. *Report of Professor M'Clune*. Philadelphia: E. C. Markley & Sons. (1869) [M'Clune Expedition]

Paine, R. T. "On the Solar Eclipse of August, 1869." *Monthly Notices of the Royal Astronomical Society.* xxix, #1. P285. (1869a)

Paine, R. T. "Solar Eclipse of 1869." *Monthly Notices of the Royal Astronomical Society.* xxx, #1. P1. (1869b) [Paine Expedition (continued)]

Peters, C. H. F. "The Litchfield Expedition." [New York, New York] *Times*. P5. (28 July 1869) [Litchfield Observatory Expedition])

III. PRINCIPAL BIOGRAPHICAL SOURCES[5]

AsheFamily.info (*accessed 29 October 2021*) [Edward Ashe]

Bailey, S. I. "Edward Charles Pickering." *Biographical Memoirs.* Washington: National Academy of Sciences. (1932)

Bigelow, F. H. "William Harkness." *Popular Astronomy.* xi, #6. P181.

Blogs.UI.edu (accessed 20 November 2021) [Gustav Hinrichs]

Brent, Joseph. *Charles Sanders Peirce: A Life.* Bloomington (Indiana): Indiana University Press. (1993)

Comstock, G. C. James Craig Watson. *Biographical Memoirs.* Washington: National Academy of Sciences. (1888)

Comstock, G. C. *John Huntington Crane Coffin. Biographical Memoirs.* Washington: National Academy of Sciences. (1913)

Comstock, G. C. *Benjamin Apthorp Gould. Biographical Memoirs.* Washington: National Academy of Sciences. (1922)

---

[5] Direct quotes are referenced in Citations.

Frost, E. B. *Charles Augustus Young. Biographical Memoirs.* Washington: National Academy of Science. P90. (1910)

Hill, G. W. Asaph Hall. *Biographical Memoirs.* Washington: National Academy of Sciences. (1908)

Kelly, H. A. & Burrage, W. L. *American Medical Biographies.* Baltimore (Maryland): Norman, Remington Company. (1920) [Edward Curtis]

Kendall, P. M. *Maria Mitchell: Life, Letters, and Journals.* Boston: Lee & Shepherd. (1896)

Lovering, J. Joseph Winlock. *Biographical Memoirs.* Washington: National Academy of Sciences. (1876)

Mayer, A. G. & Woodward, R. S. Alfred Marshall Mayer. *Biographical Memoirs.* Washington: National Academy of Sciences. (1916)

Newcomb, S. *Reminiscences of an Astronomer.* (1903) [Simon Newcomb]

Nichols, E. L. Henry Morton. *Biographical Memoirs.* Washington: National Academy of Sciences. (1915)

Potter, S. *Too Near for Dreams: The Story of Cleveland Abbe.* Boston: American Meteorological Society. 2020.

"Robert Treat Paine." *Proceedings of the American Academy of Arts and Sciences.* xxi. P532. (1886)

Sands, B. F. From Reefer to Rear Admiral: Of Nearly Half a Century of Naval Life. New York: Frederick A. Stokes Company. (1899) [Benjamin Sands]

Sheehan, W. Christian Heinrich Friedrich Peters. *Biographical Memoirs.* Washington: National Academy of Sciences. (1999)

"Sketch of General Albert J. Myer." *Popular Science Monthly.* xviii. P408. (1881)

"Stephen Alexander." *Proceedings of the American Academy of Arts and Sciences.* x. P504. (1883)

Stevens, W. B. *Missouri: The Center State.* Chicago: S. J. Clark Publishing Company. [Jacob Blickensderfer]

Vassar.edu (accessed 29 October 2021) [Mary Whitney]

Wagner, H. R. "George Davidson, Geographer of the Northwest Coast of America." *California Historical Society Quarterly.* xi, #1. P298. (1932)

Walcott, C. D. "Samuel Pierpont Langley." *Biographical Memoirs.* Washington: National Academy of Sciences. (1912)

"Winthrop S. Gilman Dead." [New York, New York] *Times.* (5 October 1884)

IV.   PRINCIPAL GEOGRAPHICAL SOURCES[6]

Bailey, D. R. *History of Minnehaha County, South Dakota*. Sioux Falls (South Dakota): Brown & Saenger. (1899)

Dixon, J. M. *Centennial History of Polk County, Iowa*. Des Moines (Iowa): State Register. (1876)

Downer, H. E. *History of Davenport and Scott County Iowa*. Chicago: S. J. Clarke Publishing Company. (1910)

Freeman, W. S. *History of Plymouth County, Iowa*. Indianapolis (Indiana): B. F. Bowen. (1917)

Goldthwait, N. E. *History of Boone County, Iowa*. Des Moines (Iowa): Union Historical Company (1880)

Gresham, J. M. *Biographical and Historical Souvenir for the Counties of Clark, Crawford, Harrison, Floyd, Jefferson, Jennings, Scott and Washington, Indiana*. Chicago: Chicago Printing Company. (1889)

History of Black Hawk County, Iowa. Chicago: Western Historical Company. (1878)

History of Des Moines County, Iowa. Chicago: Western Historical Company. (1879)

History of Henry County, Iowa. Chicago: Western Historical Company. (1879)

History of Johnson County, Iowa. Iowa City (Iowa): University of Iowa. (1883)

History of Mahaska County, Iowa. Des Moines (Iowa): Union Historical Company. (1878)

History of Sangamon County, Illinois. Chicago: Inter-state Publishing Company. (1881)

History of Wapello County, Iowa. Chicago: Western Historical Company. (1878)

McCulla, T. *History of Cherokee County Iowa*. Chicago: S. J. Clarke Publishing Company. (1914)

New History of Shelby County, Kentucky. Prospect (Kentucky): Harmony House. (2003)

Taylor, O. *Historic Sullivan: A History of Sullivan County, Tennessee*. Bristol (Tennessee): King Printing Company. (1909)

---

[6] Direct quotes are referenced in Citations.

V.   OTHER PRINCIPAL SECONDARY SOURCES[7]

Books I recommend for further reading on their respective topics.

Antia, H. M; Bhatnagar, A.; & Ulmschneider, P. *Lectures on Solar Physics.* New York: Springer. (2003)

Baron, D. *American Eclipse: A Nation's Epic Race to Catch the Shadow of the Moon and Win the Glory of the World.* New York: W. W. Norton & Company. (2017) [about the 1878 total eclipse of the Sun]

Baum, R. & Sheehan, W. *in Search of Planet Vulcan: The Ghost in* Newton's *Clockwork Universe.* New York: Plenum. (1997)

Chittenden, M. H. *History of Early Steamboat Navigation on the Missouri River.* New York: Frances P. Harper. (1903)

Clerke, A. M. *A Popular History of Astronomy during the Nineteenth Century.* Edinburgh: Adam & Charles Black. (1885)

Cottam, S. & Orchiston, W. *Eclipses, Transits, and Comets of the Nineteenth Century: How America's Perception of the Skies Changed.* New York: Springer. (2015)

Dick, S. J. *Sky and Ocean Joined: The U. S. Naval Observatory1830-2000.* Cambridge (England): University Press. (1997)

Hearnshaw, J. B. *The Analysis of Starlight: One Hundred and Fifty Years of Astronomical Spectroscopy.* Cambridge (England): Cambridge University Press. (1986)

Hollabaugh, M. *The Spirit and the Sky: Lakota Visions of the Cosmos.* Lincoln (Nebraska): University of Nebraska Press. (2017)

Hufbauer, K. *Exploring the Sun: Solar Science Since Galileo.* Baltimore (Maryland): The Johns Hopkins University Press. (1991)

Jones, B. Z. & Boyd, L. G. *The Harvard College Observatory: The First Four Directorships, 1839-1919.* Cambridge (Massachusetts): Harvard University Press (1971)

Jortner, A. *The Gods of Prophetstown.* New York: Oxford University Press. (2012)

King, H. C. *The History of the Telescope.* Buckinghamshire (England): Charles Griffin & Company. (1955)

Moyer, A. E. *A Scientist's Voice in American Culture.* Berkeley (California): University of California Press. (1992)

---

[7] Direct quotes are referenced in Citations.

Phalen, W. J. *How the Telegraph Changed the World*. Jefferson (North Carolina): McFarland & Company. (2015)

Rothenberg, Marc. *The Educational and Intellectual Background of American Astronomers, 1825–1875*. PhD Dissertation. Bryn Mawr University. (1985)

Seidelmann, P. K. & Hohonkerk, C. Y. (editors) *The History of Celestial Navigation*. New York: Springer. (2020)

Sheehan, W. & Westfall, J. *The Transits of Venus*. Amherst (New York): Prometheus Books. (2004)

Signal Service History: The Corps that Almost Wasn't. Colorado Springs (Colorado): Colorado Springs Historical Society. (2015)

Vaucouleurs, G. de. *Astronomical Photography: From Daguerreotype to the Electron Camera*. New York: The MacMillian Company. (1961)

Warner, D. J. *Alvan Clark and Sons: Artists in Optics*. Washington: Smithsonian Institution. (1968)

Westfall, J. & Sheehan, W. *Celestial Shadows: Eclipses, Transits, and Occultations*. New York: Springer. (2015)

Wolmar, C. *The Great Railroad Revolution*. New York: PublicAffairs. (2012)

VI.   GENERAL ASTRONOMY BOOKS *CIRCA* 1869[8]

These books were consulted for context. Originals were examined so that any margin notes could be read.

Airy, G. B. *Popular Astronomy*, 6th edition. London: MacMillan & Company. (1868)

Bouvier, H. M. *Familiar Astronomy*. Philadelphia (Pennsylvania): Childs & Peterson. (1857)

Dick, T. *Celestial Scenery*. Brookfield (Massachusetts): E. & I. Merrian. (1838)

Dick, T. *Sidereal Heavens*. New York: Harper & Brothers. (1840)

Flammarion, C. *Marvels of the Heavens*. London: Richard Bentley & Son. (1872)

Guillemin, A. *The Heavens*, 7th edition. New York: Scribner & Welford. (1881)

Herschel, J. F. W. *Outlines of Astronomy*. New York: P. T. Collier & Son. (1902)

Houghton, S. *Manual of Astronomy*, new edition. London: Cassell, Petter, & Galpin. (1857)

---

[8] Direct quotes are referenced in Citations.

Jackson, E. P. *Astronomical Geography*. Boston (Masssachusetts): D. C. Heath & Company. (1898)

Lardner, D. *Popular Astronomy*. London: Walton & Maberly. (1856)

Loomis, E. *Treatise on Astronomy*. New York: Harper & Brothers. (1872)

Mitchel, O. M. *Orbs of Heaven*, 6th edition. London: Nathaniel Cooke. (1856)

Mitchel, O. M. *Planetary and Stellar Worlds*. London: Partridge & Oakley. (1852)

Newcomb, S. & Holden, E. S. *Astronomy for High Schools and Colleges*, 6th edition. New York: Henry Holt & Company. (1889)

Newcomb, S. *Popular Astronomy*, 2nd edition. London: MacMillan & Company. (1898)

Olmstead, D. *Introduction to Astronomy*. New York: Collins & Brother. (1861)

Proctor, R. A. *Other Worlds Than Ours*. New York: D. Appleton & Company. (1896)

Sharpless, I. & Philips, G. M. *Astronomy for Schools and General Readers*, 3rd edition. Philadelphia (Pennsylvania): J. B. Lippincott Company. (1882)

Warren, H. W. *Recreations in Astronomy*. New York: Harper & Brothers. (1879)

Webb, T. W. *Celestial Objects for Common Telescopes*, 2nd edition. London: Longmans, Green, & Company. (1868)

Young, C. A. *Textbook of General Astronomy for Colleges and Scientific Schools*. Boston (Massachusetts): Ginn & Company. (1888)

VII. OTHER SOURCES

<u>Period Newspapers</u>

[Eddyville, Iowa] *Advertiser*. P1. (4 September 1869)

[Jefferson, Iowa] *Bee*. P1. (29 April 1869)

[Boone, Iowa] *Boone County Democrat*. P1. (4 August 1869a)

[Boone, Iowa] *Boone County Democrat*. P1. (12 August 1869b)

[Monmouth College] *Courier*. P1. (1869)

[Waterloo, Iowa] *Courier*. P1. (8 August 1869)

[Louisville, Kentucky] *Daily Express*. P1. (6 August 1869)

[Terre-Haute, Indiana] *Daily Express*. P1. (31 July 1869a)

[Terre Haute, Indiana] *Daily Express*. P1. (7 August 1869b)

[Terre Haute, Indiana] *Daily Express*. P1. (9 August 1869c)

[Keokuk, Iowa] *Daily Gate City*. P1. (8 August 1869)

[Cincinnati, Ohio] *Daily Gazette*. P1. (9 August 1869a)

[Cincinnati, Ohio] *Daily Gazette*. P1. (14 August 1869b)

[Davenport, Iowa] *Daily Gazette*. P1. (6 August 1869a)

[Davenport, Iowa] *Daily Gazette*. P1. (7 August 1869b)

[Evansville, Indiana] *Daily Journal*. P1. (9 August 1869)

[Bloomington, Illinois] *Daily Leader*. P1. (10 August 1869)

[Council Bluffs, Iowa] *Daily Nonpareil*. P1. (8 August 1869)

[Des Moines, Iowa] *Daily State Register*. P1. (8 August 1869)

[Cherokee, Iowa] *Daily Times*. P1. (20 June 1989)

[Fayetteville, North Carolina] *Eagle*. P1. (8 August 1869)

[Vinton, Iowa] *Eagle*. P1. (18 August 1869)

[Cairo, Illinois] *Evening Bulletin*. P1. (9 August 1869)

[Louisville, Kentucky] *Evening Express*. P1. (7 August 1869a)

[Louisville, Kentucky] *Evening Express*. P1. (9 August 1869b)

[Webster City, Iowa] *Freeman*. P1. (21 April 1869a)

[Webster City, Iowa] *Freeman*. P1. (25 August 1869b)

[Cedar Falls, Iowa] *Gazette*. P1. (13 August 1869)

[Burlington, Iowa] *Hawk Eye*. P1. (7 August 1869a)

[Burlington, Iowa] Hawk Eye. P1. (11 August 1869b)

[Mount Vernon, Iowa] *Hawkeye*. P1. (30 July 1869)

[Oskaloosa, Iowa] *Herald*. P1. (12 August 1869)

[London, England] *Illustrated London News*. (9 October 1869)

[Indianapolis, Indiana] *Journal*. P1. (9 August 1869)

[Indianola, Iowa] *Journal*. P1. (15 July 1869a)

[Indianola, Iowa] *Journal*. P1. (30 September 1869b)

[Macomb, Illinois] *Journal*. P1. (30 July, 1869a)

[Macomb, Illinois] *Journal*. P1. (3 September, 1869b)

[Muscatine, Iowa] *Journal*. P1. (14 August 1869)

[Sioux City, Iowa] *Journal*. P1. (12 August 1869)

Fairfield, Iowa] *Ledger*. P1. (12 August 1869)

[Marshalltown, Iowa] *Marshall* [County] *Times*. P1. (14 August 1869)

[Greensboro, North Carolina] *Patriot*. P1. (11 February 1869a)

[Greensboro, North Carolina] *Patriot*. P1. (12 August 1869b)

[Bloomington, Illinois] *Pentagraph*. P1. (11 August 1869)

[Brighton, Iowa] *Pioneer and Home Visitor*. P1. (8 August 1869a)

[Brighton, Iowa] *Pioneer and Home Visitor*. P1. (20 August 1869b)

[Wilmington, North Carolina] *Post*. P1. (8 August 1869)

[Iowa City, Iowa] *Press Citizen*. P1. (28 July 1869)

[Marion, Iowa] *Register*. P1. (28 July 1869a)

[Marion, Iowa] *Register*. P1. (8 September 1869b)

[Marion, Iowa] *Register*. P1. (15 September 1869c)

[Hampton, Iowa] *Reporter*. P1. (4 August 1869)

[Iowa City, Iowa] *Republican*. P1. (28 July 1869)

[Tipton, Indiana] *Republican*. P1. (12 August 1869)

[Greenville, Indiana] *Republican Banner*. P1. (6 August 1869)

[Shelbyville, Kentucky] *Shelby* [County] *Sentinel*. P1. (11 August 1869)

[Iowa City, Iowa] *State Democratic Press*. P1. (4 August 1869a)

[Iowa City, Iowa] *State Democratic Press*. P1. (11 August 1869b)

[Indianapolis, Indiana] *State Sentinel*. P2. (4 August 1869)

[New York, New York] *Sun*. P1. (9 August 1869)

[Toledo, Iowa] *Tama County Republican*. P1. (29 July 1869a)

[Toledo, Iowa] *Tama County Republican*. P1. (12 August 1869b)

[Rutgers University] *Targum*. P1. (October 1869)

Cedar Rapids [Iowa] *Times*. P1. (17 June 1869a)

Cedar Rapids [Iowa] *Times*. P1. (12 August 1869b)

Cedar Rapids [Iowa] *Times*. P1. (26 August 1869c)

Cedar Rapids [Iowa] *Times*. P1. (9 September 1869d)

[Fort Dodge, Iowa] *Times*. P1. (30 July 1869a)

[Fort Dodge, Iowa] *Times*. P1. (15 August 1869b)

[Paris, Kentucky] *True Kentuckian*. P1. (11 August 1869)

[Nashville, Tennessee] *Union*. P1. (7 August 1869)

[Knoxville, Iowa] *Voter*. P1. (15 July 1869a)

[Knoxville, Iowa] *Voter*. P1. (19 August 1869b)

[Raleigh, North Carolina] *Weekly North Carolina Standard.* P1. (11 August 1869)

[Vincennes, Indiana] *Weekly Western Sun.* P1. (14 August 1869)

Alphabetized by Publication

[Ann Arbor, Michigan] *Argus.* P1. (25 December 1891)

[Sidney, Iowa] *Argus Herald.* P1. (24 August 2017)

Astronomical and Meteorological Observations Made During the Year 1869. Washington: Government Printing Office. (1869)

Atlantic Almanac for 1869. (1869)

[Cherokee, Iowa] *Courier.* P1. (21 January 1965)

Harvard Gazette. P1. (5 May 2019)

[Indianapolis, Indiana] *Journal.* P2. (8 May 1899)

[Iowa City, Iowa] *Press-Citizen.* (1949)

The Monitor [of the Archdiocese of San Francisco, California]. lviii, #11. P1. (1904)

Monthly Notices of the Royal Astronomical Society. xxx. P174. (1870)

Musical Review and Musical World: A Journal of Secular and Sacred Music. xii. P189. (1862)

Publications of the Astronomical Society of the

[Des Moines, Iowa] *Register.* P1. (5 August 2019)

[Springfield, Illinois] *Register.* P1. (22 August 2017)

Smithsonian American Art Museum. Commemorative Guide. Nashville (Tennessee): Beckon Books. (2015)

The Tennessean. P1. (18 August 2017)

Transportation in Iowa. New York: Arno Press. (1981)

Alphabetized by Title

"Astronomical Phenomena and Progress." *American Annual Cyclopedia.* New York: D. Appleton & Company. P37. (1870)

"Benjamin Banneker." *Atlantic Monthly.* xi, #63. (1863)

"Centenary of William Harkness, 1837-1903." *Nature.* cxl. P1004. (1937)

"Centennial for a Spectrometer." *Dartmouth Alumni Magazine.* lxxiv, #4. P34. (1966)

"Gifts and Loans." in *Maria Mitchell Association Annual Report.* Nantucket (Massachusetts): Maria Mitchell Association. P22. (1907)

"Jonathan Ingersoll Bowditch." *Proceedings of the American Academy of Arts and Sciences.* xxiv. (1889) "The Government and the Eclipse Expedition." *Nature.* ii. P409. (1870)

"The Great Eclipse." *The Friend, A Religious and Literary Journal.* xlii, #49. P388. (1869)

"The Great Solar Eclipse." *The Ladies' Repository.* P285. (1869)

"The Late Solar Eclipse." *American Phrenological Journal.* xlix, #1. P395 (1869)

"Observations of the Eclipse as Seen at West-Port, KY." *Scientific American.* xxi, #11. P165. (1869)

"Photographs of the Eclipse." *Friends' Intelligencer.* xxxvi. P498. (1870)

Recent Solar Researches." *Maine Farmer. xxxviii*, #36. P4. (1870)

"Report of the Council." *Monthly Notices of the Royal Astronomical Society.* iii. P77. (1870)

"Scientific Intelligence." American Journal of Science and the Arts. xxx. P261. (1860)

"Solar Eclipse of July 18, 1860." *American Journal of Science and Arts.* xxx, #88. P36. (1860)

"The Total Eclipse of 1868." *Astronomical Register.* vii. P186. (1869)

"The Total Eclipse of August Seventh." *Catholic World.* x, #8. P113. (1869)

"Total Solar Eclipse in August." *Western Christian Advocate,* xxxvi, #31. P242. (1869)

Alphabetized by Author

40th Congress. "House Miscellaneous Document." in *United States Congressional Serial Set Number 1385.* Washington: Government Printing Office. (1869)

A

AAAS.org (accessed 1 April 2022)

AAIHS.org (accessed 27 March 2022)

Abbe, C. "The Approaching Eclipse." *The Mystic Star.* x. P181. (1869)

Abbe, C. "The Structure of the Corona." *Nature.* v, #123. P367. (1872)

Abbe, C. *Report on the Solar Eclipse of July, 1878.*[9] Washington: United States Printing Office. (1881)

---

[9] same as *Professional Papers of the Signal Service.* #1

Abbe, C. William Ferrel. *Biographical Memoirs.* Washington: National Academy of Sciences. (1892)

Abbe, T. *Professor Abbe and the Isobars.* New York: Vantage Press. (1955)

AbbeMuseum.org (accessed 20 March 2022)

Adams, C. "Railway Problems in 1869." *North American Review.* cx. P116. (1870)

Airy, G. "The Total Eclipse Expedition." *Astronomical Register.* ixvi, #1. P257. (1870)

Albertson, M. A. "Curator's Report." *Maria Mitchell Association Annual Report.* xi. P9. (1913)

Alexander, S. & Henry, J. *"Observations of Relative Radiation of the Solar Spots." Proceedings of the American Philosophical Society. iv. P173. (1845)*

AmPhilSoc.org (accessed 8 February 2022)

AncientPages.com (accessed 23 February 2022)

Archives.Dickinson.edu (accessed 9 March 2022)

Ashe, E. D. "On Solar Spots." *Transactions of the Literary and Historical Society of Quebec.* v. P5. (1867)

Ashe, E. D. "Solar Eclipse of August 7, 1869." *Monthly Notices of the Royal Astronomical Society.* iii. P3. (1870a)

Ashe, E. D. "On His Photographs Taken During the Total Solar Eclipse, Aug. 7, 1869." *Monthly Notices of the Royal Astronomical Society.* iii. P173. (1870b)

AutomotiveHistory.org (accessed 3 March 2022)

B

Bappu, M. K. V. "The Eclipses of the Sun." Inaugural Address given at a seminar, *The Total Eclipse of February 16, 1980.* Osmania (Australia): Center of Advanced Studies in Astronomy. (1979)

Barnard, E. E. "The Development of Photography in Astronomy." *Science.* viii, #194. P341. (1898)

Bartky, I. R. "Chicago's Dearborn Observatory." *Journal of Astronomical History and Heritage.* ii, #2. P93. (2000)

Bell, T. E. "Ingenuity in the Moon's Shadow." *The Sciences.* #14. P14. (1999)

Bell, T. E. "The Eclipse Chasers." *Griffith Observer.* lxv. P2. (2001)

Bell, T. E. "The Victorian Space Program." *The Bent of Tau Beta Pi.* #1. P15. (2003)

Blaise, C. *Time Lord.* New York: Pantheon Books. (2001)

Blogs.DavenportLibrary.com (accessed 30 November 2021)

Blogs.LoC.gov (accessed 3 March 2022)

Bray, W. "James Cremer." *Stereo World.* vi, #3. P4. (1979)

BridgesDB.com (accessed 3 March 2022)

Brothers, A. "Eclipse photography." *Monthly Notices of the Royal Astronomical Society.* xxxii. P290. (1872)

Broughton, P. *"James Craig Watson." Journal of the Royal Astronomical Society of Canada.* xc. P74. (1996)

Byrd, M. E. "Anna Winlock." *Popular Astronomy.* xii. P254. (1904)

C

Cameron, F. V.; Clarke, F. W.; Seaman, W. H. *et al.* "Robert Browne Warder." *Science.* xxiii, #579. P195. (1906)

Cameron, G. L. *Public Skies: Telescopes and the Popularization of Astronomy in the Twentieth Century.* Doctoral dissertation in the Program of History of Science and Technology. Ames (Iowa): Iowa State University. (2010)

Campbell, W. W. "The Solar Corona." *Publications of the Astronomical Society of the Pacific.* xix, #13. P71. (1907)

Campbell, W. W. "The Astronomical Activities of Professor George Davidson." *Publications of the Astronomical Society of the Pacific.* xxvi, #152. P28. (1914)

Carlson, D. *Star Gazing: Observatories at Gettysburg College, 1874-Present.* Unpublished document. (Spring 2006)

Carter, B. & Carter, M. S. *Simon Newcomb: America's Unofficial Astronomer Royal.* Saint Augustine (Florida): Mantanza Publishing. (2006)

Challis, J. "On the Indications by Phenomena of Atmospheres to the Sun, Moon, and Planets." *Memoirs of the Royal Astronomical Society.* xxiii, #8. P231. (1864)

Chamberlain, V. "Astronomical Content of Plains Indian Calendars." *Journal for the History of Astronomy.* Supplement 1. xv, #6. P1. (1984)

Cherba, C. "Dr. Asa Horr – Dubuque's Early Physician, Weatherman, and Scientist." *Julien's Journal.* xxxv, #12. P1. (2020)

Claridge, G. C. "Coronium." *Journal of the Royal Astronomical Society of Canada.* xxxi, #8. P337. (1937)

Clark, G. K. C. "Short and Somewhat Personal History of Yukon Glacier Studies in the Twentieth Century." *Arctic.* lxvii supplement, #1. P1. (2014)

Clarke, D. *Stellar Polarimetry.* Weinheim (Germany): Wiley VCH. (2010)

Coffin, J. H. C. "Suggestions for Observing the Total Eclipse of the Sun on August 7, 1869." *American Ephemeris and Nautical Almanac.* Supplement. Washington: Government Printing Office. (1869b)

Copernicus, K. "The Eclipse Party." *Nassau Literary Magazine.* xxvi, #2. P104. (1869)

Corbin, A. *The Life and Times of the Steamboat Red Cloud or How Merchants, Mounties, and the Missouri Transformed the West.* College Station (Texas): Texas A & M University Press. P15. (2006)

CosmosClub.org (accessed 27 December 2021)

Cottam, S.; Pearson, J.; Orchiston, W.; *et al.* "The Total Solar Eclipses of 7 August 1869 and 29 July 1878 and the Popularization of Astronomy in the USA as Reflected in the New York Times." in Orchiston, W. *Exploring the History of New Zealand Astronomy: Trials, Tribulations, Telescopes and Transits.* New York: Springer. (2015)

Crocker, H. *Green River Steamboating[:] a Cultural History, 1828-1923.* Masters Theses, Western Kentucky University. (1970)

Croffut, W. A. *Celestial Flirtation.* Cincinnati (Ohio): Moore, Wilstach & Moore. (1869)

Crookes, W. "The American Eclipse." *Nature.* i, #6. P170. (1869)

Crosby, J. O. "Scientific Studies of Dr. Asa Horr." *Annals of Iowa.* xii, #3. P170. (1915)

Crozier, E. S. "The Great Eclipse of 1869." *Odd Fellow's Companion.* v. P111. (1869)

Curtis, E. "The Des Moines Eclipse Expedition." *The Philadelphia Photographer.* vi. P309. (1869b)

Curtis, E. *Total Eclipse of the Sun, August 7, 1869.* Washington: Surgeon General's Office, Army Medical Museum. (1869c)

Curtis, H. D. "Eighty Years of Astronomy at the University of Michigan." *Michigan Alumnus Quarterly Review.* xli, #1. P244. (1934)

Curtis, W. *The Hill-Brown Theory of the Moon's Motion.* New York: Springer. (2010)

D

[Anchorage, Alaska] *Daily News* (2016)

Daniels, R. *Images of Rail: Sioux City Railroads.* Charleston (North Carolina): Arcadia Publishing. (2008)

Davidson, G. "An Explanation of an Indian Map." *Manzana.* i. P75. (1901)

Dawson, B. H. "The Flash Spectrum." *Popular Astronomy*. xxv. P10. (1917)

Dick, S. "Sears Cook Walker and the Philadelphia High School Observatory." *Bulletin of the American Astronomical Society*. xxii, P122. (1990)

Director's Staff Division. *Discovering Historic Iowa Transportation Milestones*. Ames (Iowa): Department of Transportation. (1974)

Donnelly, K. "On the Boredom of Science: Astronomy in the Nineteenth Century." *British Journal for the History of Science*. xlvii, #3. P479. (2013)

Donati, G. B. "Professor Donati's Lecture on Solar Phenomena." *Astronomical Register*. viii. P34. (1870)

Douglas, James. "On Recent Spectroscopic Observations of the Sun, and the Total Eclipse of the 7th August 1869." *Transactions of the Literary and Historical Society of Quebec*. vii. P56. (1870)

Downward, J. B. "The Main Question and Aims Guiding Peirces' Phenomenology." *Cognito*. xvi, #1. P87. (2015)

E

Eastman, J. *Memoirs of the Royal Astronomical Society*. xli. P191. (1879)

Edison.Rutgers.edu (accessed 3 March 2022)

Engelhardt, G. *Philadelphia Pennsylvania: The Book of Its Course & Co-operating Bodies*. Philadelphia: Lippincott Press. (1898)

ExploreNorth.com (accessed 5 February 2022)

F

Farrell, N. E. *Colorado, the Rocky Mountain Gem, as it is in 1868*. Chicago: The Western News Company. (1868)

Federal Writers' Project. *Slave Narratives: Indiana*. Washington: Work Projects Administration. (1941)

Fiala, A. D.; Dunham, David W.; & Sofia, S. "Variation of the Solar Diameter from Solar Eclipse Observations, 1715–1991." *Solar Physics*. v, #152. P97. (1994)

Fort, C. *New Lands*. Wake Forest (North Carolina): Baen Books. P18. (1923)

Fotheringham, J. K. *Historical Eclipses*. Oxford (England): Clarenden Press. (1921)

Franklin, R. S. "The Eclipse in Siberia." *The Overland Monthly*. vi. P519. (1871)

Furman, F. (editor). *Morton Memorial: A History of the Stevens Institute of Technology, with Biographies of the Trustees, Faculty, and Alumni, and a Record of the Achievements of the Stevens Family of Engineers*. Hoboken (New Jersey): Stevens Institute of Technology. (1905)

Furness, C. E. "Mary W. Whitney." *Popular Astronomy*. xxx. P597. (1922)

G

Gallarno, G. "How Iowa Cared for Orphans of Her Soldiers of the Civil War." *The Annals of Iowa.* xv, #3. P168. (1926)

GilderLehrman.org (accessed 6 February 2022)

Gillman, A. W. *Searches into the History of the Gillman or Gilman Family.* London: Elliot Stock. (1895)

Gilman, W. S. "The Anvil Protuberance." *Journal of the Franklin Institute.* lix, #6. P417. (1870)

Goodman, R. *Lakȟóta Star Knowledge.* (3rd Edition; edited by Seeger, A.). Rosebud Sioux Reservation: Sinҭe Gleška University Press. (2017)

Gould, B. "Aus einem Schreiben des Herrn. Dr. Gould an den Herausgeber." *Astronomische Nachteren.* liv, #1776. P375. (1869)

Gould, B. "The Corona." *Nature.* ii, #34. P141. (1870)

Grant, R. & Hofsommer, D. L. *Iowa's Railroads.* Bloomington (Indiana): Indiana University Press. (2009)

GraphicArts.Princeton.edu (accessed 12 March 2022)

GreatAmericanStations.com (accessed 25 March 2022)

GreeneCountyIowaHistoricalSociety.org (accessed 13 March 2022)

GreenwichMeanTime.com (accessed 12 March 2022)

H

Haig, C. T. "Spectroscopic Observations of the Total Eclipse, August 18, 1868." *Proceedings of the Royal Society of London.* xvii. P74. (1868)

Haines, Aubrey L. "A Voyage to Montana." *Montana.* l, #1. P18. (2000)

Hale, E. E. (editor). *Autobiography, Diary and Correspondence.* Boston: Houghton, Mifflin & Company. (1891)

Hall, A. "Supplementary Notes on Observations for Magnetism and Position, made in the U. S. Naval Observatory Expedition to Observe the Solar Eclipse of Aug. 7, 1869." *Astronomische Nachrichten.* lxxv. P323. (1870)

Hall, A. "On the Application of Photography to the Determination of Astronomical Data." *American Journal of Science.* ii. P25. (1871)

Hall, A. *The Astronomer's Wife.* Baltimore (Maryland): Nunn & Company. (1908)

HAO.NCAR.edu (accessed 13 December 2021)

Harkness, W. M. "Observations on Terrestrial Magnetism and on the Deviations of the Compasses of the United States Iron Clad Monadnock during Her Cruise from Philadelphia to San Francisco, in 1865 and 1866." *Smithsonian Contributions to Knowledge.* ccxxxix. P1. (1871a)

Harkness, W. M. "Report of Professor Wm. Harkness, U. S. N." in Sands, B. F. *Washington Observations for 1869, Appendix I: Reports of the Observations of the Total Solar Eclipse of December 22, 1870.* Washington: Government Printing Office. P43. (1871b)

Helmholtz, H. L. F. "The Conservation of Force." Introductory lecture to a series delivered at Karlsruhe, Baden [Germany] Winter 1862–1863. (1862)

Henry, D. L. *Across the Shaman's River.* Fairbanks (Alaska): University of Alaska Press. (2017)

Heyeraft, S. *History of Elizabethtown Kentucky and its Surroundings.* Elizabethtown (Kentucky): Women's Club of Elizabethtown. (1921)

Himes, C. F. *Some of the Methods and Results of Observation of the Total Eclipse of the Sun, August 7th, 1869.* Gettysburg (Pennsylvania): J. E. Wible. (1869a)

Himes, C. F. "The Total Eclipse of the Sun of 1869." *Evangelical Quarterly Review.* lxxxi. P124. (1869b)

Hinckley, T. C. "William H. Seward Visits His Purchase." *Oregon Historical Quarterly.* ixxii, #2. P127 (1970)

Hind, J. R. "Stellar Objects Seen During the Eclipse of 1869." *Nature.* xviii. P663. (1878)

History.com (accessed 15 December 2021)

Hoel, A. S. "Measuring the Heavens: Charles S. Peirce and Astronomical Photography." *History of Photography.* xli, #1. P49. (2016)

Hofsommer, D. L. "A Chronology of Iowa Railroads." *Railroad History.* #132. P70. (1975)

Hogg, S. H. "Early Days of Astronomy at Toronto - Part IV: The Total Solar

Eclipse of August 7, 1869." *Journal of the Royal Astronomical Society of Canada.* lxxvi, #4. P235. (1982)

Hough, G. "Eclipse Observations at Mattoon, Illinois." *Journal of the Franklin Institute.* lix. P58. (1870)

Houlette, W. *Iowa: The Pioneer Heritage.* Des Moines (Iowa): Wallace-Homestead Book Company. (1970)

Howe, C. S. "John Nelson Stockwell." *Science. lii, #1. P35. (1921)*

HumanRights.Iowa.gov (accessed 2 April 2022)

Humiston, T. "When the Eclipse Arrived 150 Years Ago." *Cecil* [County] *Whig.* (17 August 2017)

Hunter, A. "The Origin of Coronium Lines." *Nature.* cl. P756. (1942)

Hussey, T. "History of Steamboating on the Des Moines River, from 1837 to 1862." *Annals of Iowa.* iv, #5. P323. (1900)

J

Janssen, M. "The Total Solar Eclipse of 1868." *Astronomical Register*. P131. (1869)

Jarrell, R. A. "Origins of Canadian Government Astronomy." *Journal of the Royal Astronomical Society of Canada*. lxix. P75. (1975)

Jarrell, R. A. *The Cold Light of Dawn*. Toronto: University of Toronto Press. (1988)

Johnston, A. K. *School Atlas of Astronomy*. Edinburgh: Blackwood. (1869)

Johnston, G. *The Poets and Poetry of Chester County, Pennsylvania*. Philadelphia: J. B. Lippencott Company. (1870)

Jones, P. B. "The Eclipse Expedition." *The Philadelphia* [Pennsylvania] *Photographer*. vi. P271. (1869)

JuliensJournal.com (accessed 20 March 2022)

K

Kalb, A. J. "The Eclipse—A Remarkable One." *The Christian Advocate*. P1. (5 August 1869)

Kaplan, G. "Obituary: Alan Fiala." *Bulletin of the American Astronomical Society*. xliii. P9. (2011)

King, E. S. "Arthur Searle." *Proceedings of the American Academy of Arts and Sciences*. lvii, #18. P508. (1922)

Knight, H. G. "Observations of the Eclipse as Seen at West-Port, KY." *Scientific American*. xxi, #11. P165. (1869)

KUHistory.ku.edu (accessed 9 March 2022)

L

Langley, S. P. "The Eclipse." *Nature*. xvii, #461. P457. (1878)

Langley, S. P. *The New Astronomy*. Boston (Massachusetts): Houghton Mifflin & Company. (1891)

Leger, E. "Observations or Supposed Observations of the Transits of Intra-Mercurial Planets or other Bodies Across the Sun's Disk." *Observatory*. iii. P135. (1879)

Lib.UIowa.edu (accessed 25 March 2022)

LindaHall.org (accessed 11 March 2022)

Lockyer, J. N. "The Recent Total Eclipse of the Sun." Nature. i. P14. (1869a)

Lockyer, J. N. "Spectroscopic Observations of the Sun." *Astronomical Register*. #84. P253. (1869b)

Lockyer, J. N. *Contributions to Solar Physics*. London: Macmillan & Company. (1874a)

Lockyer, J. N. *A Popular Account of Inquiries into the Physical Constitution of the Sun, with Special Reference to Spectroscopic Re*searches. London: E. Clay Sons & Taylor. (1874b)

M

Mallery, G. *A Calendar of the Dakota Nation*. United States Government Printing Office. (1877)

Mallery, G. "Pictographs of the North American Indian—a Preliminary Paper" in Powell, J. W. *Fourth Annual Report of the Bureau of Ethnology to the Secretary of the Smithsonian Institution*. Washington: Government Printing Office. P13. (1886)

MathWomen.AgnesScott.org (accessed 2 April 2022)

Mayer, A. "An Abstract of Some of the Results, Measurements and Examinations of the Photographs of the Total Solar Eclipse of August 7, 1869." *Proceedings of the American Philosophical Society*. xi. P81. (1869)

McCulla, T. *History of Cherokee County*. Chicago: S. J. Clark Publishing Company. (1914)

Meadows, A. J. *Early Solar Physics*. Oxford (England): Pergamon Press. (1970)

Mericle, J. "For 1869 Solar Eclipse, Burlington Was the Best Spot." *The* [Burlington, Iowa] *Hawk Eye*. (11 August 2017)

Messier, C. "Memoire I." *The Monthly Review or, Literary Journal*. lii. P625. (1775)

Mitchell, D. G. (editor). *Atlantic Almanac for 1869*. Boston: Office of the Atlantic Almanac. (1869)

Mitchell, M. "The Total Eclipse of 1869." *Hours at Home*. P555. (October 1869)

Mitchellville.org (access 4 March 2022)

MoMA.org (accessed 25 March 2022)

Moore, E. G. "The Reverend Thomas William Webb." *Journal of the British Astronomical Association*. lxxxv. P426. (1975)

Morton, H. "Eclipse Photography." *Philadelphia* [Pennsylvania] *Photographer*. vi. P237. (1869a)

Morton, H. "Solar Eclipse—August 7, 1869." *Journal of the Franklin Institute*. lxxxviii. P200. (1869b)

N

NDStudies.gov (accessed 20 March 2022)

Newcomb, S. "On the Supposed Intra-Mercurian Planets." *Astronomical Journal*. vi. P162. (1860).

Newcomb, S. "A Proposed Arrangement for Observing the Corona and Searching for intra-Mercurial Planets during a total Eclipse of the Sun." *American Journal of Science*. xlvii, #141. P413. (1869c)

Newcomb, S. *Researches on the Motion of the Moon*. United States Printing Office. (1870)

Newcomb, S. "Report of Professor Newcomb, U. S. N." in Sands, B. F. *Washington Observations for 1869, Appendix I: Reports of the Observations of the Total Solar Eclipse of December 22, 1870*. Washington: Government Printing Office. P6. (1871)

Norberg, A. L. "Newcomb's Early Astronomical Career." *Isis*. lxix, #2. P209. (1978)

Norton, W. A. "On the Corona Seen in Total Eclipses of the Sun." *American Journal of Science*. l, #149. P250. (1870)

Now.UIowa.edu (accessed 20 November 2021)

O

Oliver, C. "Obituary Notice of Joseph Zentmayer." *Proceedings of the American Philosophical Society*. xxxi, #142. P358. (1893)

Orchiston, W.; Nakamura, T. & Strom, S. (editors) *History of Astronomy in the Asia-Pacific Region*. New York: Springer. P339. (2011)

Osterbrock, D. E. *James E. Keeler: Pioneer Astrophysicist and the Early Development of American Astrophysics*. Cambridge (England): Cambridge University Press. (1984)

P

Pang, A. S. "The Social Event of the Season: Solar Eclipse Expeditions and Victorian Culture." *Isis*. xxxxiv, #2. P252. (1993)

Peter, H. & Bhola N. D. "Discovery of the Sun's Million-Degree Hot Corona." *Frontiers of Astronomy and Space Science*. i, #2. P2. (2014)

Peters, J. "Observations of a Solar Eclipse, October 27, 1780, Made at St. John's Island, by Mess'rs. Clarke and Wright." *Memoirs of the American Academy of Arts and Sciences*. i. P143. (1783)

Petersen, W. J. "The Eclipse of 1869." *Palimpsest*. l, #2. P81. (1970)

Philosophical Society of Washington. *Bulletin*. i. Washington: Smithsonian Institution. (1874)

Pickering, E. *Observation of the Corona During the Eclipse.* P1. Unpublished manuscript. (1869)

Pickering, E. "Polarisation of the Corona." *Nature*. i. P82. (1870)

PioneerPhotographers.homestead.com (accessed 29 October 2021)

PlasticExpert.co.uk (accessed 3 March 2022)

Poole, D. *Among the Sioux of Dakota.* New York: D. Van Nostrand. (1881)

Porter, W. *Polk County Iowa and City of Des Moines.* Des Moines (Iowa): George A. Miller Printing Company. (1898)

Powell, J. W. *The Exploration of the Colorado River and its Canyons.* Washington: Smithsonian Institution. (1875)

Powell, J. W. *Fourth Annual Report of the Bureau of Ethnology to the Secretary of the Smithsonian Institution.* Washington: Government Printing Office. (1886)

Powell, W. S. (editor) "Charles Phillips." *Dictionary Of North Carolina Biography.* Chapel Hill (North Carolina): University of North Carolina Press. (1996)

Pratt, L. G. *Discovering Historic Iowa.* Des Moines (Iowa): Iowa State Department of Public Instruction. (1976)

PreCinemaHistory.net (accessed 3 March 2022)

Proctor, R. A. "Further Remarks on the Corona." *Monthly Notices of the Royal Astronomical Society*. xxx. P221. (1870)

Proctor, R. A. "Note on Oudemann's Theory of Coronal Streamers." *Monthly Notices of the Royal Astronomical Society.* xxxi. P7. (1871a)

Proctor, R. A. "Note on the Corona." *Monthly Notices of the Royal Astronomical Society.* xxxi. P153. (1871b)

Proctor, R. A. "Astronomy in America." *Popular Science Monthly.* x, #11. P75. (1876)

Purcell, W. L. "Them Was the Good Old Days" in *Davenport, Scott County Iowa.* Davenport (Iowa): Purcell Printing Company. (1922)

R

Robitaille, P-M. "Continuous Emission and Condensed Matter within the Corona." *Progress in Physics*. P1. (2013)

Rothschild, R. F. "Colonial Astronomers in Search of the Longitude of New England." *Maine History.* xxii, #4. P175. (1983)

Russell, A. J. "Discovery of the Sun's Million-Degree Hot Corona. *Frontiers in Astronomy and Space Science.* v, #9. (2018)

Russell, H. N. "Charles Augustus Young." *Monthly Notices of the Royal Astronomical Society.* lxix. P257. (1909)

S

Saal, R. "City at Center of Illinois's Last Total Eclipse." [Springfield, Illinois] *State Journal-Register.* (21 August 2017)

Safford, T. H. "Eclipse 1869." *The Western Monthly.* ii. P119. (1869)

Saint Cyr, O. C.; Young, D. E.; Pesnell, W. D. *et al.* "Recent Studies of the Behavior of the Sun's White-Light Corona Over Time." *American Geophysical Union, Fall Meeting Conference Paper.* SH44A-06. (2008)

Salisbury, D. *Elephant's Breath and London Smoke: Historical Color Names, Definitions, and Uses in Fashion, Fabric and Art.* Second Edition. Austin (Texas): Manthua-Maker. (2015)

Schaefer, B. E. "The True Identities of Professor Moriarty and Colonel Moran." *Bulletin of the American Astronomical Society.* xxiv. P1167 (1992)

Schmidt, R. E. "The Tuttles of Harvard College Observatory." *Antiquarian Astronomer.* #6. P74. (2012)

Schnorr, K; Mäckel, V., Oreshkina, S. *et al.* "Coronium in the Laboratory." *Astrophysical Journal.* dcclxxvi, #121. P1. (2013)

Scidmore, E. R. *Alaska—Its Southern Coast and the Sitkan Archipelago.* Boston (Massachusetts): D. Lothrop & Company. (1885)

scua.library.uni.edu (accessed 4 July 2022)

Seabroke, G. M. "On the Determination Whether the Corona is a Solar or Terrestrial Phenomenon." *Monthly Notices of the Royal Astronomical Society.* xxx. P193. (1870)

SeaSky.org (accessed 20 March 2022)

Seward, W. *Alaska Speech of William H. Seward at Sitka, August 12, 1869.* Washington: Philp & Solomons. (1869)

Seymour, S. *A Reminiscence of the Union Pacific Railroad, Containing Some Account of the Discovery of the Eastern Base of the Rocky Mountains; and of the Great Indian Battle of July 11, 1867.* Quebec: A. Coté & Company. (1873)

Shankland, P. D. "Nineteenth Century Astronomy at the U. S. Naval Academy." *Journal of Astronomical History and Heritage.* v, #1. P165. (2002)

ShelbyvilleHistory.org (accessed 13 November 2021)

Shellen, H. *Spectrum Analysis in its Application to Terrestrial Substances, and the Physical Constitution of Heavenly Bodies.* Lassell, J. & Lassell, C. (translators). Huggins, W. (editor). Second edition. New York: D. Appleton & Company. (1872)

Sigismondi, S. "Guidelines for Measuring Solar Radius with Baily Beads Analysis." *Science in China.* lii, #11. P1773, (2009)

Sigismonti, Constantino & Morcos, A. B. "Long Term Variations of Solar Radius." *General Relativity.* xliii. P1197. (2011)

Silverman, S. M. and Mullen, E. G. "Sky Brightness During Eclipses: A Review." *Applied Optics.* xiv. P2838. (1975)

Smith, W. "MacArthur Bridge Would Have Turned 100 this Wednesday." [Burlington, Iowa] *Hawkeye.* (26 March 2017)

Soley, J. R. *Historical Sketch of the United States Naval Academy.* Washington: United States Government Printing Office. (1876)

Space.com (accessed 28 October 2021)

Stark, R. B. "Robert Abbe: Pioneer in Plastic Surgery." *Bulletin of the New York Academy of Medicine.* xxxi, #12. P927. (1955)

Steel, D. *Eclipse: The Celestial Phenomenon that Changed the Course of History.* Washington: Joseph Henry Press. (2001)

Stevens, W. B. *Saint Louis: The Fourth City.* Saint Louis (Missouri): S. J. Clark Publishing Company. (1909)

T

Tandberg-Hanssen, E. "The History of Solar Prominence Research" in Webb, D. F.; Schmieder, B.; & Rust, D. M. (editors). *New Perspectives on Solar Prominences: Proceedings of a Meeting Held in Aussois, France.* San Francisco California: Astronomical Society of the Pacific Conference Series. P11. (1997)

Tennant, J. F. "Report on the Total Eclipse of the Sun, August 17–18, 1868." *Monthly Notices of the Royal Astronomical Society.* xxix, #10. P1. (1869)

TerraceHill.org (accessed 28 November 2021)

TheCatholicNewsArchive.org (accessed 16 February 2022)

TheFirstScout.blogspot.com (accessed 24 October 2021)

TypeWriterMuseum.org (accessed 3 March 2022)

U

UNav.edu (accessed 3 February 2022)

Uniform-reference.net (accessed 3 March 2022)

USDeadlyEvents.com (accessed 23 March 2022)

V

VacuumCleanerHistory.com (accessed 3 March 2022)

W

Warner, D. J. "Lewis M. Rutherfurd: Astronomical Photographer and Spectroscopist." *Technology and Culture*. xii, #2. P190. (1971)

Waterhouse, J. "Wet collodion in Eclipse Photography." *Observatory*, xxxix, #39, P304. (1916)

Wead, C. K.; Woodward, S.; Bryce, J. *et al.* "Simon Newcomb." *Philosophical Society of Washington Bulletin*. xv. P136. (1909)

Weaver, J. B. (editor). *The Past and Present of Jasper County Iowa*. Indianapolis (Indiana): B. F. Bowen & Company. (1912)

Webb, T. W. "American Photographs of Total Solar Eclipse of August 7, 1869." *Monthly Notices of the Royal Astronomical Society*. xxx, #4. P4. (1869)

Weber, L. L. *An industrial History of Boone County, Iowa*. Iowa City (Iowa): University of Iowa. P5 & P9. (1935)

Wendell, O. C. & Hale, G. E. "Alvan Graham Clark." *Astrophysical Journal*. vi. P136.

Whitesell, P. S. *A Creation of His Own: Tappan's Detroit Observatory*. Ann Arbor (Michigan): University of Michigan. (1998)

Whitesell, P. S. "Detroit Observatory: Nineteenth-century Training Ground for Astronomers." *Journal of Astronomical History and Heritage*. vi, #2. P69. (2003)

Wiener, P. *Evolution and the Founders of Pragmatism*. Philadelphia (Pennsylvania): University of Pennsylvania Press. (2016)

Williams, O. "On the Trail of the Corona." *Annals of Iowa*. xxix, #2. P81. (1947)

Wilson, B. H. "The Eclipse of 1869." *Palimpsest*. li, #2. P81 (1970)

Winston, I. "Charles Anthony Schott." *Science. xiv, #345. P212. (1901)*

WisconsinHistory.org (accessed 10 December 2021)

Wlasuk, P. T. "So Much for Fame*!" Quarterly Journal of the Royal Astronomical Society*. xxxvii. P683. (1996)

Wright, S. W. *Some Historic Markers in Iowa*. Iowa City: State Historical Society of Iowa. (1943)

WUStl.edu (accessed 2 April 2022)

Y

Young, C. A. "On A New Method of Observing Contacts at the Sun's Limb, and Other Spectroscopic Observations During the Recent Eclipse." *American Journal of Science and Arts.* xlviii, #144. P370. (1869a)

Young, C. A. "Spectroscopic Observations of the American Eclipse Party in Spain." *Nature.* iii. P261. (1871)

Young, C. A. "The Corona Line." *Nature.* iv. P28. (1872)

Young, C. A. *The Sun.* Akron (Ohio): The Werner Company. (1895)

Z

Zirker, J. B. & Oddbjørn, E. "Why is the Sun's Corona So Hot? Why Are the Prominences So Cool?" *Physics Today.* lxx, #8. P36. (2017)

Other

ЧЕольцсБ, А. Ф. *Извѣстія Сибирскаго Отдѣла ПипЕрскѣаго Русскаго ГЕографичЕскаго ОбщЕства.* i, #4 & #5. P1. (1871)

Citations

Roman numerals are a finding aid to the proper section within References.

I = Archives
II = Reports of Eclipse Expeditions
III = Principal Biographical Sources
IV = Principal Geographical Sources
V = Principal Secondary Sources
VI = General Astronomy Books *Circa* 1869
VII = Other Sources

<u>Chapter 1</u>

"The rush of this black wing . . ."
VII *Daily Nonpareil* 1869, Q1

"A vivid example of the manner in which events . . ."
VII Hogg 1982, Q232

Eclipse of 1780
VII Rothschild 1983

John Clarke
VII Peters 1783

Eclipse of 1860
"Solar Eclipse of July 18, 1860" 1860

The shadow of the Moon . . .
VII Space.com 2021

"I shall only say that . . ."
quoted in, *e.g.*, VII Cottam 2011, Q261

The passage of the Moon's umbra . . .
e.g., VII Kalb 1869

Chapter 2

"Newcomb was destined to become . . ."
VII Carter & Carter 2006

Learn'd Astronomer
e.g., VII Blaise 2001

Moriarity
VII Schaefer 1992

Newcomb loved to travel . . .
VII Pang 1993

Newcomb rank
VII uniform-reference.net 2022

The Ephemeris . . .
VII James 2021

"Golden era"
VII Dick 2007, Q161

"Each point of his surface . . ."
VI Haughton 1867, Q134

Between this and the next . . .
VII Tennant 1869

Motion picture projector
VII PreCinemaHistory.net 2021

Typewriter
VII TypeWriterMuseum.org 2021

Vacuum cleaner
VII VacuumCleanerHistory.com 2021

Voting machine
VII Edison.Rutgers.edu 2021

Plastic
VII PlasticExpert.co.uk 2021
Brooklyn Bridge
VII BridgesDB.com 2021

Automobile accident
VII AutomotiveHistory.org 2021

"Certainly, never before . . ."
quoted in VII Hoel 2016, Q52

"A telescope, a sextant, a chronometer . . ."
VII Blogs.LoC.gov 2021

"At no point . . ."
VII *Oskaloosa Herald* 1869, Q1

The state of Iowa is blessed . . .
VII Grant & Hofsommer 2009

A steam locomotive reached Iowa as early as 1855
VII *Transportation in Iowa* 1981

Commodities flowed east . . .
VII Director's Staff Division, Q11

Jesse James
VII Pratt 1976

The following table refers to railroad towns mentioned in this book
VII Hofsommer 1975 and other sources

"A look full of strength . . ."
VII Wead, Woodward, Bryce *et al.* 1909, Q136

This was not to be . . .
VII Bell 2001

It seems like a big jump for practitioners of the old astronomy.
e.g., VII Donnelly 2013

Chapter 3

"Some light upon this dark subject"
e.g., VII *Daily Gazette* 1869a, Q1, but datable back to at least VII Messier 1775,
Q127

"Nothing there is beyond hope . . ."
quoted in, *e.g.*, VII Fotheringham 1921, Q21

"All of our readers . . ."
VII *Daily Gazette*, Q1 (1869a)

"Everyone knows . . ."
VI Newcomb 1898, Q21

"The early inhabitants . . ."
VI Newcomb 1898, Q24

"Is that of a cone . . ."
VI Newcomb 1898, Q26

"The duration of a solar eclipse . . ."
VI Newcomb 1898, Q28

"Returning, now, to the apparent motions . . ."
VI Newcomb 1898, Q29

"Owing to the constant . . ."
VI Newcomb 1898, Q30

Using historical data
VII Curtis 2010

"Newcomb would eventually apply . . ."
e.g., Newcomb 1870

Spectroscope attention
VII *Astronomical Register* 1869

Most observers agreed . . .
VII Tennant 1869

Chapter 4

poem
VII Croffut 1869, *verses* 1 & 2

"The first condition . . ."
VII Newcomb 1890, Q63

Mitchellville
VII Mitchellville.org 2021

Population of Des Moines
VII Porter 1898

"The occupation of a station . . ."
II Sands 1869a, Q7

It would be Harkness . . .
VII Bappu 1979

He made an exhaustive study of terrestrial magnetism . . . .
VII Harkness 1871a

Breakdown
VII "Centenary of William Harkness, 1837–1903" 1937

"As the city has not built up so far north . . ."
II Harkness 1869, Q25

The spot is now in the middle of Interstate 235 . . .
VII *Register* 2019

"With a roofline parallel . . ."
II Harkness 1869, Q26

Clark 7-¾-inch-aperture achromat . . .
VII Shankland 2002

First company to manufacture . . .
Clark made many of these telescopes for the USNO . . .
VII Cameron 2021

"A remarkably fine one"
II Harkness 1869, Q28

"By Thursday, July 22 . . ."
II Harkness 1869, Q26

Chapter 5

"The vast black orb"
VII *Daily State Register* 1869, Q1

"Before the advent . . ."
"I emerged from the box . . ."
"I then took a single glance . . ."
II Newcomb 1869b, Q9

"While everyone understands . . ."
VII Crookes 1869

"I first took a general view . . ."
"An immense protuberance . . ."
"It did not seem materially brighter . . ."
"The large predominance of red . . ."
"Its structure was not uniform . . ."
"It seemed to be of a jagged outline . . ."
"A fish tail gaslight . . ."
II Newcomb 1869b, Q9

"I cannot describe the sensation . . ."
II Young 1877, Q48

The alternate reality . . .
VII Norberg 1878

"Fortunately, in order to provide against . . ."
II Curtis 1879, Q132

Curtis was using the full aperture . . .
VII Curtis 1869c

"Double the time!"
"They looked to me . . ."
II Curtis 1879, Q134

"The head of . . ."
VII Curtis 1869a, Q145

"Of a beauty and delicacy of detail . . ."
VII Curtis 1869c, Q142

This was something of an exaggeration
VII Bell 1999

As Des Moines was . . .
VII *Astronomical and Meteorological Observations Made During the Year 1869*
1869

"You have had it already."
"Never mind, give it to me again . . ."
"My view of it may have lasted five or ten seconds . . ."
II Harkness 1869, Q60

"All over."
II Harkness 1869, Q61

"The corona seemed to be composed . . ."
"The portion nearest the Sun . . ."
"Four of the star points . . ."
II Eastman 1869, Q103

"I was considerably disappointed . . ."
II Eastman 1869, Q104

Des Moines was abandoned . . .
VII Bell 2003

"The event, for the proper observation . . ."
II Harkness 1869, Q61

Even today geologists call the vicinity . . .
VII Clark 2014

"The following reasons . . ."
II Hall 1869, Q199

Russo-American *Treaty of 1824*
VII ExploreNorth.com 2021 (article 4)

Karl Neiman
VII ЧЕольцсБ 1871

"Almost the only vegetation . . ."
VII Franklin 1871, Q522

"For scientific purposes . . ."
VII Franklin 1871, Q524

Hall and Rogers
VII Hall 1870

Chapter 6

"Great Pan is dead."
e.g., Himes 1869,

"A rather dilapidated old dwelling-house . . ."
V Seidelman & Hohonkerk 2020, Q187

"For observation of the eclipse of the Sun . . ."
VII 40th Congress 1869, Q6

"For defraying expenses of observers . . ."
II Coffin 1877a, Q3

First-ever iron bridge
VII Smith 2017

"Previous to my arrival . . ."
"Besides other considerate attentions . . ."
II Coffin 1877b, Q19

Philadelphia Central High School Observatory
VII Dick 1990

Gould acted ad a sort of attaché to Coffin
VII Comstock 1922

"We were all charmed with Prof. Young . . ."
V Baron 2017, Q122

"Those whose position or engagements . . ."
II Morton 1877, Q116

"Abundant photographic records"
VII Mericle 2017, Q1

"156 feet above . . ."
II Coffin 1877b, Q18

"Thick and smoky"
II Coffin 1877b, Q19

James Clarke
VII Hale 1891

"Of the Nebula of Orion . . ."
II Coffin 1877c, Q124

"Band of rose-colored light . . ."
II Coffin 1877c, Q125

"A large prominence appeared suddenly . . ."
II Coffin 1877c, Q124

"Like streaming remnants . . ."
II Gould 1877, Q135

Charles Young's spectroscope
VII "Centennial for a Spectrometer" (1966)

"My assistant, Mr. Emerson . . ."
II Young 1877, Q46

The stone and plate were placed by . . .
VII Wright 1943

During the expedition . . .
VII Carlson 2006

"I attempted . . ."
II Eaton 1877, Q65

Chapter 7

"Athens of Iowa"
VII GreatAmericanStations.com 2021

"The marshal's tasks . . ."
IV *History of Henry County, Iowa* 1879

Some sources erroneously place Watson at Burlington
e.g., VII Whitesell 2003

"Tubby"
VII Curtis 1934, Q45

"Waited upon by . . ."
"Other prominent citizens of the place . . ."
"That the city government would provide . . ."
"Hall of fine arts . . ."
II Watson 1877, Q79

There is no indication . . .
VII Petersen 1970

"The city council issued permits . . ."
IV *History of Henry County, Iowa* 1879, Q1

George Merriman
VII Whitesell 1998

Donald McIntyre
VII [Ann Arbor, Michigan] *Argus* (1891)

"The sky was beautifully clear"
II Watson 1877, Q79

"A small group of religious fanatics . . ."
VII Petersen 1970, Q54

Pickering and polarimetry
VII Clarke 2010

"This is probably the best way . . ."
II Pickering 1877a, Q97

William Pilger
VII Wilson 1970

Chapter 8

"The hotels are making a good thing . . ."
VII *Herald* 1869, Q1

"If the halls . . ."
VII Mitchell 1869b, Q555

Jonathan Bowditch
VII "Jonathan Ingersoll Bowditch" 1889

Arthur Searle
VII King 1922

John Stockwell
VII Howe 1921

John Coffin and the Naval Academy
VII Soley 1876

"A variety of unfortunate circumstances . . ."
VII Jones 1869, Q261

Lewis Rutherfurd
VII Warner 1971

Horace Tuttle
VII Schmidt & Warner 1971

William Chauvenet
VII WUStl.edu 2021

"In person he . . ."
VII *Proceedings of the American Academy of Arts and Sciences* 1883, Q511

"Every inhabitant of Wapello County . . ."
IV *History of Wapello County* 1878, Q417

Charles Himes
VII Archives.Dickinson.edu 2021

"Ottumwa, the city of perseverance . . ."
IV *History of Wapello County* 1878, Q461

Steamboats
VII Hussey 1900

"'Och, Pat . . .'"
VII Copernicus 1869, Q7

"The ladies, as usual . . ."
VII Copernicus 1869, Q8

"Appeared not unlike pinnacled glaciers . . ."
II Alexander 1877, Q106

The first attempt . . .
e.g., VII Janssen 1869

Temperature of sunspots
VII Alexander 1845

"If the beginning of the darkness was solemn . . ."
VII Copernicus 1869, Q9

James M'Clune
VII Johnston 1890

"Oskaloosa is a quiet place"
IV *History of Mahaska County, Iowa*, Q470

"Which gave it an appearance . . ."
II M'Clune 1877, Q6

"Red flames traveled along the Moon's limb . . ."
II Austin 1877, Q75

Chapter 9

eclipse photography
VII Brothers 1872

wet collodion process
VII Waterhouse 1916

"Photography combines the powers . . ."
VII Hall 1871, Q25

"The first photograph of a total solar eclipse . . ."
II Morton 1869a, Q237

Still, the light-greedy large achromats . . .
VII Mayer 1869

Joseph Zentmeyer
VII Oliver 1893

"Mr. Zeitmeyer's office in Walnut Street . . ."
VII AmericanArtifacts 2021b

"Five furniture cart loads of material"
II Morton 1877, Q126

"Guided by the desire of securing . . ."
VII Morton 1869b, Q207

James Cremer
VII Bray 1979

Edward Wilson
VII GraphicArts.Princeton.edu 2021

"We rose on Friday morning, August 6 . . ."
II Mayer 1877, Q130

Wet collodion process
VII AlternativePhotography.com 2021

"The weather on the eventful day . . ."
VII Morton 1869b, Q208

"Five seconds was more than sufficient . . ."
II Morton 1877, Q123

"I experienced an indescribable feeling . . ."
"Our work was finished . . ."
II Mayer 1877, Q137

"The best method of examining . . ."
II Mayer 1877, Q140

"The most elevated and bright of these . . ."
II Mayer 1877, Q143

"An albatross with bill and head . . ."
VII Mayer 1869, Q204

"An eagle, with outstretched wings . . ."
II Morton 1877, Q143

"The serrated character . . ."
II Morton 1877, Q121

"Like an ear of corn"
VII Morton 1877, Q143

"It consists of a solid central mass . . ."
"To the left appears a mass of rolling cloud . . ."

"Resembles in shape a great whale . . ."
II Morton 1877, Q211

"The length of the entire mass . . ."
"To the right of this . . ."
II Morton 1877, Q212

One of these might have been . . .
VII PioneerPhotographers 2021

"Common portrait camera"
II Pickering 1877, Q156

Many astronomers would have thought . . .
VII Pickering 1869b

"The trouble and anxiety . . ."
II Morton 1877, Q120

"One of the Ottumwa pictures . . ."
II Morton 1877, Q209

Baily's Beads
VII AmericanArtifacts 2021

"The corona approached much more nearly . . ."
II Himes 1877, Q153

Thomas Webb
VII Moore 1975

"The best way to examine the glass photographs . . ."
quoted in VII Webb 1869, Q5

"Almost the first impression given me . . ."
II Gould 1877b, Q36

"Published an extraordinary number . . ."
VII *Musical Review and Musical World* 1862, Q189

Chapter 10

"The astronomers . . ."
VII *Evening Express* 1869b, Q1

The very first issue . . .
VII Lockyer 1869a

"All observations made by the navigator . . ."
II B. Peirce 1872, Q1

"Hence the examination of the Sun's disk . . ."
II Cutts & Mosman 1872, Q117

"The English government last year sent . . ."
quoted in V Jones & Boyd 1971, Q162

"It is nothing less than a duty . . ."
II B. Peirce 1872, Q5

When later Shelby College was on the brink . . .
VII *Publications of the Astronomical Society of the Pacific* 1903

Alvan G. Clark
VII Wendell & Hale 1897

County Courthouse
VII "The Total Eclipse of August Seventh" 1869, P113

George Searle
VII *Monitor* 1904

London Smoke
VII Salisbury 2015

John Whipple
VII MoMA.org 2021

Whipple given credit
e.g., VII Shellen 1872 (in German) and Barnard 1898 (in English[1])

---

[1] posthumously

"Excursion to the Sun"
VII ShelbyvilleHistory 2021, Q1

Charles Peirce
VII Downward 2015

"The observers were in continual apprehension . . ."
II C. Peirce 1872, Q126

"The crowd was placed . . ."
quoted in II C. Peirce 1872, Q127

"Two seconds more . . ."
"low, long, and yellow"
II C. Peirce 1872, Q127

"To complete the sweep the whole visible heavens . . ."
quoted in VII Schmidt 2012, Q79

"During the total phase . . . "
II Searle 1872, Q132

"Just as the total phase . . ."
II Searle 1872, Q130

Near Mammoth Cave
II Walcott 1912

"Our isolated station . . ."
"Was the resort of all the inhabitants . . ."
II Langley 1872, Q134

"The Sun went out as suddenly . . ."
VII Langley 1891, Q40

"Like suddenly kindled electric light . . . "
VII Langley 1891, Q41

Times zones
VII GreenwichMeanTime.com 2021

Illinois State Capitol
VII MakendaEclipse.com 2021

Charles Schott
VII Winston 1901

Alexander Twining
VII Archives.Lib.UConn.edu 2021

Nathan Dupuis
VII QueensU.ca 2021

Alexander Evans
VII History.House.gov 2021

So as its complement . . .
VII Humiston 2017

"People were hurrying to and fro . . ."
VII Saal 2017, Q1

"Suddenly the Sun burst out again . . ."
VII Humiston 2017, Q1

"Watched the instrument through the time . . ."
quoted in VII Saal 2017, Q1

"I saw no . . ."
II Fay 1872, Q160

"I saw one or two . . ."
II Fay 1872, Q169

Joseph Warner
VII Wiener 2016

These men, too, set up their station . . . ."
VII Williams 1947

"I gladly accepted this use of so powerful . . ."
II Lane 1872, Q167

And so it was that Assistant . . .
VII Campbell 1914

George Davidson
VII Holway 1912

"Up the Sutchitna or Kneek Rivers . . ."
II Davidson 1872, Q177

He was a Baily's Beads denier
e.g., VII Davidson 1900

"The ladies and gentlemen on board had a view . . ."
II Davidson 1872, Q180

William Seward
Seward in Alaska
VII Seward 1869

Chapter 11

This chapter quotes Ashe 1870a extensively, beginning with *"For the benefit of those who may undertake . . ."* on II Ashe 1870a, Q7. Abridgements occur at II Ashe 1870a, Q10 and II Ashe 1870a, Q18. The excerpt concludes with *". . . in the negatives."* II Ashe 1870a, Q20.

Quebec Observatory
VII Jarrell 1975

Ashe sunspot theory
VII Ashe 1867

Labrador
VII AAAS.org 2022

"The American astronomers . . ."
II Ashe 1870a, Q6

Only half of American youth were in school . . .
VII GilderLehrman.org 2021

Jefferson City
VII GreeneCountyIowaHistoricalociety.org 2021

Henry Vail
VII Houlette 1970

"To Captain Ashe . . ."
"Shortly before totality . . ."
quoted in II Ashe 1870a, Q42

"When totality took place . . ."
quoted in II Ashe 1870a, Q43

"A murmur is now heard and voices arise . . ."
quoted in II Ashe 1870a, Q44

"It remains only for me . . ."
"I have to thank you . . ."
"I remain . . ."
quoted in II Ashe 1870a, Q45

"The most remarkable photogram . . ."
II Ashe 1870a, Q37

Others were not so sure
VII Curtis 1869b

"It will be necessary to refute some opinions . . ."
"When I found that it would be many months . . ."
II Ashe 1870a, Q24

"THE OBSERVATORY, CRANFORD, MIDDLESEX . . ."
"My Dear Sir . . ."
quoted in II Ashe 1870a, Q24

Ashe's images seemed to prove . . .
VII Ashe 1869

"My Dear sir,— . . ."
quoted in II Ashe 1870a, Q25

"Crimes I am charged with"
II Ashe 1870a, Q26

Bumping into your own telescope . . .
II Ashe 1870b

"A committee appointed by the Counsel . . ."
"And also by Commander Ashe . . ."

"But they are not equal in definition . . ."
VII *Monthly Notices of the Royal Astronomical Society* 1870a, Q105

But was denied funding to attend . . ."
VII Jarrell 1988

Chapter 12

"After a night's encampment in the woods . . ."
"At about 3:00 o'clock . . ."
II Myer 1869, Q193

"Breaking into pieces"
quoted in II Myer 1869, Q196

Dudley Observatory
VII Hough 1870

Lewis Swift
VII Wlasuk 1996

"The whole population of Mattoon was gathered . . ."
VI *Targum* 1869, Q1

Then there was Robert Paine . . .
II Paine 1869b

Steamboat to the centerline
VII Paine 1869a

Few steamboats made it
VII Corbin 2006

Coal-mining town of Boonesboro
VII Weber 1935

"The corona was good . . ."
II Paine 1869b, Q1

"His brain was out of proportion to his delicate body".
III "Winthrop S. Gilman Dead" 1884, Q1

Gilman set some sort of record . . .
II Gilman 1869

"Of fine violet . . ."
II Gilman 1869, Q180

He unsuccessfully solicited money . . .
VII *Register* 1869

Jacob Blickensderfer
VII OzarksFF.com 2021

Eclipse guide
i.e., VII Coffin 1869

It may be that Blickensderfer . . .
VII Daniels 2008

"On August 7 came the eclipse of the Sun . . ."
"Everyone in town . . ."
"I was glad to have seen this phenomenon . . ."
VII *Courier* 1965, Q1

"With their noses . . ."
VII Crozier 1869b, Q111

"Lividity of death . . ."
II Crozier 1869a, Q5

"Coruscation"
VII "Observations of the Eclipse as Seen at West-Port, KY," 1869, Q165

"Drays and wagons could be seen . . ."
"Like some giant bird in a fairy-tale."
"Mountains of hydrogen . . ."
VII Crozier 1869b, Q112

"McFarland . . . selected a point in the line . . ."
"We packed our carpet-sack . . ."
"Took passage on the Louisville Mail Packet"
We took the train . . ."

"And were escorted to our hotel . . ."
"Mt. Vernon is . . ."
"Was selected for our observatory . . ."
"With no greater anxiety . . ."
"Gathered in crowds at our right and left . . ."
"While the shadow was creeping over the Sun . . ."
"The corona was seen to splendid advantage . . ."
"Prof. Christy had charge of the instrument . . ."
"The Prof. had passed through the same region . . ."
VII *Courier*, Q1

"The entire enterprise . . ."
VII Hazen 1869, Q4

"Egotistical, tactless, and mistrustful . . ."
VII Now.UIowa 2021

"The phenomenon indicated . . ."
II Irish 1869, Q27

"The eclipse on Saturday was a success."
VII *State Democratic Press* 1869b, Q1

"Because of the more extended horizon . . ."
VII *State Democratic Press* 1869a, Q1

Cupolas were popular architectural feature . . .
II B. Peirce 1872b

"But when the totality came on . . ."
VII Blog.UI 2021

Twelve colleges
VII Kalb 1872

Sometimes it was not about how to observe . . .
VII Williams 1947

Official residence of Iowa's governor"
VII TerraceHill.org 2021

Here were seven . . .
VII *Press Citizen* 1948, Q1

"In preparing for an observation . . ."
VII Mitchell 1869, Q558

Telescope was her "Little Dollond" . . .
VII Albertson 1913

"The whorl of a half-blown morning glory . . ."
II Mitchell 1877, Q56

"Was surprised at the irregularity . . ."
"Ruby-colored leaf-shaped flames . . ."
II quoted in Mitchell 1877, Q58

Mary Whitney
Sarah Glazier
Mary Reybold
VII Vassar.edu 2022

We do not know what . . .
VII *Maria Mitchell Association Annual Report* 1907

Isabella Carter
VII Vassar.edu 2022

"Roof of Mr. Foote's . . ."
quoted in II Mitchell 1877, Q57

"There were some seconds of breathless suspense . . ."
VII Mitchell 1869, Q558

"Instantly the corona burst forth . . ."
VII Mitchell 1869, Q559

After failure to obtain funding . . .
VII Stark 1955

Robert Abbe
VII Stark 1955

Robert Warder
VII Cameron, Clarke, Seaman *et al.* 1906

"By reason of his familiarity with the heavens . . ."
VII *Daily Gazette* 1869b, Q1

"Southeast near Omaha . . ."
"From Chicago to Omaha . . ."
"Our principal diversion . . ."
VII *Daily Gazette* 1869b, Q1

Treaty of Fort Laramie
VII NDStudies.gov 2021

"We soon after ascended to the low bluffs . . ."
VII *Daily Gazette* 1869b, Q1

Paris Exposition Universelle
II Abbe 1872

"The Totality has passed away like a dream . . ."
VII *Daily Gazette* 1869b, Q1

Chapter 13

"There is no premonition of a solar eclipse . . ."
VII *State Sentinel* 1869, Q2

"The day was a most beautiful one . . . "
VII *Times* 1869b, Q1

"The light grew cold and weird . . ."
II Gould 1877, Q30

"Line of silver . . ."
VII Crozier 1869a, Q3

"Persons standing near . . ."
II Watson 1877, Q84

"Dark, weird and ghastly appearance . . ."
II Eaton 1877, Q64

"Saw nothing of which to give notice"
II Myer 1869, Q195

"Mrs. Farrell distinctly saw . . ."
II Gilman 1869, Q176

"Cold dark shadow"
"It appeared to strike me bluntly . . ."
II Cutts & Mosman 1872, Q119

"Just before totality . . ."
"All nature seemed to hesitate . . ."
II Eaton 1877, Q65

"Fearful as a procession of spirits . . ."
e.g., IV *The History of Jasper County, Iowa* 1878, Q405

"Embrace thousands of acres . . ."
VII *Courier* 1869, Q1

"The feeling was as if one . . ."
II Gilman 1869, Q176

"Just as the last visible ray disappeared . . ."
"It like a flicker of candle going out"
VII *Gazette* 1869, Q1

"The sky presented the appearance . . ."
II Pickering 1877a, Q96

"The appearance of the heavens was peculiar"
II Eaton 1877, Q64

"A few minutes previous to totality . . . "
"Horizon changed color . . ."
II Gilman 1969, Q176

"Mississippi of a leaden color . . ."
quoted in II Mitchell 1877, Q57

"A bright orange-purple light . . ."
II Coffin 1877, Q25

"Pale orange color of vegetation"
II *Democrat* 1869, Q1

"Arrayed in every color of the rainbow"
"As if bands of broad ribbon . . ."
quoted in II Myers 1869, Q195

Sky brightness
VII See Silverman and Mullen 1975

"The sky was also sensibly darker . . ."
II Stockwell 1877, Q62

"The light diminished enough that the steam . . ."
VII *Sun* 1869, Q1

"Throughout the period of totality . . ."
II Gould 1877, Q80

"The darkness was at no time so great . . ."
II Mitchell 1877, Q56

"Could read print on common newspaper"
II Eaton 1877, Q64

"Of six gentlemen, all perfect vision . . ."
II McLeod 1872, Q160

Frederick Bardwell
VII KUHistory.ku.edu 2021

"Nearly the same as when the Sun is hid . . ."
II Bardwell 1877, Q189

"About equal to that on a clear moonless . . ."
VII Eastman 1879, Q191

"Sun visible to naked eye without pain . . ."
II Mitchell 1877, Q57

"Appearance of an approaching thunder storm . . ."
II Eaton 1877, Q64

"Trees could be seen basking . . . "
II Horr, Anderson, & Wormwood 1872, Q172

"The peculiar brilliance with which the stars . . ."
VII Himes 1869b, Q140

"And when the light returned . . ."
VII *Courier* 1965, Q7

"And when the light returned again . . ."
VII *Daily Times* 1989, Q11

"There was no sound . . ."
quoted in II Hall 1869, Q201

"The world seemed to breathe again . . ."
II Watson 1869, Q84

"The temperature fell"
"The suffering chill was appalling""
VII Devens 1883, Q138

"Coats were buttoned closer . . ."
VII *Ledger* 1869, Q1

"Dew was deposited on the grass . . ."
VII M'Clune 1869, Q7

"The eclipse yesterday caused the thermometer . . ."
VII *North Iowa Times* 1869, Q1

"The day had been very hot . . ."
"The heat of the Sun was no longer oppressive . . ."
II Austin 1877, Q75

"A low moaning wind now spring . . ."
II Mayer 1877, Q137

"A tornado . . ."
II Eaton 1877, Q65

"Deranged in transportation . . ."
II Coffin 1877b, Q20

"The cocks all began to crow . . ."
"Cattle were evidently much alarmed . . ."
quoted in II C. Peirce 1872, Q127

"Doves flew to their cotes . . ."
VII *Times* 1869b, Q1

"Turkeys were surprised . . ."
VII *Pioneer* 1869, Q1

"Pigs squealed in resentment . . ."
VII Williams 1947, Q85

"Farmers reported that sheep . . ."
VII Blogs.DavenportLibrary.com 2021

"Horses continued to feed quietly . . ."
II Myers 1869, Q196

"Growing so restive and alarmed . . ."
VII *Daily State Register* 1869, Q1

"Someone noticed a bat flying about"
VII Blogs.IU.edu 2021

"Even the everlastingly howling curs . . ."
II Davidson 1872, Q179

"Before the Sun was entirely obscured . . ."
II Hall 1869, Q202

"The birds ceased their music . . ."
VII *Daily State Register* 1869, Q1

"One of them one of our party . . ."
VII *Ledger* 1869, Q1

"The swallows that nest in the cornice . . ."
VII *State Democratic Press* 1869b, Q1

"Jaybirds became very boisterous . . ."
VII Williams 1947, Q104

"Flies nested on ceilings"
VII Blogs.DavenportLibrary.com 2021

"Fireflies twinkled in the foliage"
II Mitchell 1869, Q559

"Bees came swarming to the hives . . ."
II M'Clune 1869, Q7

"Katydids chirped their nocturnal notes . . ."
VII "Astronomical Phenomena and Progress" 1869, Q40

"Shower of bright specks"
quoted in II Myer 1969, Q193

"Bodies like meteors crossing . . ."
II Himes 1877, Q154

"Flowers opened again their petals . . ."
"The eclipse was gone . . ."
VII *Times* 1869b, Q1

"Few people trouble themselves . . ."
VII *Journal* 1869, Q1

"Mr. Goodfellow's sight . . ."
"But, limited as was the extent of my view . . ."
II B. Peirce 1872c, Q164

"I turned my head . . ."
II Young 1877, Q48

"I was completely surprised . . ."
II Eaton 1877, Q65

"I was completely surprised . . ."
II Eaton 1877, Q65

"Each individual fact . . ."
e.g., VII Helmholtz 1862

Chapter 14

Above Benjamin Sand's signature
e.g., VII Costantino 2008, Q1773

Haphazard attempts . . .
VII Newcomb 1871

"Eighty-five rods northwest . . ."
quoted in II Newcomb 1869b, Q17

"Original record not sent . . ."
II Newcomb 1869b, Q18

"1,985 feet east and 416 feet north . . ."
quoted in II Newcomb, Q18

"Four squares from court-house . . ."
quoted in II Newcomb 1869b, Q20

"The beginning of the total phase . . ."
quoted in II Newcomb 1869b, Q21

"The original record is forwarded . . ."
II Newcomb 1869b, Q22

Charles Phillips
VII Powell 1996

Asa Horr
VII Cherba 2020; Crosby 2015

Civil War Orphans' Home
VII Gallarno 1926

Photographs show the building
VII scua.library.uni.edu 2021

"Recently returned from Germany."
VII *Gazette* 1869, Q1

Alan Fiala
VII Kaplan 2001

Solar radius change . . .
e.g., VII Sigismoni & Morcos 2011

Henry Ristine
II Horr, Anderson, & Wormwood 1872, Q171

William Pulsifer
VII Stevens 1909

Charles Marsh
VII Marsh 1910

James Keeler
VII Osterbrock 1984

Matthew Maury
VII SeaSky.org 2021

Chapter 15

"Paris, KY., Aug 7"
VII *True Kentuckian 1869*, Q1

'Discovery' of Vulcan
VII *Leger 1879*

"Not go on a wild goose chase . . ."
III Sheehan 1999, Q14

"As if herald and handmaids . . ."
VII "The Great Eclipse" 1869, Q286

One or more Vulcans
VII Newcomb 1860

Indeed, Newcomb could narrow the search . . .
VII Newcomb 1869c

"The main object I kept in view . . ."
II Newcomb 1869b, Q8

"Was disappointed in not finding . . ."
I Newcomb Diary 1869a

"I devoted 25 seconds . . ."
"A slight motion of the unclamped telescope . . ."
"Had I not in advance made myself familiar . . ."
II Gould 1869, Q34

"Far enough from the public highway . . ."
II Bardwell 1869, Q189

"A rosette of bright purple . . ."

"With regard to the search . . ."
II Bardwell 1969, Q190

"I had no time to examine the space . . ."
II Cutts & Mosman 1872, Q123

N. E. Farrell
VII Farrell 1868

"At no time during the eclipse . . ."
II Gilman 1869, Q181

"A few moments after the corona . . ."
"Each of the observers felt quite positive . . ."
VII Hind 1878, Q663

"Exclaiming that he and able men saw a miniature . . ."
"During totality the object was forgotten . . ."
II Gilman 1869, Q176

"It may be that out from the floating . . ."
VII Fort 1923, Q18

"How was the photography?" [and rest of dialog]
VII Mitchell 1869, Q560

"In giving to the clouds . . ."
"Having no astronomical knowledge . . ."
quoted in II Myer 1869, Q196

Corona too bright in 1869?
VII "The Total Eclipse of August Seventh" 1869

"One of the most energetic and able men . . ."
quoted in VII Whitesell 1998, Q146

Chapter 16

"Well, it proved one thing . . . "
VII *Journal* 1869b, Q1

"Those astronomers who have come . . ."
VII *Journal* 1869a, Q1

American Phrenological Journal
VII *American Phrenological Journal* 1869

"Murderous riot on the steamship . . ."
VII *Tama County Republican* 1869b, Q1

"Personal habits of . . .
VII *Tama County Republican* 1869a, Q1

"If you would be beautiful . . ."
VII *Daily Gazette* 1869a, Q1

"Afraid it would never 'go off' . . ."
VII *State Democratic Press* 1869a, Q1

"Thousands of people . . ."
VII *Herald* 1869, Q1

"The proposed eclipse of tomorrow . . ."
VII *Daily Gazette* 1869a, Q1

"The best we have seen"
"Prof. White, of New York"
VII *Bee* 1869, Q1

"This exhibition will come off promptly . . ."
VII Reporter 1869, Q1

Christopher Columbus predicted a solar eclipse
VII *Times* 1869a, Q1

The only two total eclipses of the Sun . . .
VII *State Democratic Press* 1869a, Q1

"The Sun will rise eclipsed . . ."
VII *Boone County Democrat* 1869, Q1

"The shadows of Earth will commence . . ."
VII *Freemason* 1869a, Q1

Large stars will be observable . . .
VII *Boone County Democrat* 1869, Q1

It will be the last total solar eclipse . . .
VII *Daily Gazette* 1869b, Q1

"We learned that on that day of the eclipse . . ."
VII "Total Solar Eclipse in August" 1869, Q242

"People who are too penurious to subscribe . . ."
VII *Register* 1869b, Q1

"We have seen no general statement . . ."
VII *Republican* 1869, Q1

The requisite information was to be found . . .
VII e.g., "The Great Eclipse" 1869

"Cumberland Almanac"
VII *Tennessean* 2017, Q1

"The scientific inclined . . ."
VII *Daily Gate City* 1869b, Q1

"Sunday evening Rev. Mr. Percival . . ."
VII *Marshall Times* 1869, Q1

"A total eclipse, of the Sun, is so rare . . ."
VII Safford 1869, Q123

"So it happened that a general enthusiasm . . ."
VII *Eagle* 1869, Q1

*"One old gentlema*n . . ."
"There was not stuff enough to cover . . ."
VII *Evening Bulletin* 1869, Q1

"We know of no better place . . ."
VII *Evening Express* 1869a, Q1

"The editor of the . . ."
VII *Republican Banner* 1869, Q1

" . . . at exactly the time calculated . . ."
VII *Weekly North Carolina Standard* 1869, Q1

"Party of Cincinnati gentlemen . . ."
VII Crocker 1970, Q105

Steamboat General Lytle
VII USDeadlyEvents.com 2021

"And as the hour for the great obscuration . . ."
VII *Courier* 1869a, Q1

"The entire population was in the streets . . ."
VII WisconsinHistory.org 2021

"The solar eclipse . . . appeared promptly . . ."
VII *Daily Express* 1869b, Q1

"Of course, eclipse watchers in different places saw . . ."
e.g., *Patriot* 1869b

"The sun sets on the 7th at 58 minutes past 6 . . .
VII *Eagle* 1869, Q1

"Great preparations have been made . . ."
VII *Daily* Gazette 1869a, Q1

"All persons wishing to view the eclipse . . ."
VII *Hawk Eye* 1869a, Q1

"Observed one enthusiastic eclipser . . ."
VII *Evening Express* 1869b, Q1

"We regret to announce that Alderman Riggs . . ."
"Quite a number of enthusiastic gazers . . ."
VII *Republican* 1869, Q1

"In every parlor . . ."
VII *Eagle* 1869, Q1

"The Sun looked cheerfully down . . ."
VII *Union* 1869, Q1

"The whole population became astronomers . . ."
VII *Pioneer and Home Visitor* 1869b, Q1

"All business for the time suspended . . ."
"Every person of any considerable age . . ."
VII Weaver 1912, Q370

"Men walked more softly and spoke in lower tones . . ."
VII *Hawk Eye* 1869b, Q1

"Very pious . . . lady . . ."
"I would just like to know why . . ."
VII *State Democratic Press* 1869a, Q1

"Product of the McGuffy *Reader . . ."*
"I have thought a thousand times . . ."
VII Williams 1947, Q81

"All quiet this week on . . ."
VII *Gazette* 1869, Q1

"A minute passed as if an hour"
"We were all back on Earth again . . ."
VII Williams 1947, Q86

As little as 100 words
e.g., VII *Weekly Western Sun* 1869

"At 5 minutes before 4 o'clock . . ."
VII *Ledger* 1869, Q1

"A dark shadow hid the Sun . . ."
VII *Tama County Republican* 1869b, Q1

"It seemed to speak directly to our spirits . . ."
VII *Gazette* 1869, Q1

"This great occurrence on Saturday last . . ."
VII *Boone County Advocate* 1869b, Q1

"The observations made in Brighton . . ."
VII *Pioneer and Home Visitor* 1869a, Q1

"The succession of phenomenon . . ."
VII *Daily Courier* 1869, Q1

"There are in Sioux City . . ."
VII *Journal* 1869, Q1

"On Vine Street, below Third . . ."
VII *Daily Gazette* 1869a, Q1

"A couple . . . who insisted . . ."
VII *North Iowa Times* 1869, Q1

"Mr. J. S. Moore took a photographic view . . ."
VII *Tama County Republican* 1869b, Q1

"Last week two or three parties bought . . ."
VII *Freeman* 1869, Q1

"The hole in the Moon observed . . ."
VII *Register* 1869c, Q1

"Before the next . . ."
VII *Evening Express* 1869a, Q1

"As the Moon began to move away . . ."
VII *Journal* 1869, Q1

"At Shakopee, Minn. Two girls aged 15 and 16 . . ."
VII *Daily Journal* 1869, Q1

"Hon. O. A. Allen died to-day . . ."
VII *Daily Journal* 1869, Q1

"Funeral procession found itself . . ."
VII *Pentagraph* 1869, Q1

"Our Republican friends in the county . . ."
"Last evening we were shown an egg . . ."
VII *Journal* 1869a, Q1

"Two notable things visited . . ."
VII *Journal,* Q1

"We have been enjoying splendid October weather . . ."
VII *Times* 1869a, Q1

"His wheat was all . . ."
VII *State Democratic Press* 1869c, Q1

"That the Sun would never shine clearly after . . ."
VII *Eagle* 1869, Q1

"The great eclipse of the Sun occurred . . ."
VII Heyeraft 1921, Q87

"A Mrs. Gifford of Marion County. . ."
VII *Voter* 1869b, Q1

"A young woman . . . lost her life . . ."
VII *Advertiser* 1869, Q1

"When trying to make his way to his brother's . . ."
VII McCulla 1914, Q507

"Gaze of all humanity"
VII *Post* 1869, Q1

"State papers that have recovered . . ."
VII *Times* 1869d, Q1

"The Comet is Here"
"For more than ten years past . . ."
VII *Pioneer and Home Visitor* 1869a, Q1

"The journal says that the eclipse . . ."
VII *Weekly Express* 1869, Q1

"Children resumed their play . . ."
VII *Times* 1869b, Q1

"Ever heard about the big eclipse . . ."
VII Purcell 1922, Q67

"The gas company will not light . . ."
VII *Weekly Express* 1869, Q1

"It's something different . . ."
VII *Argus Herald* 2017, Q1

Chapter 17

"Even thoughtless women and children . . ."
"Why . . ." [and rest of dialog]
VII *Sentinel* 1869, Q1

American Philosophical Society
VII AmPhilSoc.org 2021

"The beaux at Des Moines who plumed themselves . . ."
quoted in VII Petersen 1970, Q107

"We should meet . . ."
VII *Voter* 1869a, Q1

"By the way, a lady . . ."
VII Hind 1878, Q663

"Conversed intelligently on the matter"
B. Peirce editorially inserted into II Horr, Anderson & Wormwood 1872, Q172

"Some five or six intelligent observers . . ."
II Eaton 1877, Q65

"Our party now consisted of . . ."
VII *Courier* 1869, Q1

Anna Winlock
VII Byrd 1904

Saint-Mary-of-the-Woods College
VII *Daily Express*, 1869b

*University of Iow*a
VII Lib.UIowa.edu 2021

"My assistants, a party of young students . . ."
VII Mitchell 1869, Q560

A large number of prominent citizens . . ."
VII *Daily Express* 1869c, Q1

Arabella Mansfield
VII HumanRights.Iowa.gov 2021

Mary Whiting
VII Furness 1922

Mary Newsom
VII MathWomen.AgnesScott.org 2021

"At that moment flashed a light . . ."
VII "The Great Solar Eclipse" 1869, Q286

"All this equipment, scores of experts . . ."
VII *Hawkeye* 2021, Q1

"Why . . ." [and rest of dialog]
VII *Sentinel* 1869, Q1

"It appears to me that the wisest men . . ."
VII "Benjamin Banneker" 1863, Q1

Harvard would not appoint . . .
VII *Harvard Gazette* 2019

Civil War
See Gould 1869

"Asaph Hall, who was outside the pale . . ."
Angeline Stickney
VII Hall 1908

"Original, though not outspoken, abolitionist"
III "Winthrop S. Gilman Dead" 1884, Q1

American Colonization Society
VII AAIHS.org 2022

"Gotten up for the purpose . . ."
VII *Progress* 1869, Q2

"An old . . ."
VII *Evening Express* 1869a, Q1

"Men's faces looked like . . . "
VII "Photographs of the Eclipse" 1870, Q498

"The old woman remembers the Big Eclipse is. . ."
"And everyone was scared to death"
"Lived down in Kentucky . . ."
VII Federal Writers' Project 1941, Q202

"In 1888 Langley published . . ."
VIII HAO.UCAR.edu 2021

"In this part of Kentucky . . ."
VII Langley 1888, Q42

"The African . . ."
II Ashe 1870a, Q3

Chapter 18

Symbol
VII Goodman 2017

Sarah Washburn
VII Haines 2000, Q17

Abbe Museum
VII AbbeMuseum.org 2021

Here is an example of skewed reporting . . .
e.g., VII History.com 2021 or AncientPages.com 2021

"Less anxious about . . ."
VII Henry 2017, Q61

"The chief was a man of commanding presence . . ."
"He fulfilled in spirit and letter . . ."
VII Davidson 1901, Q76

"About the time the Sun was half obscured . . ."
II Davidson 1872, Q179

"Delighted with the great trick of his friend . . ."
VII Scidmore 1885

Davidson painting
VII AncientPages.com 2021

Tglit Map
VII Penn.Museum.edu 2021

"William Seward's experience with Alaska . . ."
VII Hinckley 1971

"Mr. Seward and his son . . ."
"The chiefs wanted to know more . . ."
VII Hinckley 1971, Q141

"The Indians huddled together in awe"
II Hall 1869, Q200

"The Indians exhibited a good deal . . ."
VII Franklin 1871, Q524

"The Esquimaux seemed to be . . ."
VII Franklin 1871, Q519

The Sisseton
VII TheFirstScout.Blogspot.com 2021

The Yanktonai
VII Mallery 1886

"So everyone . . . yelled very loudly . . ."
VII TheFirstScout.Blogspot.com 2021

Two Bears
VII American-Tribes.com 2021

Winter Count
VII Chamberlain 1984

Lone Dog
VII Mallery 1886

Colonel Garrick Mallery published . . ."
VII Mallery 1877

The Oglála
VII TheFirstScout.Blogspot.com 2021

"For Indians in the Sioux District . . ."
VII Poole 1881, Q11

The Brulé
VII Poole 1881

"Some days before the great eclipse . . ."
VII Poole 1881, Q76

"The Indians were impassive lookers on . . ."
VII Poole 1881, Q77

Chapter 19

"The observatories must have been left . . ."
VII Mitchell 1869, Q556

Newcomb after the eclipse
Carter & Carter 2006

"Suggested the name . . ."
VII Copernicus 1869, Q106

"All eyes were fixed upon our proceedings . . ."
VII Copernicus 1869, Q107

"An accident occurred . . ."
"Two of us . . ."
VII Copernicus 1869, Q108

"Some person caught his foot in the wire . . ."
"The fever heat of excitement . . ."
VII Copernicus 1869, Q111

Stereophotographs were a popular fad
VII NPS.gov 2021

"The body of the Moon . . ."
VII *Daily Journal* 1869, Q1

"Anvil Protuberance"
VII Gilman 1870, Q418

"The semblance of a horse's tail . . ."
VII Knight 1869, Q165

History of prominences
VII Tandberg-Hanssen 1997

"Any correct observation of color . . ."
VII Mitchell 1869, Q559

A trip to the chemist's laboratory . . .
VII Lockyer 1869b

1868 eclipse
VII Haig 1868

Were there any other chemical elements present . . .
VII Douglas 1870

"Perhaps the most thrilling effect in nature . . ."
VII *Atlantic Almanac for 1869* 1869, Q62

Indeed, 1869 solar-eclipse observers . . .
VII e.g., Haig, C. T. 1868

1869 estimates of the corona's extent . . .
VII Proctor 1871b

"Instead of the gorgeous spectacle I witnessed . . ."
VII Newcomb 1871, Q11

Review of *theories of the corona* appears in VII Lockyer 1874

Eventually, it would be realized . . .
e.g., VII Meadows 1970

For centuries, people had longed . . .
VII Campbell 1907

Stephen Alexander and a lunar atmosphere
VII Challis 1864

Theory of the corona before 1869
VII Seabroke 1870

He resurrected the idea that the solar corona . . .
VII Gould 1870

Coronal polarization
VII Pickering 1870

Iowa-based astronomers noted . . .
VII Proctor 1870

There was some attempt to correlate . . .
VII Proctor 1871b

Harkness/Young priority
e.g., VII Lynn 1904 *versus* Meadows 1970, P26

Was the corona behaving like an aurora . . .
VII Norton 1870

"During the Civil War, the auroras . . ."
VII *Smithsonian American Art Museum* 2015, Q1

Later he disavowed it
VII Young 1871

Breaking it all down, did the corona change . . .
VII Proctor 1871a

The correlation of number of sunspots . . .
VII Saint Cyr, O. C.; Young, D. E.; Pesnell, W. D. *et al.* 2008

Since the 1930s, we know that this is because . . .
VII Hunter 1942

The absence of hundreds of other . . ."
VII Young 1872, Q28

As an aside, it is interesting that Young . . .
VII Russell 1909; and VII Dawson 1917

Coronium
VII Claridge 1937

Dmitri Mendeleev
VII Langley 1878

Periodic Table
VII Alarez, Sales, & Seco 2008

Yet Young and Harkness confirmed the presence . . .
VII Harkness 1871b

Meanwhile, geologists and chemists found . . .
VII Young 1895

Walter Grotrian
Hannes Alfvén
Bengt Edlén
VII Peter & Bhola 2014; and VII Russell 2018

Ionized iron
VII Hunter 1942

The solar corona must be much hotter . . .
VII Schnorr, Mäckel, & Oreshkina 2013

"As recently as the 1850s . . ."
VII "Recent Solar Researches" 1870

Homer Lane
VII SNAC.Cooperative.org 2021

"Odd looking and odd mannered little man . . ."
III Newcomb 1903

The Sun does not do this
VII Fiala, Dunham & Sofia 1994

State of the Sun
e.g., Donati 1870

"Always impressive . . ."
VII Himes 1869a, Q4

Chapter 20

"Providence has made . . ."
VII Robitaille 2013, Q1

"Never before was an eclipsed Sun . . ."
"The Government, the Railway and other companies . . ."
VII Lockyer 1874a, Q246

"Its unblanching eye . . ."
VII Himes 1869a, Q2

John Wesley Powell
VII Powell 1875

The white-water journey was perilous . . .
VII "The Colorado River Region and John Wesley Powell" 1969, Q1

Charles Peirce, much later back at Harvard . . .
VII Hoel 2016

"The Survey's greatest computer"
VII *Cloud (circa)* 2010, Q35

"About half an hour before totality . . ."
VII Young 1869a, Q370

"Genius in this direction . . ."
II Newcomb 1903

The herculean effort of 1869 . . .
VII "The Government and the Eclipse" 1870, Q409

To them, government funding meant supply of a ship.
e.g., VII Airy 1870, Q257

"The railroads and telegraph were supported by . . ."
e.g., VII Adams 1870, Q116

"In the preceding report . ."
II Cutts 1872, Q122

"Accidental . . ."
VII *Journal* 1899, Q2

They conspired to produce another trans-American . . ."
VII Abbe 1881

Vulcan search telescope
VII Hussey 1912

Postlog

"Was obliged to make an inclement passage . . ."
III "Robert Treat Paine" 1886, Q534

"Sick and nearly blind"
"On reaching his destination . . ."
VII Wolba.ch 2022

Appendix I

II Newcomb 1869b, Q15

Appendix II

"Bits of glass of every description . . ."
VII *Daily Gazette* 1869b, Q1

"With due precaution . . ."
VI Webb 1868

Illustration Notes

I tried to choose illustrations that are not commonly seen. When a traditional subject is called for, I strove to select an image different from those that are regularly used. For readers interested in learning more about the illustrations, I provide the notes below.

Introduction: "The Rush of This Black Wing of Night"

Fig. 2 drawing by contemporary Middlebury student, Samuel Mosley
Fig. 5 Campbell Hoffman (programmer); John DeGroote, Director
Fig. 6 Campbell Hoffman (programmer); John DeGroote, Director

A Trans-American Eclipse

Fig. 1 framed print located at Renaissance Des Moines Savery Hotel
Fig. 2 digitized by Morgan Aaronson
Fig. 4 digitized by Rosemary Meany
Fig. 5 Charles Peirce (cartographer)

"Some Light Upon This Dark Subject"

Fig. 1 two-page chromolithograph. From Johnston 1869, plate V. Digitized by Rosemary Meany
Fig. 3 Romania, 1999
Fig. 4 available from the NASA Center for AeroSpace Information

T. Hockey, *America's First Eclipse Chasers*, Springer Praxis Books,
https://doi.org/10.1007/978-3-031-24124-6

Navy Astronomers 2,000 Kilometers Ashore

Fig. 2 from a postcard
Fig. 3 from a photogravure
Fig. 4 digitized by Rosemary Meany
Fig. 5 P. Marvin (photographer); from a *carte-de-viste*
Fig. 6 from Lockyer 1874b. Digitized by Rosemary Meany

"The Vast Black Orb"

Fig. 1 digitized by Morgan Aaronson
Fig. 2 digitized by Rosemary Meany
Fig. 3 digitized by Rosemary Meany
Fig. 6 digitized by Morgan Aaronson

New Astronomy in the Old West

Fig. 1 Matthew Brady[1] (photographer)
Fig. 2 H. Wellge (cartographer); Milwaukee (Wisconsin): American Publishing
  Company. 44 centimeters X 82 centimeters.
Fig. 3 from Engelhardt 1898-1899, P55
Fig. 8 digitized by Rosemary Meany
Fig. 9 digitized by Morgan Aaronson
Fig. 10 digitized by Morgan Aaronson

Observing in Style

Fig. 7 from Whitesell 2003, Figure 10. Digitized by Joe Tenn

Meeting of the Grayhairs

Fig. 2 from a *carte-de-viste*
Fig. 3 John Browne (photographer); gift of Mary Himes Vale and Sarah Vale
  Rush
Fig. 7 M'Clune 1869, P

"Overhanging Monster Wings": The Philadelphia Photographic Corp

Fig. 1 from Oliver 1893, plate 12. digitized by Rosemary Meany
Fig. 2 from a *carte-de-viste*
Fig. 3 digitized by Rosemary Meany
Fig. 4 digitized by Morgan Aaronson
Fig. 5 from Furman 1905, P
Fig. 6 digitized by Rosemary Meany
Fig. 7 T. Sinclair (illustration artist). 29.6 centimeters X 39.1 centimeters.

---

[1] famous for documenting the Civil War

Surveying a Solar Eclipse

Fig. 4 from a lantern slide
Fig. 9 from Schmidt 2012, Figure 9
Fig. 16 archived at the California Department of Parks and Recreation. Oil on
masonite. 46.5 centimeters X 32.4 centimeters.

The Canadians: Toques on the Frontier

Fig. 2 from *Illustrated London News* 1869

Chasing the Umbra through Time and Space

Fig. 1 Luthers Boots (painter).
Fig. 2 from a *carte-de-viste*
Fig. 3 Chicago Lithographic Company. Not to scale. 50 cm X 66 cm.
Fig. 4 from Gilman 1895. A. H. Ritchie (photographer). Etching on steel
Fig. 7 framed print located at Sanford Museum
Fig. 9 from Collection of Wes and Shelley Cowan
Fig. 16 from Garson 2001. Digitized by the author.

"A Darkness That Can Be Felt"

Fig. 1 College of Saint Rose Image Collection
Fig. 4 from Purcell 1922. Watercolor original to this book

Standing of the Edge Looking Up

Fig. 2 digitized by Morgan Aaronson

Vulcan

Fig. 1 Lithograph by E. Jones and G. W. Newman (1846). 51 cm X 50 cm.
Fig. 4 Recruiting poster titled, *United States Soldiers at Camp "William Penn"
Philadelphia*, Pa. 33 X 40 cm. Commissioned by Supervisory Committee
for Recruiting Colored Regiments.

Americans in Totality

Fig. 1 digitized by Rosemary Meany
Fig. 2 Wilstach, Baldwin & Company
Fig. 5 Registered at United States Patent Office by L. Sternberger for Eclipse
Brand Shirts

**In the Shadow of Benjamin Banneker and "Even Thoughtless Women and
Children Hush …"**

Fig. 1 Also pictured are Helen Storke (standing) and Clara Glover (reclining).
Fig. 4 J. A. Whipple (photographer). Gift of Larry J. West.

Fig. 3 F. Graham (cartographer)
Fig. 6 from Cottam and Orchiston (2015)

Fire Cloud

Fig. 2 Alexander Gardner (photographer)
Fig. 5 artifact archived at the Plains Indian Museum
Fig. 6 Alexander Gardner (photographer)
Fig. 8 from Poole 1881, cover photograph

What Did It All Mean?

Fig. 1 almost certainly by James Cremer (photographer)
Fig. 2 from Gillman, 1870 (plate 1)
Fig. 5 digitized by Rosemary Meany
Fig. 6 oil on canvas. 142.3 centimeters x 212.2 centimeters. 1911 gift to the museum from Eleanor Blodgett.

… And What Happened After That?

Fig. 1 based on NASA data
Fig. 2 J. Hillers (photographer).
Fig. 6 digitized by Rosemary Meany
Fig. 7 T. R. Burnham (photographer)
Fig. 8 *Jones, J. and* Butterfield, C. *Eclipse*. Summit Records. 2019. digitized by Rosemary Meany

Index

A

Abbe, Cleveland
    Robert, 191, 268, 316, 332, 402
Abington (Virginia), 347
abolition, 257
achromat, 23, 48–50, 59, 91, 110, 118, 137, 174,
    312, 384, 391
A Connecticut Yankee in King Arthur's Court, 275
actinometer, 16, 20, 24, 102, 193
Adair (Iowa), 25
Adams, John Quincy, 16
Africa, 257
African-American, 86, 226, 256, 257, 259–261,
    263
Agnew, F.H., 137, 345
Alaska, 6, 69, 153, 155, 203, 268, 269, 316, 331,
    397, 421
Alaskan Native, 267, 268
Albany (New York), 106, 172
albedo, 223
Aldrin, Buzz, 10
Aleutian Islands, 69
Alexander, Stephen, 76, 106, 108, 111, 152,
    158, 200, 211, 244, 257, 258, 287,
    316, 424
Alfvén, Hannes, 295, 425
Alkaid, 200
Allen, Benjamin, 187
Altair, 200
altitude sickness, 321
Alton (Illinois), 210, 340
Alvan Clark & Sons, 83, 88, 311
American Academy of Arts and Science, 133

Association for the Advancement of Science,
    191
Astronomical Society, 226, 319, 394
Colonization Society, 257, 420
American Journal of Science
    Nautical Almanac, 50, 74, 93, 305
    Philosophical Society, 107, 130, 252, 418
Anchor Bay, 155
Anderson, William, 214
angular size, *see* apparent size
animal, 134, 202, 272, 284
Anthony, Susan, 255
Antigua, 1
aperture, 23, 45, 55, 56, 61, 83, 96, 117, 122, 171,
    183, 189, 194, 210, 286, 327, 384
apogee, 36
Appalachian Mountains, 6
apparent size, 33, 35, 36, 230, 236, 281, 306, 307
Arabic, 265
Arago, François, 100
Archilochus, 39
Arctic Circle, 70
Arcturus, 200, 229
Argentina, 1, 317
Aristotle, 39
Armstrong, Neil, 10
Arnold, W.T., 146
artificial horizon, 20, 50, 51
Ashe, Edward, 157–164, 166–169, 263, 284, 316,
    333, 397–398, 420
Asia, 3, 36, 220, 265
Association for the Advancement of Women, 255
asteroid, 158, 186, 222, 224, 232

Astronomy, 50
astrophysics, 16, 262, 263, 292, 319
Atlantic Ocean, 6, 36, 236
Atlas of Astronomy, 243
atmosphere, 18, 101, 121, 180, 196, 205, 276,
281, 288–290, 292, 293, 295, 297, 315,
325, 424
atmospheric pressure, 202
aurora, 290, 291, 424
Austin, Edward, 105, 116
autobiography, 26, 217
automobile, 8, 19, 381
A.W. & F. Voodry's Carriage Factory, 44

B

Bache, Alexander, 104, 106
Baconian tradition, 315
Baily, Francis, 43
Baily's Beads, 84, 97, 129, 147, 171, 175, 189,
228, 309, 311, 393, 397
Bald Mountain, 71
bandit, 184
Banneker, Benjamin, xii, 252–264, 419, 430
Bardwell, Frederick, 106, 200, 226, 227, 316, 347,
405, 411
barnyard, 243
Bedford (Indiana), 178, 343
Bell, James, 208
Trudy, 48, 60, 188
Beloit (Iowa), 18
Bering Sea, 6, 70, 72
Bermuda, 6
Bethlehem (Pennsylvania), 122
Bible, 36, 37, 99
Big Island (Hawaii), 1
Black Bear, 273
Black Drop Effect, 149, 211
Black Friday, 76
Black Hawk War, 77
Black, James, 149
Blickensderfer, Jacob, 176–178, 193, 315, 316,
332, 400
Bloomington (Illinois), 152, 309, 341
Bloomington (Indiana), 186, 343
Board of Visitors, 172
bolometer, 318
Boone (Iowa), 25, 333
Boone County (Iowa), 173
Boone, Daniel, 113
Boris, W.S., 209
Boston (Massachusetts), 83, 135, 184
Boston Latin School, 104

Bowditch Comet Seeker, 143
Bowditch, Jonathan, 104, 136
Nathaniel, 137
Bowling Green (Kentucky), 147
Bowman, Thomas, 248
Brazelton House, 99
Bristol (Tennessee), 135, 144, 145, 226, 309, 347
Britain, 172, 314
Brookline (Massachusetts), 321
Brooklyn Bridge, 19, 381
Brulé, 275, 422
Brünnow, Franz, 106
Buffalo (New York), 210
Bunsen burner, 285
Burksville (Kentucky), 345
Burlington (Iowa), 25, 76, 77, 90, 99, 173, 187,
202, 240, 252, 292, 337
Burlington and Missouri Railroad, 78
Burlington, Cedar Rapids and Northern Railway,
216
Burlington City Park Board, 102
Burlington Collegiate Institute, 189
Burnham, Sherburne, 106
Burton-Speke Expedition, 92
Byrd, Max, 18, 19, 419

C

Cambridge (Massachusetts), 49, 100, 104, 139
camera, 20, 22, 47, 59–62, 81, 117–122, 128, 130,
139, 148, 166, 179, 181, 246, 317, 320
Canada, 4, 9, 12, 26, 157, 159, 169, 267, 271
canoe, 26, 269
Carlisle (Kentucky), 209, 347
Carlisle (Pennsylvania), 120
Carlyle (Illinois), 210, 341
Carte du Ciel, 319
Carter, George, 196, 340
Isabella, 190, 337, 402
Case, Zophar, 210, 340
Cecil, Sackville, 46, 152, 333
Cedar Falls (Iowa), xvi, xvii, 25, 212, 214, 334
Cedar Falls Civil War Orphan Home, 212
Cedar Rapids (Iowa), 249
Celestial Meridian, 18
Pole, 24
Sphere, 14, 18, 24, 29, 35, 38, 196, 319
cemetery, 102, 208
census, 6, 14, 44, 77, 93, 110, 113, 158, 222, 259
centerline, 19, 33, 44, 47, 81, 113, 125, 142, 145,
154, 158, 173, 178, 186, 207, 208, 217,
282, 399
Central Meridian, *see* Celestial Meridian